Lecture Notes in Computer Science 6945

Commenced Publication in 1973
Founding and Former Series Editors:
Gerhard Goos, Juris Hartmanis, and Jan van Leeuwen

Giorgio Delzanno   Igor Potapov (Eds.)

Reachability Problems

5th International Workshop, RP 2011
Genoa, Italy, September 28-30, 2011
Proceedings

 Springer

Volume Editors

Giorgio Delzanno
Università di Genova
Dipartimento di Informatica e Scienze dell'Informazione
via Dodecaneso 35, 16146 Genoa, Italy
E-mail: delzanno@disi.unige.it

Igor Potapov
University of Liverpool
Department of Computer Science
Ashton Building, Ashton St, Liverpool, L69 3BX, UK
E-mail: potapov@liverpool.ac.uk

ISSN 0302-9743   e-ISSN 1611-3349
ISBN 978-3-642-24287-8   ISBN 978-3-642-24288-5 (eBook)
DOI 10.1007/978-3-642-24288-5
Springer Heidelberg Dordrecht London New York

Library of Congress Control Number: 2011936633

CR Subject Classification (1998): F.3, D.2, F.2, D.3, F.4, F.4.1, F.1

LNCS Sublibrary: SL 1 – Theoretical Computer Science and General Issues

Typesetting: Camera-ready by author, data conversion by Scientific Publishing Services, Chennai, India

Printed on acid-free paper

Springer is part of Springer Science+Business Media (www.springer.com)

Preface

This volume contains the papers presented at the 5th Workshop on Reachability Problems RP 2011 during September 28–30, 2011 in the Department of Informatics and Computer Science, University of Genoa, Italy. RP 2011 was the fifth in the series of workshops following four successful meetings at Masaryk University of Brno, Czech Republic, in 2010, Ecole Polytechnique, France, in 2009, at the University of Liverpool, UK, in 2008 and at Turku University, Finland, in 2007.

The Reachability Workshop is specifically aimed at gathering together scholars from diverse disciplines and backgrounds interested in reachability problems that appear in algebraic structures, computational models, hybrid systems, logic, and verification.

Reachability is a fundamental problem that appears in several different contexts: finite- and infinite-state concurrent systems, computational models like cellular automata and Petri nets, decision procedures for classical, modal and temporal logic, program analysis, discrete and continuous systems, time critical systems, hybrid systems, rewriting systems, probabilistic and parametric systems, and open systems modelled as games.

Typically, for a fixed system description given in some form (reduction rules, systems of equations, logical formulas, etc.) a reachability problem consists in checking whether a given set of target states can be reached starting from a fixed set of initial states. The set of target states can be represented explicitly or via some implicit representation (e.g., a system of equations, a set of minimal elements with respect to some ordering on the states). Sophisticated quantitative and qualitative properties can often be reduced to basic reachability questions. Decidability and complexity boundaries, algorithmic solutions, and efficient heuristics are all important aspects to be considered in this context. Algorithmic solutions are often based on different combinations of exploration strategies, symbolic manipulations of sets of states, decomposition properties, reduction to linear programming problems, and they often benefit from approximations, abstractions, accelerations and extrapolation heurisitics. Ad hoc solutions as well as solutions based on general purpose constraint solvers and deduction engines are often combined in order to balance efficiency and flexibility.

The purpose of the conference is to promote exploration of new approaches for the predictability of computational processes by merging mathematical, algorithmic and computational techniques. Topics of interest include (but are not limited to): reachability for infinite state systems, rewriting systems; reachability analysis in counter/timed/cellular/communicating automata; Petri-nets; computational aspects of semigroups, groups and rings; reachability in dynamical and hybrid systems; frontiers between decidable and undecidable reachability problems; complexity and decidability aspects; predictability in iterative maps and new computational paradigms.

All these aspects were discussed in the 20 presentations of the fifth edition of the RP workshop.

The proceedings of the previous editions of the workshop appeared in the following volumes:

Mika Hirvensalo, Vesa Halava, Igor Potapov, Jarkko Kari (Eds.): Proceedings of the Satellite Workshops of DLT 2007. TUCS General Publication No 45, June 2007. ISBN: 978-952-12-1921-4.

Vesa Halava and Igor Potapov (Eds.): Proceedings of the Second Workshop on Reachability Problems in Computational Models (RP 2008). Electronic Notes in Theoretical Computer Science. Volume 223, Pages 1-264 (26 December 2008).

Olivier Bournez and Igor Potapov (Eds.): Reachability Problems, Third International Workshop, RP 2009, Palaiseau, France, September 23–25, 2009, Lecture Notes in Computer Science, 5797, Springer 2009.

Antonin Kucera and Igor Potapov (Eds.): Reachability Problems, Fourth International Workshop, RP 2010, Brno, Czech Republic, August 28–29, 2010, Lecture Notes in Computer Science, 6227, Springer 2010.

The four keynote speakers at the 2011 edition of the conference were:

- Krishnendu Chatterjee, IST Austria, "Graph Games with Reachability Objectives: Mixing Chess, Soccer and Poker"
- Bruno Courcelle, Labri, Université Bordeaux 1, "Automata for Monadic Second-Order Model-Checking"
- Joost-Pieter Katoen, RWTH Aachen, "Timed Automata as Observers of Stochastic Processes"
- Jean-Francois Raskin, CFV, Université Libre de Bruxelles, "Reachability Problems for Hybrid Automata"

There were 24 submissions. Each submission was reviewed by at least three Program Committee members. The full list of the members of the Program Committee and the list of external reviewers can be found on the next two pages. The Program Committee is grateful for the highly appreciated and high-quality work produced by these external reviewers. Based on these reviews, the Program Committee decided finally to accept 16 papers, in addition to the four invited talks.

We gratefully acknowledge the financial support from the Games for Design and Verification initiative of the European Science Foundation that helped us invite keynote speakers of exceptionally high scientific level to Genoa.

We also gratefully acknowledge the support of the University of Genoa, of the Department of Informatics and Computer Science, and of the ASAP team for the help in the organization of the workshop.

It is also a great pleasure to acknowledge the team of the EasyChair system, and the fine cooperation with the *Lecture Notes in Computer Science* team of Springer, which made possible the production of this volume in time for the conference. Finally, we thank all the authors for their high quality contributions, and the participants for making this edition of RP 2011 a success.

September 2011
Giorgio Delzanno
Igor Potapov

Organization

Program Committee

Parosh Abdulla	Uppsala University, Sweden
Davide Ancona	University of Genoa, Italy
Bernard Boigelot	University of Liege, Belgium
Olivier Bournez	LIX - Ecole Polytechnique, France
Cristian Calude	University of Auckland, New Zealand
Giorgio Delzanno	University of Genoa, Italy
Stephane Demri	LSV - ENS Cachan - CNRS, France
Javier Esparza	Technische Universität München, Germany
Laurent Fribourg	LSV - ENS Cachan, France
Vesa Halava	University of Turku, Finland
Juhani Karhumaki	University of Turku, Finland
Antonin Kucera	Masaryk University, Czech Rebublic
Alexander Kurz	University of Leicester, UK
Jerome Leroux	CNRS LABRI Bordeaux, France
Alexei Lisitsa	University of Liverpool, UK
Igor Potapov	University of Liverpool, UK
Arnaud Sangnier	LIAFA - University Paris 7 - CNRS, France
Hsu-Chun Yen	National Taiwan University, Taiwan
Gianluigi Zavattaro	University of Bologna, Italy

Additional Reviewers

André, Étienne	Jancar, Petr
Atig, Mohamed Faouzi	Jha, Sumit
Bardin, Sebastien	Knapp, Alexander
Bell, Paul	Lazic, Ranko
Czerwiński, Wojciech	Rezine, Ahmed
Doyen, Laurent	Schmitz, Sylvain
Guan, Nan	Soulat, Romain
Göller, Stefan	Trtik, Marek
Habermehl, Peter	
Holik, Lukas	

Table of Contents

Graph Games with Reachability Objectives (Invited Talk) 1
  Krishnendu Chatterjee

Observing Continuous-Time MDPs by 1-Clock Timed Automata
(Invited Talk) 2
  Taolue Chen, Tingting Han, Joost-Pieter Katoen, and
  Alexandru Mereacre

Automata for Monadic Second-Order Model-Checking (Invited Talk) . . . 26
  Bruno Courcelle

Reachability Problems for Hybrid Automata (Invited Talk) 28
  Jean-François Raskin

Synthesis of Timing Parameters Satisfying Safety Properties 31
  Étienne André and Romain Soulat

Formal Language Constrained Reachability and Model Checking
Propositional Dynamic Logics 45
  Roland Axelsson and Martin Lange

Completeness of the Bounded Satisfiability Problem for Constraint
LTL 58
  Marcello M. Bersani, Achille Frigeri, Matteo Rossi, and
  Pierluigi San Pietro

Characterizing Conclusive Approximations by Logical Formulae 72
  Yohan Boichut, Thi-Bich-Hanh Dao, and Valérie Murat

Decidability of LTL for Vector Addition Systems with One Zero-Test . . . 85
  Rémi Bonnet

Complexity Analysis of the Backward Coverability Algorithm for
VASS 96
  Laura Bozzelli and Pierre Ganty

Automated Termination in Model Checking Modulo Theories 110
  Alessandro Carioni, Silvio Ghilardi, and Silvio Ranise

Monotonic Abstraction for Programs with Multiply-Linked
Structures 125
  Parosh Aziz Abdulla, Jonathan Cederberg, and Tomáš Vojnar

Efficient Bounded Reachability Computation for Rectangular
Automata.. 139
    Xin Chen, Erika Ábrahám, and Goran Frehse

Reachability and Deadlocking Problems in Multi-stage Scheduling...... 153
    Christian E.J. Eggermont and Gerhard J. Woeginger

Improving Reachability Analysis of Infinite State Systems by
Specialization 165
    *Fabio Fioravanti, Alberto Pettorossi, Maurizio Proietti, and
    Valerio Senni*

Lower Bounds for the Length of Reset Words in Eulerian Automata 180
    Vladimir V. Gusev

Parametric Verification and Test Coverage for Hybrid Automata Using
the Inverse Method 191
    Laurent Fribourg and Ulrich Kühne

A New Weakly Universal Cellular Automaton in the 3D Hyperbolic
Space with Two States 205
    Maurice Margenstern

A Fully Symbolic Bisimulation Algorithm 218
    Malcolm Mumme and Gianfranco Ciardo

Reachability for Finite-State Process Algebras Using Static Analysis.... 231
    Nataliya Skrypnyuk and Flemming Nielson

Author Index... 245

Graph Games with Reachability Objectives
(Invited Talk)

Krishnendu Chatterjee

IST Austria (Institute of Science and Technology Austria)

Abstract. Games played on graphs provide the mathematical framework to analyze several important problems in computer science as well as mathematics,
such as the synthesis problem of Church, model checking of open reactive systems and many others. On the basis of mode of interaction of the players these
games can be classified as follows: (a) *turn-based* (players make moves in turns);
and (b) *concurrent* (players make moves simultaneously). On the basis of the
information available to the players these games can be classified as follows:
(a) *perfect-information* (players have perfect view of the game); and (b) *partial-
information* (players have partial view of the game). In this talk we will consider
all these classes of games with *reachability* objectives, where the goal of one
player is to reach a set of target vertices of the graph, and the goal of the op-
ponent player is to prevent the player from reaching the target. We will survey
the results for various classes of games, and the results range from linear time
decision algorithms to EXPTIME-complete problems to undecidable problems.

G. Delzanno and I. Potapov (Eds.): RP 2011, LNCS 6945, p. 1, 2011.

Observing Continuous-Time MDPs
by 1-Clock Timed Automata*

Taolue Chen[1], Tingting Han[1], Joost-Pieter Katoen[2], and Alexandru Mereacre[1]

[1] Department of Computer Science, University of Oxford, United Kingdom
[2] Software Modeling and Verification Group, RWTH Aachen University, Germany

Abstract. This paper considers the verification of continuous-time Markov decision process (CTMDPs) against single-clock deterministic timed automata (DTA) specifications. The central issue is to compute the maximum probability of the set of timed paths of a CTMDP $\mathcal{C}$ that are accepted by a DTA $\mathcal{A}$. We show that this problem can be reduced to a linear programming problem whose coefficients are maximum timed reachability probabilities in a set of CTMDPs, which are obtained via a graph decomposition of the product of the CTMDP $\mathcal{C}$ and the region graph of the DTA $\mathcal{A}$.

1 Introduction

Markov decision processes (MDPs) are a prominent mathematical system model for modeling decision-making—modeled as nondeterministic choices—in situations where outcomes are partly random and partly under the control of a decision maker [24]. MDPs, also referred to as turn-based $1\frac{1}{2}$-player games, are intensively used in decision making and planning with a focus on optimization problems which are typically solved via dynamic programming. They are a discrete-time stochastic control process where at each time step, the decision maker (i.e., the scheduler) may select any action α that is enabled in the current state s. The MDP reacts on this choice by probabilistically moving to state s' with probability $\mathbf{P}(s, \alpha, s')$. A discrete-time Markov chain (DTMC) is an MDP where for each state only a single action is enabled. Since the mid-eighties, MDPs (and DTMCs as special subclass) have been the active subject of applying model checking. Whereas the initial focus was on qualitative properties (e.g., "can a state be reached almost surely, i.e., with probability one?"), the emphasis soon shifted towards *quantitative* properties. Several specification formalisms have been adopted, such as LTL [34,19], probabilistic versions of CTL [9,6], as well as automata [19,21]. The key issue in the quantitative verification of MDPs is to determine the maximum, or dually, minimum probability of a certain event of interest, such as $\Diamond G$, $\Box \Diamond G$, and so forth, where G is a set of states which is either given explicitly or as a state formula. For finite-state MDPs, it is well-known that e.g., extremum reachability probabilities can be obtained by solving linear

---

* This research is partially supported by the EU FP7 Project MoVeS and the ERC Advanced Grant VERIWARE.

programming (LP) problem and that memoryless schedulers suffice to obtain such extrema. If the reachability event is constrained by the maximum number of allowed transitions, one has to resort to finite-memory schedulers, but still a simple value iteration technique suffices to compute the extremum probabilities with the required accuracy. Such techniques have been implemented in model checkers such as PRISM[1] and LiQUOR [18] and successfully applied to several practical case studies such as randomized distributed protocols.

Continuous-time Markov decision processes (CTMDPs) [32] extend MDPs by associating a random delay in state s on selecting action α by the scheduler. Choosing action α in state s yields a random delay in s by the CTMDP which is governed according to an exponential distribution with rate $r^\alpha(s)$. Thus, the probability to wait at most d time units in state s on choosing α is $1 - e^{-r^\alpha(s)\cdot d}$. After delaying, a CTMDP evolves like an MDP probabilistically to state s' with probability $\mathbf{P}(s, \alpha, s')$. A continuous-time Markov chain (CTMC) is a CTMDP where for each state only a single action is enabled. The state residence time in a CTMC is thus independent of the action chosen. CTMCs have received quite some attention by the verification community since the late nineties. This work has primarily focused on CSL (Continuous Stochastic Logic), a timed probabilistic version of the branching-time temporal logic CTL. The key issue in CSL model checking is to compute the probability of the event $\Diamond^{\leq T} G$ where $T \in \mathbb{R}_{\geq 0}$ acts as a time bound. It has been shown that such probabilities can be characterized as least solution of Volterra integral equation systems and can be computed in a numerically stable and efficient way by reducing the problem to transient analysis of CTMCs [4]. This has been implemented in model checkers such as MRMC [25][2] and PRISM, and has been applied successfully to several cases from systems biology and queueing theory, to mention a few.

Recently, the verification of CTMCs has been enriched by considering linear-time properties equipped with timing constraints. In particular, [15,16] treat linear real-time specifications that are given as *deterministic timed automata* (DTA) [2]. DTA are automata equipped with clock variables that can be used to measure the elapse of time, can be reset to zero, and whose value can be inspected in transition guards. The fact that these automata are deterministic means that for any clock valuation and state, the successor state is uniquely determined. Whereas timed automata are typically used as system models describing the possible system behaviors, we use them—in analogy to [1]—as objectives that need to be fulfilled by the system. In our context, DTA specifications include properties of the form "what is the probability to reach a given target state within the deadline, while avoiding unsafe states and not staying too long in any of the dangerous states on the way?". DTA have recently also been adopted as specification language for generalized semi-Markov processes (and their game extensions) in [11,12]. The central issue in checking a DTA specification is computing the probability of the set of paths in a CTMC that are accepted by the DTA. This can be reduced to computing the (simple) reachability probability

---

[1] http://www.prismmodelchecker.org/
[2] http://www.mrmc-tool.org/trac/

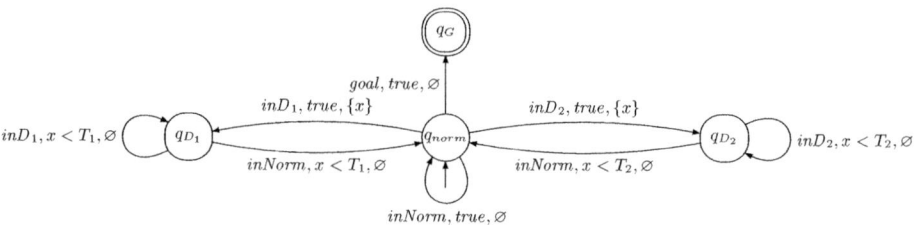

Fig. 1. An example 1-clock DTA that goes beyond timed reachability

in a (somewhat simplified variant of) *piecewise deterministic Markov process* (PDP, [20]), basically a stochastic hybrid model which is obtained by a synchronous product construction between the CTMC and the region graph of the DTA [16]. A prototypical implementation of this technique has recently been presented [7] and has led to the efficient verification of CTMCs of several hundreds of thousands of states against one-clock DTA specifications. The appealing properties of this algorithm are that it resorts to standard computational procedures, i.e., graph analysis, region graph construction, solving systems of linear equations, and transient analysis of CTMCs for which efficient algorithms exist.

In contrast to MDPs, CTMDPs have received far less attention by the verification community; in fact, the presence of nondeterminism and continuous time makes their analysis non-trivial. CTMDPs have originated as continuous-time variants of finite-state probabilistic automata [26], and have been used for, among others, the control of queueing systems, epidemic, and manufacturing processes. Their analysis is mainly focused on determining optimal schedulers for criteria such as expected total reward and expected (long-run) average reward, cf. the survey [23]. The formal verification of CTMDPs has mostly concentrated on computing extremum probabilities for the event $\Diamond^{\leq T} G$ with time bound $T \in \mathbb{R}_{\geq 0}$. Whereas memoryless schedulers suffice for extremum reachability probabilities in MDPs, maximizing (or minimizing) timed reachability probabilities requires *timed* schedulers, i.e., schedulers that "know" how much time has elapsed so far [29,28,8]. As these schedulers are infinite objects, most work has concentrated on obtaining ϵ-optimal schedulers—mostly piecewise-constant schedulers that only change finitely often in the interval $[0, T]$—that approximate the extremum probability obtained by a timed scheduler up to a given accuracy $\epsilon > 0$ [31,33]. Recently, the use of adaptive uniformization has been proposed as an alternative numerical approach to obtain such ϵ-optimal schedulers [13]. Another approach is to concentrate on sub-optimal schedulers, and consider the optimal *time-abstract* scheduler [5,10]. This is a much simpler and efficient procedure that does not rely on discretization, and in several cases suffices. Some of the techniques for both timed and time-abstract schedulers have recently been added to the model checker MRMC [25].

In this paper, we concentrate on a larger class of properties and consider the verification of CTMDPs against linear real-time specifications given as *single-clock* DTA. Note that single-clock DTA cover a whole range of safety and liveness objectives and naturally include timed reachability objectives such as $\diamond^{\leq T} G$. We believe that DTA are a very natural specification formalism that captures a rich set of practically interesting properties. For instance, Fig. 1 presents an example 1-clock DTA that goes beyond timed reachability properties. It asserts "reach a given target G (modeled by state q_G) while not staying too long (at most T_1 and T_2 time units in respective zones D_1 and D_2) in any of the two "dangerous zones on the way". For simplicity, we assume the dangerous zones D_1 and D_2 are not adjacent. In case the system stays too long in one of the dangerous zones, it resides in either location q_{D_1} or q_{D_2} forever, and will never reach the goal state. This property can neither be expressed in CSL nor in one of its existing dialects [3,22]. The central issue now in checking such a DTA specification is computing the extremum probability of the set of paths in a CTMDP $\mathcal{C}$ that are accepted by the DTA $\mathcal{A}$. We show that the approach in [15,16,7] can be adapted to this problem in the following way. We first establish that the extremum probability of CTMDP $\mathcal{C}$ satisfying DTA $\mathcal{A}$ can be characterized as the extremum reachability probability in the product of $\mathcal{C}$ and the region graph of $\mathcal{A}$. Here, the region graph is based on a variant of the standard region construction for timed automata [2]. The product $\mathcal{C} \otimes \mathcal{G}(\mathcal{A})$ is in fact a simple instance of a piecewise deterministic Markov *decision* process (PDDP, [20]). The extremum reachability probabilities in $\mathcal{C} \otimes \mathcal{G}(\mathcal{A})$ are then characterized by a Bellman equation. These results so far are also applicable to DTA with an arbitrary number of clocks (although formulated in this paper for single-clock DTA only). For 1-clock DTA, we then show that solving this Bellman equation can be reduced to an LP problem whose coefficients are extremum timed reachability probabilities in the CTMDP $\mathcal{C}$, i.e., events of the form $\diamond^{\leq T} G$. The size of the obtained LP problem is in $\mathcal{O}(|S| \cdot |Q| \cdot m)$, where S is the state space of CTMDP $\mathcal{C}$, Q is the state space of DTA $\mathcal{A}$, and m is the number of distinct constants appearing in the guards of $\mathcal{A}$.

To put in a nutshell, this paper shows that the verification of CTMDPs against 1-clock DTA objectives can be done by a region graph construction, a product construction, and finally solving an LP problem whose coefficients are extremum timed reachability probabilities in CTMDPs. 1-clock DTA objectives model a rich class of interesting properties in a natural manner and include timed reachability. To the best of our knowledge, this is the first work towards treating linear real-time objectives of CTMDPs. The main appealing implication of our result is that CTMDPs can be verified against 1-clock DTA objectives using rather standard means. The availability of the first practical implementations for timed reachability of CTMDPs paves the way to a realization of our approach in a prototypical tool.

Organization of this paper. Section 2 defines the basic concepts for this paper: CTMDPs, DTA, and formalizes the problem tackled in this paper. Section 3 shortly recapitulates a mathematical characterization of maximum timed reachability probabilities in CTMDPs. Section 4 introduces the product $\mathcal{C} \otimes \mathcal{G}(\mathcal{A})$ and provides a Bellman equation for reachability events in this product. Section 5 is the core of this paper, and shows that for 1-clock DTA, the solution of the Bellman equation can be obtained by solving an LP problem whose coefficients are extremum timed reachability probabilities in CTMDPs obtained from $\mathcal{C} \otimes \mathcal{G}(\mathcal{A})$. Section 6 concludes the paper. The proof of Theorem 2 is included in the appendix.

2   Preliminaries

Given a set H, let $\Pr : \mathcal{F}(H) \to [0, 1]$ be a probability measure on the measurable space $(H, \mathcal{F}(H))$, where $\mathcal{F}(H)$ is a σ-algebra over H.

2.1   CTMDP

Let AP be a fixed, finite set of atomic propositions.

Definition 1 (CTMDP). *A continuous-time Markov decision process is a tuple $\mathcal{C} = (S, s_0, Act, \mathbf{P}, r, L)$, where*

- *S is a finite set of* states;
- *s_0 is the* initial state;
- *Act is a finite set of* actions;
- *$\mathbf{P} : S \times Act \times S \to [0, 1]$ is a* transition probability matrix, *such that for any state $s \in S$ and action $\alpha \in Act$, $\sum_{s' \in S} \mathbf{P}(s, \alpha, s') \in \{0, 1\}$;*
- *$r : S \times Act \to \mathbb{R}_{\geq 0}$ is an* exit rate function; *and*
- *$L : S \to 2^{\mathrm{AP}}$ is a* labeling function.

The set of actions that are enabled in state s is denoted $Act(s) = \{\, \alpha \in Act \mid r^\alpha(s) > 0 \,\}$ where $r^\alpha(s)$ is a shorthand for $r(s, \alpha)$. The operational behavior of a CTMDP is as follows. On entering state s, an action α, say, in $Act(s)$ is non-deterministically selected. The CTMDP now evolves probabilistically as follows. Given that action α has been chosen, the residence time in state s is exponentially distributed with rate $r^\alpha(s)$. Hence, the probability to leave state s via action α in the time interval $[l, u]$ is given by $\int_l^u r^\alpha(s) \cdot e^{-r^\alpha(s) \cdot t} \, dt$ and the average sojourn time in s is given by $\frac{1}{r^\alpha(s)}$. We say that there is an α-transition from s to s' whenever $\mathbf{P}^\alpha(s, s') \cdot r_\alpha(s) > 0$ where $\mathbf{P}^\alpha(s, s')$ is shorthand of $\mathbf{P}(s, \alpha, s')$. If multiple outgoing α-transitions exist, they compete: the probability that transition $s \xrightarrow{\alpha} s'$ is taken is $\mathbf{P}^\alpha(s, s')$. Putting the pieces together, this means that the CTMDP transits from state s to s' on selecting α in s in the time interval $[l, u]$ with a likelihood that is given by:

$$\mathbf{P}^\alpha(s, s') \cdot \int_l^u r^\alpha(s) \cdot e^{-r^\alpha(s) \cdot t} \, dt.$$

Note that the probabilistic behavior of a CTMDP conforms to that of a CTMC; indeed, if $Act(s)$ is a singleton set in each state $s \in S$, the CTMDP is in fact a CTMC. In this case, the selection of actions is uniquely determined, and the function $\mathbf{P}$ can be projected to an $(S \times S)$-matrix, the transition probability matrix. If we abstract from the exponential state residence times, we obtain a classical MDP. For CTMDP $\mathcal{C} = (S, s_0, Act, \mathbf{P}, r, L)$, its *embedded* MDP is given by $emb(\mathcal{C}) = (S, s_0, Act, \mathbf{P}, L)$.

Example 1. Fig. 2 shows an example CTMDP with AP $= \{a, b\}$ and initial state s_0. The state-labelings are indicated at the states, whereas the transition probabilities are attached to the edges. Rates are omitted from the figure and are defined as: $r^\alpha(s_0) = 10$, $r^\beta(s_0) = 5$, and $r^\beta(s_3) = r^\beta(s_1) = r^\gamma(s_2) = 1$. In s_0, there is a nondeterministic choice between the actions α and β.

Definition 2 (CTMDP paths). *A sequence* $\pi = s_0 \xrightarrow{\alpha_0, t_0} s_1 \xrightarrow{\alpha_1, t_1} \cdots$ *is an infinite path in a* CTMDP $\mathcal{C} = (S, s_0, Act, \mathbf{P}, r, L)$, *where for each* $i \geq 0$, $s_i \in S$ *is a state,* $\alpha_i \in Act$ *is an action, and* $t_i \in \mathbb{R}_{>0}$ *is the sojourn time in state* s_i. *A finite path is a fragment of an infinite path ending in a state.*

The length of an infinite path π, denoted $|\pi|$, is ∞; the length of finite path π with $n+1$ states is n. For a finite path $\pi = s_0 \xrightarrow{\alpha_0, t_0} s_1 \xrightarrow{\alpha_1, t_1} \cdots \xrightarrow{\alpha_{n-1}, t_{n-1}} s_n$, let $\pi\downarrow = s_n$ be the last state of π. Let $Paths(\mathcal{C})$ (respectively $Paths_s(\mathcal{C})$) denote the set of infinite paths (respectively starting in state s) in $\mathcal{C}$; let $Paths^n(\mathcal{C})$ (respectively $Paths_s^n(\mathcal{C})$) denote the set of finite paths of length n (respectively starting in state s). To simplify notation, we omit the reference to $\mathcal{C}$ whenever possible. An example path in the CTMDP of Fig. 2 is $\pi = s_0 \xrightarrow{\alpha, 2.5} s_2 \xrightarrow{\gamma, 1.4} s_0 \xrightarrow{\alpha, \sqrt{2}} s_1 \xrightarrow{\beta, 2.8} s_1 \cdots$.

In order to construct a measurable space over $Paths(\mathcal{C})$, we define the following sets: $\Omega = Act \times \mathbb{R}_{>0} \times S$ and the σ-field $\mathcal{J} = \sigma(2^{Act} \times \mathcal{J}_R \times 2^S)$, where $\mathcal{J}_R$ is the Borel σ-field over $\mathbb{R}_{\geq 0}$. The σ-field over $Paths^n$ is defined as $\mathcal{J}_{Paths^n} = \sigma(\{S_0 \times M_0 \times S_1 \times \cdots \times M_{n-1} \mid S_i \subseteq S, M_i \in \mathcal{J}\})$. A set $B \in \mathcal{J}_{Paths^n}$ is a base of a *cylinder set* C if $C = Cyl(B) =$

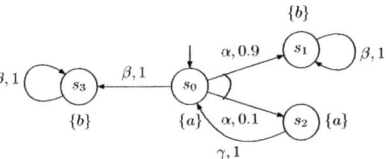

Fig. 2. An example CTMDP

$\{\pi \in Paths \mid \pi[0 \ldots n] \in B\}$, where $\pi[0 \ldots n]$ is the prefix of length n of the path π. The σ-field $\mathcal{J}_{Paths}$ of measurable subsets of $Paths(\mathcal{C})$ is defined as $\mathcal{J}_{Paths} = \sigma(\cup_{n=0}^{\infty} \{Cyl(B) \mid B \in \mathcal{J}_{Paths^n}\})$. Hence we obtain a measurable space $(Paths(\mathcal{C}), \mathcal{J}_{Paths})$.

Schedulers. Nondeterminism in a CTMDP is resolved by a *scheduler*. In the literature, schedulers are sometimes also referred to as adversaries, policies, or strategies. For deciding which of the next actions to take, a scheduler may "have access" to the current state only or to the path from the initial to the current state (either with all or with partial information). Schedulers may select the next action either *deterministically*, i.e., depending on the available information, the next action is chosen in a deterministic way, or *randomly*, i.e., depending on the available information, the next action is chosen probabilistically. In our setting, deterministic schedulers suffice to achieve extremum probabilities and can base their decision on a complete information of the current path so far. Moreover, it is not evident how to define the probability measure for randomized schedulers, as exit rates depend on the actions. Hence we only consider deterministic rather than randomized schedulers in this paper. Furthermore, like in [35], we consider measurable functions as schedulers. Formally,

Definition 3 (Schedulers). *A scheduler for CTMDP $\mathcal{C} = (S, s_0, Act, \mathbf{P}, r, L)$ is a measurable function $D : Paths(\mathcal{C}) \to Act$ such that for $n \in \mathbb{N}$,*

$$D(s_0 \xrightarrow{\alpha_0, t_0} s_1 \xrightarrow{\alpha_1, t_1} \cdots \xrightarrow{\alpha_{n-1}, t_{n-1}} s_n) \in Act(s_n). \tag{1}$$

We denote the set of all schedulers of $\mathcal{C}$ as $\mathcal{D}_{\mathcal{C}}$.

Remark 1. According to the above definition, we consider schedulers that make a decision as soon as a state is entered. In particular, the sojourn time in the current state s_n is not considered for selecting the next action. Such schedulers are called *early* schedulers in [30]. In contrast, a *late* scheduler will choose an action upon leaving a state, i.e., besides the history $s_0 \xrightarrow{\alpha_0, t_0} \cdots \xrightarrow{\alpha_{n-1}, t_{n-1}} s_n$, it will consider also the elapsed time so far in state s_n. Late schedulers suffice for determining extremum reachability probabilities for a certain class of CTMDPs, the so-called locally uniform ones, i.e., CTMDPs in which the exit rate for any enabled action in a state is the same [30].

Probability measure. For a path $\pi \in Paths(\mathcal{C})$ and $m \in \Omega = Act \times \mathbb{R}_{\geq 0} \times S$, we define the *concatenation* of π and m as the path $\pi' = \pi \circ m$. Below we define a probability measure over the measurable space $(Paths(\mathcal{C}), \mathcal{J}_{Paths})$ under the scheduler D.

Definition 4 (Probability measure). *Let $\mathcal{C} = (S, s_0, Act, \mathbf{P}, r, L)$ be a CT-MDP, $n \in \mathbb{N}$ and D a scheduler in $\mathcal{D}_{\mathcal{C}}$. The probability $\mathrm{Pr}^n_{s,D} : \mathcal{J}_{Paths^n} \to [0, 1]$ of sets of paths of length $n > 0$ starting in s is defined inductively by:*

$$\mathrm{Pr}^{n+1}_{s,D}(B) = \int_{Paths^n} \mathrm{Pr}^n_{s,D}(d\pi) \int_{\Omega} \mathbf{1}_B(\pi \circ m) \int_{\mathbb{R}_{\geq 0}} r^\alpha(\pi\!\downarrow) \cdot e^{-r^\alpha(\pi\!\downarrow) \cdot \tau}$$
$$\cdot \sum_{s' \in S} \mathbf{1}_m(\alpha, \tau, s') \cdot \mathbf{P}^\alpha(\pi\!\downarrow, s') \, dm \, d\tau,$$

where

- $\alpha = D(\pi)$, *the action selected by scheduler D on the path π of length n,*
- $B \in Paths^{n+1}$ *and for $n = 0$ we define $\mathrm{Pr}^0_s(B) = 1$ if $s \in B$, and 0 otherwise,*
- $\mathbf{1}_B(\pi \circ m) = 1$ *when $\pi \circ m \in B$, and 0 otherwise,*
- $\mathbf{1}_m(\alpha, \tau, s') = 1$ *when $m = (\alpha, \tau, s')$, and 0 otherwise.*

Intuitively, $\mathrm{Pr}^{n+1}_{s,D}(B)$ is the probability of the set of paths $\pi' = \pi \circ m$ of length $n+1$ defined as a product between the probability of the set of paths π of length n and the one-step transition probability to go from state $\pi\!\downarrow$ to state $\pi'\!\downarrow$ by the action α as selected by the scheduler D. For a measurable base $B \in \mathcal{J}_{Paths^n_s}$ and cylinder set $C = Cyl(B)$, let $\mathrm{Pr}_{s,D}(C) = \mathrm{Pr}^n_{s,D}(B)$ as the probability of subsets of paths from $Paths_s$. Sometimes we write $\mathrm{Pr}_D(C)$ to when the starting state s is clear from the context.

2.2 Single-Clock DTA

Let x be a *clock*, which is a variable in $\mathbb{R}_{\geq 0}$[3]. A *clock valuation* is a function η assigning to x the value $\eta(x) \in \mathbb{R}_{\geq 0}$. A *clock constraint* on x is a conjunction of expressions of the form $x \bowtie c$, where $\bowtie \in \{<, \leq, >, \geq\}$ is a binary comparison operator and $c \in \mathbb{N}$. Let $\mathcal{B}_x$ denote the set of clock constraints over x and let g range over $\mathcal{B}_x$.

Definition 5 (DTA). *A single-clock deterministic timed automaton (DTA) is a tuple $\mathcal{A} = (\Sigma, Q, q_0, Q_F, \rightarrow)$ where*

- Σ *is a finite alphabet;*
- Q *is a nonempty finite set of locations;*
- $q_0 \in Q$ *is the initial location;*
- $Q_F \subseteq Q$ *is a set of accepting locations; and*
- $\rightarrow \in (Q \setminus Q_F) \times \Sigma \times \mathcal{B}_x \times \{\varnothing, \{x\}\} \times Q$ *is an edge relation satisfying:*
  $$q \xrightarrow{a,g,X} q' \text{ and } q \xrightarrow{a,g',X'} q'' \text{ with } g \neq g' \text{ implies } g \wedge g' \equiv \text{FALSE}.$$

We refer to $q \xrightarrow{a,g,X} q'$ as an *edge*, where $a \in \Sigma$ is an input symbol, the *guard* g is a clock constraint on x, $X = \{\varnothing, \{x\}\}$ is the set of clocks that are to be reset and q' is the successor location. Intuitively, the edge $q \xrightarrow{a,g,X} q'$ asserts that the DTA $\mathcal{A}$ can move from location q to q' when the input symbol is a and the guard g on clock x holds, while the clocks in X should be reset when entering q'. DTA are *deterministic* as they have a single initial location, and outgoing edges of a location labeled with the same input symbol are required to have disjoint guards. In this way, the next location is uniquely determined for a given location and a given clock valuation, together with an action. In case no guard is satisfied in a location for a given clock valuation, time can progress. If the advance of time will never reach a situation in which a guard holds, the DTA will stay in that location ad infinitum. Note that DTA do not have location invariants, as in safety timed automata. However, all the results presented in this paper can be adapted to DTA with invariants without any difficulties.

Runs of a DTA are timed paths. In order to define these formally, we need the following notions on clock valuations. A clock valuation η *satisfies* clock constraint $x \bowtie c$, denoted $\eta \models x \bowtie c$, if and only if $\eta(x) \bowtie c$; it satisfies a conjunction of such expressions if and only if η satisfies all of them. Let $\mathbf{0}$ denote the valuation that assigns 0 to x. The reset of x, denoted $\eta[x := 0]$, is the valuation $\mathbf{0}$. For $\delta \in \mathbb{R}_{\geq 0}$ and η, $\eta + \delta$ is the clock-valuation η'' such that $\eta''(x) := \eta(x) + \delta$.

Definition 6 (Finite DTA path). *A finite timed path in DTA $\mathcal{A}$ is of the form $\theta = q_0 \xrightarrow{a_0, t_0} q_1 \xrightarrow{a_1, t_1} \cdots \xrightarrow{a_n, t_n} q_{n+1}$, such that for all $0 \leqslant i \leq n$, it holds $t_i > 0$, $x_0 = 0$, $x_j + t_j \models g_j$ and $x_{j+1} = (x_j + t_j)[X_j := 0]$, where x_j is the clock evaluation[4] on entering q_j, g_j is the guard on the uniquely enabled edge in*

---

[3] Throughout this paper, we use x for the clock variable of the 1-clock DTA under consideration.

[4] As there is only a single clock we sometimes write x for the value of clock x as shorthand for $\eta(x)$.

the DTA *leading from q_j to q_{j+1} when $x_j+t_j \models g_j$, and X_j is the set of clocks on that edge that needs to be reset. Path θ is* accepted *whenever $q_{n+1} \in Q_F$.*

The concepts defined on CTMDP paths, such as $|\theta|$, will be applied to timed DTA paths without modification.

Regions. We consider a variant of the standard region construction for timed automata [2] to DTA. As we consider single-clock DTA, the region construction is rather simple. We basically follow the definition and terminology of [27]. Let $\{c_0, \ldots, c_m\}$ be the set of constants appearing in the guards of DTA $\mathcal{A}$ with $c_0 = 0$. W.l.o.g. we assume $0 = c_0 < c_1 < \cdots < c_m$. Regions can thus be represented by the intervals: $[c_0, c_0], (c_0, c_1), \ldots, [c_m, c_m]$ and (c_m, ∞). (In fact, these regions are also sometimes called zones.) In the continuous probabilistic setting of this paper, the probability of the CTMC taking a transition in a point interval is zero. We therefore combine a region of the form $[c_i, c_i]$ with a region of the form (c_i, c_{i+1}) yielding $[c_i, c_{i+1})$. In the rest of the paper, we slight abuse nomenclature and refer to $[c_i, c_{i+1})$ as a region. As a result, we obtain the regions: $\Theta_0 = [c_0, c_1), \ldots, \Theta_m = [c_m, \infty)$. Let $\Delta c_i = c_{i+1} - c_i$ for $0 \leqslant i < m$ and let $\mathcal{R}_\mathcal{A}$ be the set of regions of DTA $\mathcal{A}$, i.e., $\mathcal{R}_\mathcal{A} = \{\Theta_i \mid 0 \leq i \leq m\}$. The region Θ satisfies a guard g, denoted $\Theta \models g$, iff for all $\eta \in \Theta$ we have $\eta \models g$.

Definition 7 (Region graph). *The region graph of DTA $\mathcal{A} = (\Sigma, Q, q_0, Q_F, \rightarrow)$, denoted $\mathcal{G}(\mathcal{A})$, is the tuple $(\Sigma, W, w_0, W_F, \dashrightarrow)$ with $W = Q \times \mathcal{R}_\mathcal{A}$ the set of states; $w_0 = (q_0, \mathbf{0})$ the initial state; $W_F = Q_F \times \mathcal{R}_\mathcal{A}$ the set of final states; and $\dashrightarrow \subset W \times ((\Sigma \times \{\varnothing, \{x\}\}) \uplus \{\delta\}) \times W$ the smallest relation such that:*

- $(q, \Theta_i) \xdashrightarrow{\delta} (q, \Theta_{i+1})$ *for $0 \leq i < m$;*
- $(q, \Theta_i) \xdashrightarrow{a, \{x\}} (q', \Theta_0)$ *if $\exists g \in \mathcal{B}_x$ such that $q \xrightarrow{a, g, \{x\}} q'$ with $\Theta_i \models g$; and*
- $(q, \Theta_i) \xdashrightarrow{a, \varnothing} (q', \Theta_i)$ *if $\exists g \in \mathcal{B}_x$ such that $q \xrightarrow{a, g, \varnothing} q'$ with $\Theta_i \models g$.*

States in $\mathcal{G}(\mathcal{A})$ are thus pairs of locations (of the DTA $\mathcal{A}$) and a region on clock x. The initial state is the initial location in which clock x equals zero. The transition relation of $\mathcal{G}(\mathcal{A})$ is defined using two cases: (1) a delay transition in which the location stays the same, and the region Θ_i is exchanged by its direct successor Θ_{i+1}, (2) a transition that corresponds to taking an enabled edge in the DTA $\mathcal{A}$. The latter corresponds to the last two items in the above definition distinguishing the case in which x is reset (second item) or not (third item).

Example 2. Fig. 3(a) depicts an example DTA, where q_0 is the initial state and q_1 is the only accepting state. In q_0, the guards of the two a-actions are disjoint, so this TA is indeed deterministic. The part of the region graph of the DTA that is reachable from $(q_0, \mathbf{0})$ is depicted in Fig. 3(b).

2.3   Problem Statement

We now are settled to formalize the problem of interest in this paper. Recall that our focus is on using DTA as specification objectives and CTMDPs as system

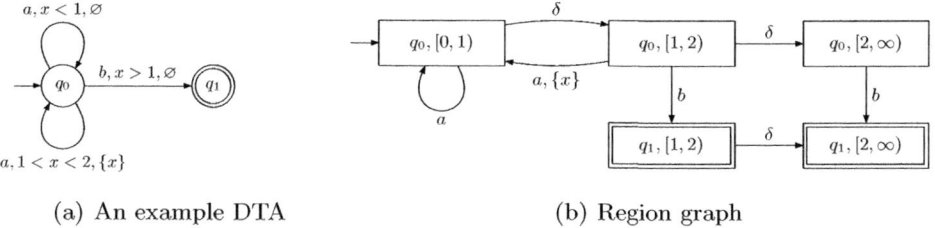

(a) An example DTA   (b) Region graph

Fig. 3. Example DTA and its region graph

models, and our aim is to determine the probability of the set of timed paths of the CTMDP $\mathcal{C}$ that are accepted by $\mathcal{A}$. Let us first define what it means for a CTMDP path to be accepted by DTA $\mathcal{A}$.

Definition 8 (Acceptance). *Given a CTMDP $\mathcal{C} = (S, s_0, Act, \mathbf{P}, r, L)$ and a single-clock DTA $\mathcal{A} = (\Sigma, Q, q_0, Q_F, \rightarrow)$, we say that an infinite timed path $\pi = s_0 \xrightarrow{\alpha_0, t_0} s_1 \xrightarrow{\alpha_1, t_1} \cdots$ in $\mathcal{C}$ is accepted by $\mathcal{A}$ if there exists some $n \in \mathbb{N}$ such that the finite fragment of π up to n, i.e., $s_0 \xrightarrow{\alpha_0, t_0} s_1 \cdots s_{n-1} \xrightarrow{\alpha_0, t_{n-1}} s_n$, gives rise to an "augmented" timed path $\theta = q_0 \xrightarrow{L(s_0), t_0} q_1 \cdots q_{n-1} \xrightarrow{L(s_{n-1}), t_{n-1}} q_n$ of $\mathcal{A}$ with $q_n \in Q_F$. Let $Paths_{s_0}(\mathcal{C} \models \mathcal{A})$ denote the set of paths in CTMDP $\mathcal{C}$ that start in s_0 and are accepted by $\mathcal{A}$.*

Note that the labels of the states that are visited along the CTMDP path π are used as input symbols for the associated timed path in the DTA. Thus, the alphabet of the DTA will be the powerset of AP, the set of atomic propositions. The aim of this paper is to determine the maximum probability of $Paths_{s_0}(\mathcal{C} \models \mathcal{A})$ over all possible schedulers, i.e.,

$$\sup_{D \in \mathcal{D}_{\mathcal{C}}} \mathrm{Pr}_{s_0, D}(Paths_{s_0}(\mathcal{C} \models \mathcal{A})).$$

In the remainder of this paper, we will show that these maximum probabilities can be characterized as a solution of an LP problem, whose coefficients are given as timed reachability probabilities in a set of CTMDPs. Let us first briefly recall such reachability probabilities.

3   Timed Reachability in CTMDP

Given a CTMDP $\mathcal{C} = (S, s_0, Act, \mathbf{P}, r, L)$, a set of goal states $G \subseteq S$, and a time bound $T \in \mathbb{R}_{\geq 0}$, let $Paths_{s_0}(\Diamond^{\leq T} G)$ denote the set of timed paths reaching G from the initial state s_0 within T time units. Formally,

$$Paths_{s_0}(\Diamond^{\leq T} G) \;=\; \{\pi \in Paths(s_0) \mid \exists t \leq T. \, \pi@t \in G\}$$

where $\pi@t$ denotes the state occupied by π at time t, i.e., $\pi@t = \pi[i]$ where i is the smallest index i such that $\sum_{j=0}^{i} t_j > t$. The timed reachability problem amounts to computing

$$\sup_{D \in \mathcal{D}_C} \mathrm{Pr}_{s_0, D}(Paths_{s_0}(\lozenge^{\leq T} G)).$$

This problem has been solved, to a large extent, forty years ago by Miller [29], and has recently been revisited in the setting of formal verification by, amongst others, [5,31]. We briefly recapitulate the main results. Let $\Psi(s, x)$ be the maximum probability to reach G, within T time units, starting from state s given that x time units have passed so far. It follows that $\Psi(s, x)$ can be characterized by the following set of Bellman equations:

$$\Psi(s, x) = \max_{\alpha \in Act(s)} \left\{ \int_0^{T-x} \sum_{s' \in S} r^\alpha(s) \cdot e^{-r^\alpha(s) \cdot \tau} \cdot \mathbf{P}^\alpha(s, s') \cdot \Psi(s', x+\tau) \, d\tau \right\},$$

if $s \notin G$ and $x \leq T$; and 1 if $s \in G$ and $x \leq T$; and 0, otherwise. The term on the right-hand side takes the action that maximizes the probability to reach G in the remaining $T-x$ time units from s by first moving to s' after a delay of τ time units in s and then proceeding from s' to reach G with elapsed time $x+\tau$.

There are different ways to solve this Bellman equation. One straightforward way is by applying discretization [28,31,17]. An alternative approach is to reduce it to a system of ordinary differential equations (ODEs) with decisions. To that end, let $P_{i,j}(t)$ be the maximum probability to reach state s_j at time t starting from state s_i at time 0. For any two states s_i and s_j we obtain the ODE [8]:

$$\frac{dP_{i,j}(t)}{dt} = \max_{\alpha \in Act(s_i)} \left\{ r^\alpha(s_i) \cdot \sum_{s_k \in S} \mathbf{P}^\alpha(s_i, s_k) \cdot (P_{k,j}(t) - P_{i,j}(t)) \right\}.$$

which using $\mathbf{R}^\alpha(s, s') = r^\alpha(s) \cdot \mathbf{P}^\alpha(s, s')$ can be simplified to:

$$\frac{dP_{i,j}(t)}{dt} = \max_{\alpha \in Act(s_i)} \left\{ \sum_{s_k \in S} \mathbf{R}^\alpha(s_i, s_k) \cdot (P_{k,j}(t) - P_{i,j}(t)) \right\}.$$

For $t \leqslant T$, we obtain the following system of ODEs in matrix form:

$$\frac{d\mathbf{\Pi}(t)}{dt} = \max_{\alpha \in Act} \left\{ \mathbf{\Pi}(t) \cdot \mathbf{Q}^\alpha \right\},$$

where $\mathbf{\Pi}(t)$ is the transition probability matrix at time t, i.e., the element (i, j) of $\mathbf{\Pi}(t)$ equals $P_{i,j}(t)$, $\mathbf{\Pi}(0) = \mathbf{I}$, the identity matrix, $\mathbf{Q}^\alpha = \mathbf{R}^\alpha - \mathbf{r}^\alpha$ is the infinitesimal generator matrix for action α where $\mathbf{R}^\alpha$ is the transition rate matrix, i.e., the element (i, j) is $r^\alpha(s_i) \cdot \mathbf{P}^\alpha(s_i, s_j)$, and $\mathbf{r}^\alpha$ is the exit rate matrix in which all diagonal elements are the exit rates, i.e., $\mathbf{r}^\alpha(i, i) = r^\alpha(s_i)$ and its off-diagonal elements are all zero. Recently, [13] showed that the above system of ODEs can be solved by adopting a technique known as adaptive uniformization.

4   Product Construction

Recall that our aim is to compute the maximum probability of the set of paths of CTMDP $\mathcal{C}$ accepted by the DTA $\mathcal{A}$, that is,

$$\sup_{D \in \mathcal{D}_C} \mathrm{Pr}_{s_0, D}(Paths_{s_0}(\mathcal{C} \models \mathcal{A})).$$

In this section, we show that this can be accomplished by computing maximum reachability probabilities in $\mathcal{C} \otimes \mathcal{G}(\mathcal{A})$, i.e., the product between $\mathcal{C}$ and the region graph of $\mathcal{A}$.

Definition 9 (Product). *The product of CTMDP $\mathcal{C} = (S, s_0, Act, \mathbf{P}, r, L)$ and DTA region graph $\mathcal{G}(\mathcal{A}) = (\Sigma, W, w_0, W_F, \dashrightarrow)$, denoted $\mathcal{C} \otimes \mathcal{G}(\mathcal{A})$, is the tuple $(Act, V, v_0, V_F, \Lambda, \hookrightarrow)$ with $V = S \times W$, $v_0 = (s_0, w_0)$, $V_F = S \times W_F$, and*

- $\hookrightarrow \subseteq V \times ((Act \times [0,1] \times \{\varnothing, \{x\}\}) \uplus \{\delta\}) \times V$ *is the smallest relation s.t.:*
  - $(s, w) \overset{\delta}{\hookrightarrow} (s, w')$ *iff* $w \overset{\delta}{\dashrightarrow} w'$; *and*
  - $(s, w) \overset{\alpha,p,X}{\hookrightarrow} (s', w')$ *iff* $p = \mathbf{P}^\alpha(s, s')$ *with* $p > 0$, *and* $w \overset{L(s),X}{\dashrightarrow} w'$.
- $\Lambda : V \times Act \to \mathbb{R}_{\geqslant 0}$ *is the* exit rate function *where:*

$$\Lambda(s, w, \alpha) = \begin{cases} r^\alpha(s) & \text{if } (s, w) \overset{\alpha,p,X}{\hookrightarrow} (s', w') \text{ for some } (s', w') \in V \\ 0 & \text{otherwise.} \end{cases}$$

Example 3. The product of the CTMDP in Fig. 2 and the DTA region graph in Fig. 3(b) is depicted in Fig. 4.

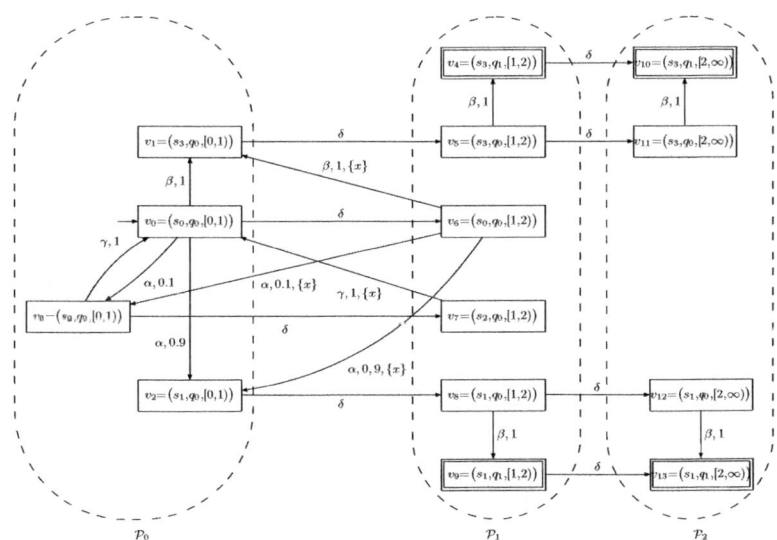

Fig. 4. The product of CTMC and DTA region graph (the reachable part)

Vertex v in the product $\mathcal{C} \otimes \mathcal{G}(\mathcal{A})$ is a triple consisting of a CTMDP state s, a DTA state q and a region Θ. Let $v|_i$ denote the i-th component of the triple v; e.g., if $v = (s, q, \Theta)$, then $v|_2 = q$. Furthermore, let $Act(v)$ be the set of enabled actions in vertex v, i.e., $Act(v) = Act(v|_1)$. Edges of the form $v \overset{\delta}{\hookrightarrow} v'$ are called *delay edges*, whereas those of the form $v \overset{\alpha,p,X}{\hookrightarrow} v'$ are called *Markovian edges*.

The product $\mathcal{C} \otimes \mathcal{G}(\mathcal{A})$ is essentially a (simple variant of a) PDDP, i.e., a decision variant of a PDP. The notions of paths and schedulers for a PDDP can be defined in a similar way as for CTMDP in Section 2; we do not dwell upon the details here. For the sake of brevity, let $\mathcal{P} = \mathcal{C} \otimes \mathcal{G}(\mathcal{A})$. In the sequel, let $\mathcal{D}_{\mathcal{P}}$ denote the set of all schedulers of the product $\mathcal{P}$. A scheduler $D \in \mathcal{D}_{\mathcal{P}}$ on the product $\mathcal{P}$ induces a PDP which is equipped with a probability measure $\mathrm{Pr}_{v_0,D}^{\mathcal{P}}$ over its infinite paths in a standard way; for details, we refer to [20]. Let $Paths_{v_0}(\Diamond V_F)$ denote the set of timed paths in $\mathcal{P}$ that reach some vertex in V_F from vertex v_0 starting with clock-valuation $0 \in \Theta_0$.

Lemma 1. *Given CTMDP $\mathcal{C}$ and DTA $\mathcal{A}$,*

$$\sup_{D \in \mathcal{D}_{\mathcal{C}}} \mathrm{Pr}_{s_0,D}(Paths_{s_0}(\mathcal{C} \models \mathcal{A})) \;=\; \sup_{D \in \mathcal{D}_{\mathcal{P}}} \mathrm{Pr}_{v_0,D}^{\mathcal{P}}(Paths_{v_0}(\Diamond V_F)).$$

Proof (Sketch). We first show that there is a one-to-one correspondence between $Paths_{s_0}(\mathcal{C} \models \mathcal{A})$ and $Paths_{s_0}^{\mathcal{P}}(\Diamond V_F)$.

($\Longrightarrow$) Let $\pi = s_0 \xrightarrow{\alpha_0,t_0} s_1 \cdots s_{n-1} \xrightarrow{\alpha_{n-1},t_{n-1}} s_n$ with $\pi \in Paths_{s_0}(\mathcal{C} \models \mathcal{A})$. We prove that there exists a path $\rho \in Paths_{s_0}^{\mathcal{P}}(\Diamond V_F)$ with $\pi = \rho|_1$. We have $x_0 = 0$ and for $0 \le i < n$, $x_i + t_i \models g_i$ with $x_{i+1} = (x_i + t_1)[X_i := 0]$. Here x_i is the clock valuation in $\mathcal{A}$ on entering state s_i in $\mathcal{C}$. We now construct a timed path θ in $\mathcal{A}$ from π such that $\theta = q_0 \xrightarrow{L(s_0),t_0} q_1 \cdots q_{n-1} \xrightarrow{L(s_{n-1}),t_{n-1}} q_n$, where the clock valuation on entering s_i and q_i coincides. Combining timed paths π and θ yields:

$$\rho = \langle s_0, q_0 \rangle \xrightarrow{t_0} \langle s_1, q_1 \rangle \cdots \langle s_{n-1}, q_{n-1} \rangle \xrightarrow{t_{n-1}} \langle s_n, q_n \rangle,$$

where $\langle s_n, q_n \rangle \in Loc_F$. It follows that $\rho \in Paths_{s_0}^{\mathcal{P}}(\Diamond V_F)$ and $\pi = \rho|_1$.

($\Longleftarrow$) Let $\rho = \langle s_0, q_0 \rangle \xrightarrow{\alpha_0,t_0} \cdots \xrightarrow{\alpha_{n-1},t_{n-1}} \langle s_n, q_n \rangle \in Paths_{s_0}^{\mathcal{P}}(\Diamond V_F)$. We prove that $\rho|_1 \in Paths_{s_0}(\mathcal{C} \models \mathcal{A})$. Clearly, we have that $\langle s_n, q_n \rangle \in Loc_F$, $x_0 = 0$, and for $0 \le i < n$, $x_i + t_i \models g_i$ and $x_{i+1} = (x_i + t_i)[X_i := 0]$, where x_i is the clock valuation when entering location $\langle s_i, q_i \rangle$. It then directly follows that $q_n \in Q_F$ and $\rho|_1 \in Paths_{s_0}(\mathcal{C} \models \mathcal{A})$, given the entering clock valuation x_i of state s_i.

Following this path correspondence, it is not difficult to show that for each scheduler D of the CTMDP $\mathcal{C}$, one can construct a scheduler D' of the product $\mathcal{P}$, such that the induced probability measures $\mathrm{Pr}_{s_0,D}$ and $\mathrm{Pr}_{v_0,D'}$ on the corresponding paths coincide. The detailed arguments are quite similar to (and actually simpler than) those of [16, Thm. 4.3]. $\square$

Thanks to this lemma, it suffices to concentrate on determining maximum reachability probabilities in the product $\mathcal{P} = \mathcal{C} \otimes \mathcal{G}(\mathcal{A})$. It is well-known [20] that in this case, memoryless schedulers suffice. This basically stems from the fact that the elapsed time is "encoded" in the state space of the product $\mathcal{P}$; recall that any vertex in $\mathcal{P}$ is of the form (s, q, Θ) where Θ is the current region of the single clock x. Namely, the decision solely depends on (s, q, Θ, x) where (s, q, Θ) is a vertex in $\mathcal{P}$, and x is the actual clock value.

Now we introduce the Bellman equation on the product $\mathcal{P}$ that characterizes $\sup_{D \in \mathcal{D}_{\mathcal{P}}} \mathrm{Pr}_{v_0,D}^{\mathcal{P}}(Paths_{v_0}(\Diamond V_F))$. The following auxiliary definition turns out to be helpful. For a vertex $v \in V$ with $v\lfloor_3 = \Theta_i$ and clock value x, we define the boundary function $\flat(v, x) = c_{i+1} - x$ if $i < m$; and ∞ if $i = m$. Intuitively, $\flat(v, x)$ is the minimum time (if it exists) to "hit" the boundary of the region of vertex v starting from a clock value x. Let $\Psi(v, x)$ be the maximum probability to reach V_F starting from vertex v given clock value x. Then it follows from [20] that $\Psi(v, x) = 1$ if $v \in V_F$, and otherwise:

$$\Psi(v, x) = \max_{\alpha \in Act(v)} \left\{ \sum_{v \overset{\alpha, p, X}{\hookrightarrow} v'} \int_0^{\flat(v,x)} \underbrace{\Lambda^\alpha(v) \cdot e^{-\Lambda^\alpha(v) \cdot \tau} \cdot p \cdot \Psi(v', (x+\tau)[X := 0])}_{(\star)} \, d\tau \right.$$

$$\left. + \sum_{v \overset{\delta}{\hookrightarrow} v'} \underbrace{e^{-\Lambda^\alpha(v) \cdot \flat(v,x)} \cdot \Psi(v', x + \flat(v, x))}_{(\star\star)} \right\}, \tag{2}$$

The term $(\star)$ represents the probability to take the Markovian edge $v \overset{\alpha, p, X}{\hookrightarrow} v'$ while the term $(\star\star)$ denotes the probability to take the delay edge $v \overset{\delta}{\hookrightarrow} v'$. (Note that there is only a single such delay edge, i.e., the second summation ranges over a single delay edge.)

Theorem 1. *For* $\mathcal{P} = (Act, V, v_0, V_F, \Lambda, \hookrightarrow)$ *we have:*

$$\Psi(v_0, \mathbf{0}) = \sup_{D \in \mathcal{D}_\mathcal{P}} \Pr^\mathcal{P}_{v_0, D}(Paths_{v_0}(\Diamond V_F)).$$

Together with Lemma 1, we thus conclude that our problem—determining the maximum probability that CTMDP $\mathcal{C}$ satisfies the DTA specification $\mathcal{A}$— reduces to determining $\Psi(v_0, \mathbf{0})$ for the Bellman equation (2) on the product $\mathcal{P} = \mathcal{C} \otimes \mathcal{G}(\mathcal{A})$.

5   Reduction to a Linear Programming Problem

In this section, we show that the solution $\Psi(v_0, \mathbf{0})$ of the Bellman equation (2) coincides with the solution of an LP problem whose coefficients are maximum timed reachability probabilities in a set of CTMDPs that are obtained by a graph decomposition of the product $\mathcal{P} = \mathcal{C} \otimes \mathcal{G}(\mathcal{A})$. Let us first define the graph decomposition of the product $\mathcal{P}$. The operational intuition can best be explained by means of our running example, see Fig. 4. The idea is to group all vertices with the same region, i.e., we group the vertices in a column-wise manner. In the example this yields three sub-graphs $\mathcal{P}_0$ through $\mathcal{P}_2$. A delay in $\mathcal{P}_i$ (with $i = 0, 1$) yields a vertex in $\mathcal{P}_{i+1}$, taking an edge in the DTA in which clock x is unaffected (i.e., not reset) yields a vertex in $\mathcal{P}_i$ (for all i), whereas in case clock x is reset, a vertex in $\mathcal{P}_0$ is obtained. This is formalized below as follows.

Definition 10 (Graph decomposition). *The* graph decomposition *of* $\mathcal{P} = (Act, V, v_0, V_F, \Lambda, \hookrightarrow)$ *yields the set of graphs* $\{\mathcal{P}_i \mid 0 \leq i \leq m\}$ *where* $\mathcal{P}_i = (Act, V_i, V_{F_i}, \Lambda_i, \hookrightarrow_i)$ *with:*

– $V_i = \{(s, q, \Theta_i) \in V\}$ and $V_{F_i} = V_i \cap V_F$,
– $\Lambda_i^\alpha(v) = \Lambda^\alpha(v)$, for $v \in V_i$, and
– $\hookrightarrow_i = \left(\bigcup_{\alpha \in Act} \{M_i^\alpha \cup B_i^\alpha\}\right) \cup F_i$ where:
  • M_i^α is the set of Markovian edges (without reset) within $\mathcal{P}_i$ labeled by α,
  • B_i^α (backward) is the set of Markovian edges (with reset) from $\mathcal{P}_i$ to $\mathcal{P}_0$,
  • F_i (forward) is the set of delay edges from the vertices in $\mathcal{P}_i$ to $\mathcal{P}_{i+1}$.

As the graph $\mathcal{P}_m$ only involves unbounded regions, it has no outgoing delay transitions.

Example 4. The product $\mathcal{P}$ in Fig. 4 is decomposed into the graphs $\mathcal{P}_0, \mathcal{P}_1, \mathcal{P}_2$ as indicated by the dashed ovals. For $\mathcal{P}_1$, e.g., we have $M_1^\beta = \{v_5 \hookrightarrow v_4, v_8 \hookrightarrow v_9\}$; $B_1^\alpha = \{v_6 \hookrightarrow v_3, v_6 \hookrightarrow v_2\}$, $B_1^\beta = \{v_6 \hookrightarrow v_1\}$, and $B_1^\gamma = \{v_7 \hookrightarrow v_0\}$. Its delay transitions are $F_1 = \{v_4 \hookrightarrow v_{10}, v_5 \hookrightarrow v_{11}, v_8 \hookrightarrow v_{12}, v_9 \hookrightarrow v_{13}\}$.

For graph $\mathcal{P}_i$ $(0 \le i \le m)$ with $|V_i| = k_i$, define the probability vector

$$\vec{U}_i(x) = [u_i^1(x), \ldots, u_i^{k_i}(x)]^\mathsf{T} \in \mathbb{R}(x)^{k_i \times 1},$$

where $u_i^j(x)$ is the maximum probability to go from vertex $v_i^j \in V_i$ to some vertex in the goal set V_F (in $\mathcal{M}$) at time point x. Our aim is to determine $\vec{U}_0(0)$. In the sequel, we aim to establish a relationship between $\vec{U}_i(0)$ and $\vec{U}_j(0)$ for $i \ne j$. To that end, we distinguish two cases:

CASE $0 \le i < m$. We first introduce some definitions.

– $\mathbf{P}_i^{\alpha, \mathsf{M}} \in [0, 1]^{k_i \times k_i}$ and $\mathbf{P}_i^{\alpha, \mathsf{B}} \in [0, 1]^{k_i \times k_0}$ are probability transition matrices for Markovian and backward transitions respectively, parameterized by action α. For $\alpha \in Act(v)$, let $\mathbf{P}_i^{\alpha, \mathsf{M}}[v, v'] = p$, if $v \overset{\alpha, p, \varnothing}{\hookrightarrow} v'$; and 0 otherwise. Similarly $\mathbf{P}_i^{\alpha, \mathsf{B}}[v, v'] = p$ if $v \overset{\alpha, p, \{x\}}{\hookrightarrow} v'$; and 0 otherwise. Moreover, let $\mathbf{P}_i^\alpha = \left(\mathbf{P}_i^{\alpha, \mathsf{M}} | \mathbf{P}_i^{\alpha, \mathsf{B}}\right)$. Note that $\mathbf{P}_i^\alpha$ is a stochastic matrix, as:

$$\sum_{v'} \mathbf{P}_i^{\alpha, \mathsf{M}}[v, v'] + \sum_{v''} \mathbf{P}_i^{\alpha, \mathsf{B}}[v, v''] = 1.$$

– $\mathbf{D}_i^\alpha(x) \in \mathbb{R}^{k_i \times k_i}$ is the delay probability matrix, i.e., for any $1 \le j \le k_i$, $\mathbf{D}_i^\alpha(x)[j, j] = e^{-r^\alpha(v_i^j)x}$. Its off-diagonal elements are zero;
– $\mathbf{E}_i^\alpha \in \mathbb{R}^{k_i \times k_i}$ is the exit rate matrix, i.e., for any $1 \le j \le k_i$, $\mathbf{E}_i^\alpha[j, j] = r^\alpha(v_i^j)$. Its off-diagonal elements are zero;
– $\mathbf{M}_i^\alpha(x) = \mathbf{E}_i^\alpha \cdot \mathbf{D}_i^\alpha(x) \cdot \mathbf{P}_i^{\alpha, \mathsf{M}} \in \mathbb{R}^{k_i \times k_i}$ is the probability density matrix for Markovian transitions inside $\mathcal{P}_i$. Namely, $\mathbf{M}_i^\alpha(x)[j, j']$ indicates the pdf to take the α-labelled Markovian edge without reset from the j-th vertex to the j'-th vertex in $\mathcal{P}_i$;
– $\mathbf{B}_i^\alpha(x) = \mathbf{E}_i^\alpha \cdot \mathbf{D}_i^\alpha(x) \cdot \mathbf{P}_i^{\alpha, \mathsf{B}} \in \mathbb{R}^{k_i \times k_0}$ is the probability density matrix for the reset edges B_i^α. Namely, $\mathbf{B}_i^\alpha(x)[j, j']$ indicates the pdf to take the Markovian edge with reset from the j-th vertex in $\mathcal{P}_i$ to the j'-th vertex in $\mathcal{P}_0$;

- $\mathbf{F}_i \in \mathbb{R}^{k_i \times k_{i+1}}$ is the incidence matrix for delay edges F_i. Thus, $\mathbf{F}_i[j, j'] = 1$ iff there is a delay edge from the j-th vertex in $\mathcal{P}_i$ to the j'-th vertex in $\mathcal{P}_{i+1}$.

Example 5 (Continuing Example 4). According to the definitions, we have the following matrices for $\mathcal{P}_0$ and $\mathcal{P}_1$. Let r_i^α be a shorthand of the exit rate $r^\alpha(s_i)$:

$$\mathbf{M}_0^\alpha(x) = \underbrace{\begin{pmatrix} r_0^\alpha & 0 & 0 & 0 \\ 0 & 0 & 0 & 0 \\ 0 & 0 & 0 & 0 \\ 0 & 0 & 0 & 0 \end{pmatrix}}_{\mathbf{E}_0} \underbrace{\begin{pmatrix} e^{-r_0^\alpha x} & 0 & 0 & 0 \\ 0 & 1 & 0 & 0 \\ 0 & 0 & 1 & 0 \\ 0 & 0 & 0 & 1 \end{pmatrix}}_{\mathbf{D}_0(x)} \underbrace{\begin{pmatrix} 0 & 0 & 0.9 & 0.1 \\ 0 & 0 & 0 & 0 \\ 0 & 0 & 0 & 0 \\ 0 & 0 & 0 & 0 \end{pmatrix}}_{\mathbf{P}_0^{\alpha,M}} = \begin{pmatrix} 0 & 0 & 0.9 r_0^\alpha e^{-r_0^\alpha x} & 0.1 r_0^\alpha e^{-r_0^\alpha x} \\ 0 & 0 & 0 & 0 \\ 0 & 0 & 0 & 0 \\ 0 & 0 & 0 & 0 \end{pmatrix}$$

Similarly,

$$\mathbf{B}_1^\beta(x) = \begin{pmatrix} 0 & 0 & 0 & 0 \\ 0 & 0 & 0 & 0 \\ 0 & 1 \cdot e^{-r_0^\beta x} & 0 & 0 \\ 0 & 0 & 0 & 0 \\ 0 & 0 & 0 & 0 \\ 0 & 0 & 0 & 0 \end{pmatrix} \quad \text{and} \quad \mathbf{F}_0 = \begin{pmatrix} 0 & 0 & 1 & 0 & 0 & 0 \\ 1 & 0 & 0 & 0 & 0 & 0 \\ 0 & 0 & 0 & 0 & 1 & 0 \\ 0 & 0 & 0 & 1 & 0 & 0 \end{pmatrix}$$

By instantiating (2), we obtain the following for $0 \le i < m$:

$$\vec{U}_i(x) = \max_{\alpha \in Act} \left\{ \underbrace{\int_0^{\Delta c_i - x} \mathbf{M}_i^\alpha(\tau) \cdot \vec{U}_i(x+\tau) d\tau + \int_0^{\Delta c_i - x} \mathbf{B}_i^\alpha(\tau) d\tau \cdot \vec{U}_0(0)}_{(\star)} \right. \tag{3}$$
$$\left. {}_{} \qquad\qquad\qquad (\star\star) \right.$$
$$\left. + \mathbf{D}_i^\alpha(\Delta c_i - x) \cdot \mathbf{F}_i \vec{U}_{i+1}(0) \right\},$$

Let us explain the above equation. First of all, recall that $\flat(v, x) = \Delta c_i - x$ for each state $v \in V_i$ with $i < m$. Term $(\star)$ (resp. $(\star\star)$) reflects the case where clock x is not reset (resp. is reset and returned to $\mathcal{P}_0$). Note that $\mathbf{M}_i^\alpha(\tau)$ and $\mathbf{B}_i^\alpha(\tau)$ are the matrix forms of the density function $(\star)$ in (2). The matrix $\mathbf{D}_i^\alpha(\Delta c_i - x)$ indicates the probability to delay until the "end" of region i, and $\mathbf{F}_i \cdot \vec{U}_{i+1}(0)$ denotes the probability to continue in $\mathcal{P}_{i+1}$ (at relative time point 0), and $\mathbf{D}_i^\alpha \cdot (\Delta c_i - x) \cdot \mathbf{F}_i$ is the matrix form of the term $(\star\star)$ in (2).

CASE $i = m$. In this case, $\vec{U}_m(x)$ is simplified as follows:

$$\vec{U}_m(x) = \max_{\alpha \in Act} \left\{ \int_0^\infty \widehat{\mathbf{M}}_m^\alpha(\tau) \cdot \vec{U}_m(x+\tau) d\tau + \tilde{\mathbf{1}}_F + \int_0^\infty \mathbf{B}_m^\alpha(\tau) d\tau \cdot \vec{U}_0(0) \right\}, \tag{4}$$

where $\widehat{\mathbf{M}}_m^\alpha(\tau)[v, \cdot] = \mathbf{M}_m^\alpha(\tau)[v, \cdot]$ for $v \notin V_F$, 0 otherwise. $\tilde{\mathbf{1}}_F$ is a characteristic vector such that $\mathbf{1}_F[v] = 1$ iff $v \in V_F$.

Our remaining task now is to solve the system of integral equations given by equations (3) and (4). First observe that, due to the fact that $\mathcal{P}_i$ only contains

Markovian edges, the struc-
ture (V_i, Λ_i, M_i) forms a CT-
MDP, referred to as C_i. For
each $\mathcal{P}_i$, we define the *augmented* CTMDP $C_i^\star$ with
state space $V_i \cup V_0$ such that
all V_0-vertices are made ab-
sorbing (i.e., their outgoing
edges are replaced by a self-
loop) and all edges connecting
V_i to V_0 are kept. The aug-

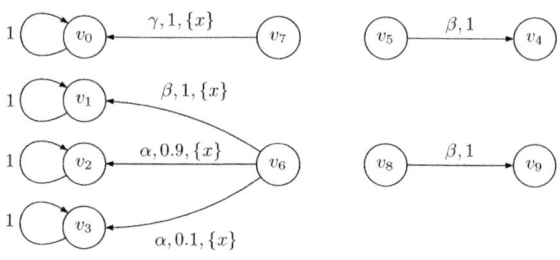

Fig. 5. Augmented CTMDP $C_1^\star$

mented CTMDP $C_1^\star$ for $\mathcal{P}_1$ in Fig. 4 is shown in Fig. 5.

By instantiating (2), we have the following equation (in the matrix form) for
the transition probability:

$$\mathbf{\Pi}(x) \;=\; \max_{\alpha \in Act} \left\{ \int_0^x \widetilde{\mathbf{M}}^\alpha(\tau)\cdot\mathbf{\Pi}(x-\tau)\,d\tau \right\} + \mathbf{D}^\alpha(x), \tag{5}$$

where $\widetilde{\mathbf{M}}^\alpha(\tau)[v,v'] = r^\alpha(v)\cdot e^{-r^\alpha(v)\cdot\tau}\cdot p$ if there is a Markovian edge $v \overset{\alpha,p,\varnothing}{\hookrightarrow} v'$; 0
otherwise. In fact, the characterization of $\Psi(s,x)$ in Section 4 is an equivalent
formulation of Eq.(5). For augmented CTMDP $C_i^\star$, $\widetilde{\mathbf{M}}^\alpha(\tau)$ we have:

$$\widetilde{\mathbf{M}}^\alpha(\tau) = \left(\begin{array}{c|c} \mathbf{M}_i^\alpha(\tau) & \mathbf{B}_i^\alpha(\tau) \\ \hline \mathbf{0} & \mathbf{I} \end{array} \right),$$

where $\mathbf{0} \in \mathbb{R}^{k_0 \times k_i}$ is the matrix with all 0's and $\mathbf{I} \in \mathbb{R}^{k_0 \times k_0}$ is the identity matrix.

Now given any CTMDP C_i (resp. augmented CTMDP $C_i^\star$) corresponding to
$\mathcal{P}_i$, we obtain Eq.(5), and write its solution as $\mathbf{\Pi}_i(x)$ (resp. $\mathbf{\Pi}_i^\star(x)$). We then
define $\bar{\mathbf{\Pi}}_i^\star \in \mathbb{R}^{k_i \times k_0}$ for an augmented CTMDP $C_i^\star$ to be part of $\mathbf{\Pi}_i^\star$, where $\bar{\mathbf{\Pi}}_i^\star$
only keeps the probabilities starting from V_i and ending in V_0. As a matter of
fact,

$$\mathbf{\Pi}_i^\star(x) = \left(\begin{array}{c|c} \mathbf{\Pi}_i(x) & \bar{\mathbf{\Pi}}_i^\star(x) \\ \hline \mathbf{0} & \mathbf{I} \end{array} \right).$$

The following theorem is the key result of this section. Its proof is technically
involved and is given in the Appendix.

Theorem 2. *For sub-graph $\mathcal{P}_i$ of $\mathcal{P}$, it holds:*

$$\vec{U}_i(0) \;=\; \mathbf{\Pi}_i(\Delta c_i) \cdot \mathbf{F}_i \cdot \vec{U}_{i+1}(0) \;+\; \bar{\mathbf{\Pi}}_i^\star(\Delta c_i) \cdot \vec{U}_0(0), \qquad \textit{if } 0 \le i < m \tag{6}$$

where $\mathbf{\Pi}_i(\cdot)$ and $\bar{\mathbf{\Pi}}_i^\star(\cdot)$ are defined on CTMDP C_i and (augmented) $C_i^\star$ as above.

$$\vec{U}_m(0) \;=\; \max_{\alpha \in Act} \left\{ \widehat{\mathbf{P}}_m^\alpha \cdot \vec{U}_m(0) + \vec{1}_F + \widehat{\mathbf{B}}_m^\alpha \cdot \vec{U}_0(0) \right\}, \qquad \textit{if } i = m \tag{7}$$

with $\widehat{\mathbf{P}}_m^\alpha(v,v') = \mathbf{P}_m^\alpha(v,v')$ if $v \notin V_{F_m}$; 0 otherwise, and $\widehat{\mathbf{B}}_m^\alpha = \int_0^\infty \mathbf{B}_m^\alpha(\tau)\,d\tau$.

Recall that we intend to solve the system of integral equations given by the equations (3) and (4) so as to obtain the vectors $\vec{U}_i(0)$ for $0 \le i \le m$. Theorem 2 entails that instead of accomplishing this directly, one alternatively can exploit equations 6 and 7, where $\vec{U}_i(0)$ $(0 \le i \le m)$ can be regarded as a family of variables and the coefficients $\Pi_i(\cdot)$ can be obtained by computing the corresponding maximum timed reachability probabilities of CTMDPs $\mathcal{C}_i^\star$. It is not difficult to see that the set of equations in Theorem 2 can be easily reduced to an LP problem, see, e.g., [9].

6   Conclusion

We showed that the verification of CTMDPs against 1-clock DTA objectives can be reduced to solving an LP problem whose coefficients are extremum timed reachability probabilities in CTMDPs. This extends the class of timed reachability properties to an interesting and practically relevant set of properties. The main ingredients of our approach are a region graph and a product construction, computing timed reachability probabilities in a set of CTMDPs, and finally solving an LP problem. The availability of the first practical implementations for timed reachability of CTMDPs paves the way to a realization of our approach in a prototypical tool. Like in [7], our approach facilitates optimizations such as parallelization and bisimulation minimization. Such implementation and experimentation is essential to show the practical feasibility of our approach and is left for further work.

Another interesting research direction is to consider other acceptance criteria for DTA, such as Muller acceptance. We claim that this can basically be done along the lines of [16] for CTMCs; the main technical difficulty is that one needs to resort to either finite memory schedulers or randomized schedulers, see e.g. [14].

References

1. Aceto, L., Bouyer, P., Burgueño, A., Larsen, K.G.: The power of reachability testing for timed automata. Theor. Comput. Sci. 300(1-3), 411–475 (2003)
2. Alur, R., Dill, D.L.: A theory of timed automata. Theor. Comput. Sci. 126(2), 183–235 (1994)
3. Baier, C., Cloth, L., Haverkort, B.R., Kuntz, M., Siegle, M.: Model checking Markov chains with actions and state labels. IEEE Trans. Software Eng. 33(4), 209–224 (2007)
4. Baier, C., Haverkort, B.R., Hermanns, H., Katoen, J.-P.: Model-checking algorithms for continuous-time Markov chains. IEEE Trans. Software Eng. 29(6), 524–541 (2003)
5. Baier, C., Hermanns, H., Katoen, J.-P., Haverkort, B.: Efficient computation of time-bounded reachability probabilities in uniform continuous-time Markov decision processes. Theor. Comput. Sci. 345(1), 2–26 (2005)
6. Baier, C., Kwiatkowska, M.: Model checking for a probabilistic branching time logic with fairness. Distrib. Comput. 11, 125–155 (1998)

7. Barbot, B., Chen, T., Han, T., Katoen, J.-P., Mereacre, A.: Efficient CTMC model checking of linear real-time objectives. In: Abdulla, P.A., Leino, K.R.M. (eds.) TACAS 2011. LNCS, vol. 6605, pp. 128–142. Springer, Heidelberg (2011)
8. Bellman, R.: Dynamic Programming. Princeton University Press, Princeton (1957)
9. Bianco, A., de Alfaro, L.: Model checking of probabilistic and nondeterministic systems. In: Thiagarajan, P.S. (ed.) FSTTCS 1995. LNCS, vol. 1026, pp. 499–513. Springer, Heidelberg (1995)
10. Brázdil, T., Forejt, V., Krcál, J., Kretínský, J., Kucera, A.: Continuous-time stochastic games with time-bounded reachability. In: FSTTCS, pp. 61–72 (2009)
11. Brázdil, T., Krcál, J., Kretínský, J., Kucera, A., Rehák, V.: Stochastic real-time games with qualitative timed automata objectives. In: Gastin, P., Laroussinie, F. (eds.) CONCUR 2010. LNCS, vol. 6269, pp. 207–221. Springer, Heidelberg (2010)
12. Brázdil, T., Krcál, J., Kretínský, J., Kucera, A., Rehák, V.: Measuring performance of continuous-time stochastic processes using timed automata. In: HSCC, pp. 33–42. ACM Press, New York (2011)
13. Buchholz, P., Schulz, I.: Numerical analysis of continuous time Markov decision processes over finite horizons. Computers & OR 38(3), 651–659 (2011)
14. Chatterjee, K., de Alfaro, L., Henzinger, T.A.: Trading memory for randomness. In: Quantitative Evaluation of Systems (QEST), pp. 206–217. IEEE Computer Society, Los Alamitos (2004)
15. Chen, T., Han, T., Katoen, J.-P., Mereacre, A.: Quantitative model checking of continuous-time Markov chains against timed automata specifications. In: LICS, pp. 309–318 (2009)
16. Chen, T., Han, T., Katoen, J.-P., Mereacre, A.: Model checking of continuous-time Markov chains against timed automata specifications. Logical Methods in Computer Science 7(1–2), 1–34 (2011)
17. Chen, T., Han, T., Katoen, J.-P., Mereacre, A.: Reachability probabilities in Markovian timed automata. Technical report, RR-11-02, OUCL (2011)
18. Ciesinski, F., Baier, C.: Liquor: A tool for qualitative and quantitative linear time analysis of reactive systems. In: Quantitative Evaluation of Systems (QEST), pp. 131–132. IEEE Computer Society, Los Alamitos (2006)
19. Courcoubetis, C., Yannakakis, M.: The complexity of probabilistic verification. J. ACM 42(4), 857–907 (1995)
20. Davis, M.H.A.: Markov Models and Optimization. Chapman and Hall, Boca Raton (1993)
21. de Alfaro, L.: How to specify and verify the long-run average behavior of probabilistic systems. In: LICS, pp. 454–465 (1998)
22. Donatelli, S., Haddad, S., Sproston, J.: Model checking timed and stochastic properties with CSL^{TA}. IEEE Trans. Software Eng. 35(2), 224–240 (2009)
23. Guo, X., Hernández-Lerma, O., Prieto-Rumeau, T.: A survey of recent results on continuous-time Markov decision processes. TOP 14(2), 177–257 (2006)
24. Howard, R.A.: Dynamic Programming and Markov Processes. MIT Press, Cambridge (1960)
25. Katoen, J.-P., Zapreev, I., Hahn, E.M., Hermanns, H., Jansen, D.N.: The ins and outs of the probabilistic model checker MRMC. Perf. Ev. 68(2), 90–104 (2011)
26. Knast, R.: Continuous-time probabilistic automata. Information and Control 15(4), 335–352 (1969)
27. Laroussinie, F., Markey, N., Schnoebelen, P.: Model checking timed automata with one or two clocks. In: Gardner, P., Yoshida, N. (eds.) CONCUR 2004. LNCS, vol. 3170, pp. 387–401. Springer, Heidelberg (2004)

28. Martin-Löfs, A.: Optimal control of a continuous-time Markov chain with periodic transition probabilities. Operations Research 15, 872–881 (1967)
29. Miller, B.L.: Finite state continuous time Markov decision processes with a finite planning horizon. SIAM Journal on Control 6(2), 266–280 (1968)
30. Neuhäußer, M.R., Stoelinga, M., Katoen, J.-P.: Delayed nondeterminism in continuous-time Markov decision processes. In: de Alfaro, L. (ed.) FOSSACS 2009. LNCS, vol. 5504, pp. 364–379. Springer, Heidelberg (2009)
31. Neuhäußer, M.R., Zhang, L.: Time-bounded reachability probabilities in continuous-time Markov decision processes. In: Quantitative Evaluation of Systems (QEST), pp. 209–218. IEEE Computer Society, Los Alamitos (2010)
32. Puterman, M.L.: Markov Decision Processes. Wiley, Chichester (1994)
33. Rabe, M., Schewe, S.: Finite optimal control for time-bounded reachability in CTMDPs and continuous-time Markov games. CoRR, abs/1004.4005 (2010)
34. Vardi, M.Y.: Automatic verification of probabilistic concurrent finite-state programs. In: FOCS, pp. 327–338. IEEE Computer Society, Los Alamitos (1985)
35. Wolovick, N., Johr, S.: A characterization of meaningful schedulers for continuous-time Markov decision processes. In: Asarin, E., Bouyer, P. (eds.) FORMATS 2006. LNCS, vol. 4202, pp. 352–367. Springer, Heidelberg (2006)

A Proof of Theorem 2

Theorem 2. For subgraph $\mathcal{P}_i$ of $\mathcal{M}$ with k_i states, it holds:

- For $0 \leq i < m$,

$$\vec{U}_i(0) = \mathbf{\Pi}_i(\Delta c_i) \cdot \mathbf{F}_i \vec{U}_{i+1}(0) + \bar{\mathbf{\Pi}}_i^\star(\Delta c_i) \cdot \vec{U}_0(0), \tag{8}$$

where $\mathbf{\Pi}_i(\Delta c_i)$ and $\bar{\mathbf{\Pi}}_i^\star(\Delta c_i)$ are for CTMDP $\mathcal{C}_i$ and (augmented) $\mathcal{C}_i^\star$, respectively.

- For $i = m$,

$$\vec{U}_m(0) - \max_{\alpha \in Act} \left\{ \widehat{\mathbf{P}}_m^\alpha \cdot \vec{U}_m(0) + \vec{1}_F + \widehat{\mathbf{B}}_m^\alpha \cdot \vec{U}_0(0) \right\}, \tag{9}$$

where $\widehat{\mathbf{P}}_m^\alpha(v, v') = \mathbf{P}_m^\alpha(v, v')$ if $v \notin V_{F_m}$; 0 otherwise, and $\widehat{\mathbf{B}}_m^\alpha = \int_0^\infty \mathbf{B}_m^\alpha(\tau) d\tau$.

Proof. We first deal with the case $i < m$. If in $\mathcal{P}_i$, for some action α there exists some backward edge, namely, for some j, j', $\mathbf{B}_i^\alpha(x)[j, j'] \neq 0$, then we shall consider the *augmented* CTMDP $\mathcal{C}_i^\star$ with $k_i^\star = k_i + k_0$ states. In view of this, the augmented version of the integral equation $\vec{V}_i(x)$ is defined as:

$$\vec{V}_i^\star(x) = \max_{\alpha \in Act} \left\{ \int_0^{\Delta c_i - x} \mathbf{M}_i^{\alpha,\star}(\tau) \cdot \vec{V}_i^\star(x+\tau) d\tau + \mathbf{D}_i^{\alpha,\star}(\Delta c_i - x) \cdot \mathbf{F}_i^\star \cdot \vec{V}_i(0) \right\},$$

where

- $\vec{V}_i^\star(x) = \begin{pmatrix} \vec{V}_i(x) \\ \overline{\vec{V}_i'(x)} \end{pmatrix} \in \mathbb{R}^{k_i^\star \times 1}$, where $\vec{V}_i'(x) \in \mathbb{R}^{k_0 \times 1}$ is the vector representing reachability probabilities for the augmented states in $\mathcal{P}_i$;

$-\mathbf{M}_i^{\alpha,\star}(\tau) = \left(\begin{array}{c|c} \mathbf{M}_i^{\alpha}(\tau) & \mathbf{B}_i^{\alpha}(\tau) \\ \hline \mathbf{0} & \mathbf{0} \end{array}\right) \in \mathbb{R}^{k_i^\star \times k_i^\star}$. The exit rate of augmented states is 0 for all actions.

$-\mathbf{D}_i^{\alpha,\star}(\tau) = \left(\begin{array}{c|c} \mathbf{D}_i^{\alpha}(\tau) & \mathbf{0} \\ \hline \mathbf{0} & \mathbf{I} \end{array}\right) \in \mathbb{R}^{k_i^\star \times k_i^\star}$.

$-\mathbf{F}_i^\star = \left(\mathbf{F}_i' | \mathbf{B}_i'\right) \in \mathbb{R}^{k_i^\star \times (k_{i+1}+k_0)}$ such that

   • $\mathbf{F}_i' = \left(\begin{array}{c} \mathbf{F}_i \\ \hline \mathbf{0} \end{array}\right) \in \mathbb{R}^{k_i^\star \times k_{i+1}}$ is the incidence matrix for delay edges and

   • $\mathbf{B}_i' = \left(\begin{array}{c} \mathbf{0} \\ \hline \mathbf{I} \end{array}\right) \in \mathbb{R}^{k_i^\star \times k_0}$.

$-\vec{V}_i(0) = \left(\begin{array}{c} \vec{U}_{i+1}(0) \\ \hline \vec{U}_0(0) \end{array}\right) \in \mathbb{R}^{(k_{i+1}+k_0)\times 1}$.

In the sequel, we prove two claims:

Claim 1. For each $0 \leq j \leq k_i$, $\vec{U}_i[j] = \vec{V}_i^\star[j]$.

Proof of Claim 1. According to the definition, we have that

$$\vec{V}_i^\star(x) = \max_{\alpha \in Act} \left\{ \int_0^{\Delta c_i - x} \left(\begin{array}{c|c} \mathbf{M}_i^{\alpha}(\tau) & \mathbf{B}_i^{\alpha}(\tau) \\ \hline \mathbf{0} & \mathbf{0} \end{array}\right) \cdot \vec{V}_i^\star(x+\tau)d\tau \right.$$
$$\left. + \left(\begin{array}{c|c} \mathbf{D}_i^{\alpha}(\Delta c_i - x) & \mathbf{0} \\ \hline \mathbf{0} & \mathbf{I} \end{array}\right) \cdot \left(\begin{array}{c|c} \mathbf{F}_i & \mathbf{0} \\ \hline \mathbf{0} & \mathbf{I} \end{array}\right) \cdot \left(\begin{array}{c} \vec{U}_{i+1}(0) \\ \hline \vec{U}_0(0) \end{array}\right) \right\} .$$

It follows immediately that $\vec{V}_i'(x) = \vec{U}_0(0)$. For $\vec{V}_i(x)$, we have that

$$\vec{V}_i(x)$$
$$= \max_{\alpha \in Act} \left\{ \int_0^{\Delta c_i - x} \mathbf{M}_i^{\alpha}(\tau)\vec{V}_i(x+\tau)d\tau + \int_0^{\Delta c_i - x} \mathbf{B}_i^{\alpha}(\tau)\vec{V}_i'(x+\tau)d\tau \right.$$
$$\left. + \mathbf{D}_i^{\alpha}(\Delta c_i - x)\cdot\mathbf{F}_i\cdot\vec{U}_{i+1}(0)\right\}$$
$$= \max_{\alpha \in Act} \left\{ \int_0^{\Delta c_i - x} \mathbf{M}_i^{\alpha}(\tau)\vec{V}_i(x+\tau)d\tau + \int_0^{\Delta c_i - x} \mathbf{B}_i^{\alpha}(\tau)d\tau \cdot \vec{U}_0(0) \right.$$
$$\left. + \mathbf{D}_i^{\alpha}(\Delta c_i - x) \cdot \mathbf{F}_i \cdot \vec{U}_{i+1}(0)\right\}$$
$$= \vec{U}_i(x) .$$

♣

Claim 2.
$$\vec{V}_i^\star(x) = \mathbf{\Pi}_i^\star(\Delta c_i - x) \cdot \mathbf{F}_i^\star \vec{V}_i(0) ,$$

where
$$\mathbf{\Pi}_i^\star(x) = \max_{\alpha \in Act} \left\{ \int_0^x \mathbf{M}_i^{\alpha,\star}(\tau)\mathbf{\Pi}_i^\star(x - \tau)d\tau + \mathbf{D}_i^{\alpha,\star}(x)\right\} .$$

Proof of Claim 2. Standard arguments yield that the optimal probability corresponds to the least fixpoint of a functional and can be computed iteratively. Let $c_{i,x} = \Delta c_i - x$. We define

$$\vec{V}_i^{\star,(0)}(x) = \vec{0}$$
$$\vec{V}_i^{\star,(j+1)}(x) = \max_{\alpha \in Act} \left\{ \int_0^{c_{i,x}} \mathbf{M}_i^{\alpha}(\tau) \vec{V}_i^{\star,(j)}(x+\tau) d\tau + \mathbf{D}_i^{\alpha,\star}(c_{i,x}) \cdot \mathbf{F}_i^{\star} \vec{V}_i(0) \right\} .$$

and

$$\mathbf{\Pi}_i^{\star,(0)}(c_{i,x}) = \mathbf{0}$$
$$\mathbf{\Pi}_i^{\star,(j+1)}(c_{i,x}) = \max_{\alpha \in Act} \left\{ \int_0^{c_{i,x}} \mathbf{M}_i^{\star}(\tau) \mathbf{\Pi}_i^{\star,(j)}(c_{i,x}-\tau) d\tau + \mathbf{D}_i^{\alpha,\star}(c_{i,x}) \right\} .$$

By induction on j, we prove the following relation:

$$\vec{V}_i^{\star,(j)}(x) = \mathbf{\Pi}_i^{\star,(j)}(c_{i,x}) \cdot \mathbf{F}_i \vec{V}_i(0) .$$

- Base case. $\vec{V}_i^{\star,(0)}(x) = \vec{0}$ and $\mathbf{\Pi}_i^{\star,(0)}(c_{i,x}) = \mathbf{0}$.
- Induction hypothesis.

$$\vec{V}_i^{\star,(j)}(x) = \mathbf{\Pi}_i^{\star,(j)}(c_{i,x}) \cdot \mathbf{F}_i^{\star} \vec{\hat{U}}_i(0) .$$

- Induction step. We have that

$$\vec{V}_i^{\star,(j+1)}(x) = \max_{\alpha \in Act} \left\{ \int_0^{c_{i,x}} \mathbf{M}_i^{\star,\alpha}(\tau) \vec{V}_i^{\star,(j)}(x+\tau) d\tau + \mathbf{D}_i^{\alpha,\star}(c_{i,x}) \cdot \mathbf{F}_i^{\star} \vec{\hat{U}}_i(0) \right\} .$$

It follows that

$$\vec{V}_i^{\star,(j+1)}(x)$$
$$= \max_{\alpha \in Act} \left\{ \int_0^{c_{i,x}} \mathbf{M}_i^{\star,\alpha}(\cdot) \vec{V}_i^{\star,(j)}(x+\tau) d\tau \mid \mathbf{D}_i^{\alpha,\star}(c_{i,x}) \cdot \mathbf{F}_i^{\star} \vec{V}_i(0) \right\}$$
$$\stackrel{\text{I.H.}}{=} \max_{\alpha \in Act} \left\{ \int_0^{c_{i,x}} \mathbf{M}_i^{\star,\alpha}(\tau) \cdot \mathbf{\Pi}_i^{\star,(j)}(c_{i,x}-\tau) \cdot \mathbf{F}_i^{\star} \vec{V}_i(0) d\tau + \mathbf{D}_i^{\alpha,\star}(c_{i,x}) \cdot \mathbf{F}_i^{\star} \vec{V}_i(0) \right\}$$
$$= \max_{\alpha \in Act} \left\{ \left(\int_0^{c_{i,x}} \mathbf{M}_i^{\star,\alpha}(\tau) \mathbf{\Pi}_i^{\star,(j)}(c_{i,x}-\tau) d\tau + \mathbf{D}_i^{\star}(c_{i,x}) \right) \cdot \mathbf{F}_i^{\star} \vec{V}_i(0) \right\}$$
$$= \max_{\alpha \in Act} \left\{ \int_0^{c_{i,x}} \mathbf{M}_i^{\star,\alpha}(\tau) \mathbf{\Pi}_i^{\star,(j)}(c_{i,x}-\tau) d\tau + \mathbf{D}_i^{\alpha} \star (c_{i,x}) \right\} \cdot \mathbf{F}_i^{\star} \vec{V}_i(0)$$
$$= \mathbf{\Pi}_i^{\alpha,(j+1)}(c_{i,x}) \cdot \mathbf{F}_i^{\star} \vec{V}_i(0) .$$

Clearly,

$$\mathbf{\Pi}_i^{\star}(c_{i,x}) = \lim_{j \to \infty} \mathbf{\Pi}_i^{\star,(j)}(c_{i,x}) ,$$

and

$$\vec{V}_i^{\star}(x) = \lim_{j \to \infty} \vec{V}_i^{\star,(j)}(x) .$$

It follows the conclusion.   ♣

We now proceed with the main proof. Let $x = 0$ and we obtain

$$\vec{V}_i^\star(0) = \mathbf{\Pi}_i^\star(c_{i,0}) \cdot \mathbf{F}_i \vec{V}_i(0) \ .$$

We can also write the above relation for $x = 0$ as:

$$\begin{pmatrix} \vec{V}_i(0) \\ \vec{V}_i'(0) \end{pmatrix} = \mathbf{\Pi}_i^\star(\Delta c_i) \left(\mathbf{F}_i' | \mathbf{B}_i' \right) \begin{pmatrix} \vec{U}_{i+1}(0) \\ \vec{U}_0(0) \end{pmatrix}$$

$$= \left(\begin{array}{c|c} \mathbf{\Pi}_i(\Delta c_i) & \bar{\mathbf{\Pi}}_i^\star(\Delta c_i) \\ \hline \mathbf{0} & \mathbf{I} \end{array} \right) \left(\begin{array}{c|c} \mathbf{F}_i & \mathbf{0} \\ \hline \mathbf{0} & \mathbf{I} \end{array} \right) \begin{pmatrix} \vec{U}_{i+1}(0) \\ \vec{U}_0(0) \end{pmatrix}$$

$$= \left(\begin{array}{c|c} \mathbf{\Pi}_i(\Delta c_i)\mathbf{F}_i & \bar{\mathbf{\Pi}}_i^\star(\Delta c_i) \\ \hline \mathbf{0} & \mathbf{I} \end{array} \right) \begin{pmatrix} \vec{U}_{i+1}(0) \\ \vec{U}_0(0) \end{pmatrix}$$

$$= \begin{pmatrix} \mathbf{\Pi}_i(\Delta c_i)\mathbf{F}_i\vec{U}_{i+1}(0) + \bar{\mathbf{\Pi}}_i^\star(\Delta c_i)\vec{U}_0(0) \\ \vec{U}_0(0) \end{pmatrix} .$$

As a result we can represent $\vec{V}_i(0)$ in the following matrix form

$$\vec{V}_i(0) = \mathbf{\Pi}_i(\Delta c_i)\mathbf{F}_i\vec{U}_{i+1}(0) + \bar{\mathbf{\Pi}}_i^a(\Delta c_i)\vec{U}_0(0) \ ,$$

by noting that $\mathbf{\Pi}_i$ is formed by the first k_i rows and columns of matrix $\mathbf{\Pi}_i^\star$ and $\bar{\mathbf{\Pi}}_i^\star$ is formed by the first k_i rows and the last $k_i^\star - k_i = k_0$ columns of $\mathbf{\Pi}_i^\star$. (8) follows from *Claim 1*.

For the case $i = m$, i.e., the last graph $\mathcal{P}_m$, the region size is infinite, therefore delay transitions do not exist. Recall that

$$\vec{U}_m(x) = \max_{\alpha \in Act} \left\{ \int_0^\infty \widehat{\mathbf{M}}_m^\alpha(\tau)\vec{U}_m(x+\tau)d\tau + \vec{1}_F + \int_0^\infty \mathbf{B}_m^\alpha(\tau)d\tau \cdot \vec{U}_0(0) \right\} \ .$$

We first prove the following claim:

Claim 3. For any $x \in \mathbb{R}_{\geq 0}$, $\vec{U}_m(x)$ is a constant vector function.

Proof of Claim 3. We define

$$\vec{U}_m^{(0)}(x) = \vec{0}$$

$$\vec{U}_m^{(j+1)}(x) = \max_{\alpha \in Act} \left\{ \int_0^\infty \widehat{\mathbf{M}}_m^\alpha(\tau)\vec{U}_m^{(j)}(x+\tau)d\tau + \vec{1}_F + \int_0^\infty \mathbf{B}_m^\alpha(\tau)d\tau \cdot \vec{U}_0(0) \right\} \ .$$

It is not difficult to see that $\vec{U}_m(x) = \lim_{j \to \infty} \vec{U}_m^{(j)}(x)$. We shall show, by induction on j, that $\vec{U}_m^{(j)}(x)$ is a constant vector function.

- Base case. $\vec{U}_m^{(0)}(x) = \vec{0}$, which is clearly constant.
- Induction Hypothesis. $\vec{U}_m^{(j)}(x)$ is a constant vector function.

– Induction step.

$$\vec{U}_m^{(j+1)}(x)$$

$$= \max_{\alpha \in Act} \left\{ \int_0^\infty \widehat{\mathbf{M}}_m^a(\tau) \vec{U}_m^{(j)}(x + \tau) d\tau + \vec{1}_F + \int_0^\infty \mathbf{B}_m^\alpha(\tau) d\tau \cdot \vec{U}_0(0) \right\}$$

$$\overset{\text{I.H.}}{=} \max_{\alpha \in Act} \left\{ \int_0^\infty \widehat{\mathbf{M}}_m^a(\tau) \cdot \vec{U}_m^{(j)}(x) d\tau + \vec{1}_F + \int_0^\infty \mathbf{B}_m^\alpha(\tau) d\tau \cdot \vec{U}_0(0) \right\}$$

$$= \max_{\alpha \in Act} \left\{ \int_0^\infty \widehat{\mathbf{M}}_m^a(\tau) d\tau \cdot \vec{U}_m^{(j)}(x) + \vec{1}_F + \int_0^\infty \mathbf{B}_m^\alpha(\tau) d\tau \cdot \vec{U}_0(0) \right\} \ .$$

The conclusion follows trivially.   ♣

Since $\vec{U}_m(x)$ is constant vector function, we have that

$$\vec{U}_m(x) = \max_{\alpha \in Act} \left\{ \int_0^\infty \widehat{\mathbf{M}}_m^\alpha(\tau) d\tau \cdot \vec{U}_m(x) + \vec{1}_F + \int_0^\infty \mathbf{B}_m^\alpha(\tau) d\tau \cdot \vec{U}_0(0) \right\} \ .$$

Moreover, it is easy to see that $\int_0^\infty \widehat{\mathbf{M}}_m^\alpha(\tau) d\tau$ boils down to $\widehat{\mathbf{P}}_m^\alpha$ and $\int_0^\infty \mathbf{B}_m^a(\tau) d\tau$ boils down to $\widehat{\mathbf{B}}_m^\alpha$. Also we add the vector $\vec{1}_F$ to ensure that the probability to start from a state in G_F is one. Hence, (9) follows trivially.   □

Automata for Monadic Second-Order Model-Checking

Bruno Courcelle

Université Bordeaux-1, LaBRI, CNRS
351, Cours de la Libération
33405, Talence, France
courcell@labri.fr

We describe the construction of finite automata on terms establishing that the model-checking problem for every monadic second-order graph property is fixed-parameter linear for tree-width and clique-width (Chapter 6 of [6]).

In this approach, input graphs of small tree-width and clique-width are denoted by terms over finite signatures. These terms reflect the corresponding hierarchical decompositions and monadic second-order sentences are translated into automata intended to run on them. For the case of clique-width, this translation is a straightforward extension of that for finite words. It is a bit more complicated in the case of tree-width ([3], [4]).

In both cases, the practical use of these constructions faces the problem that the automata are huge. (The number of states is typically an h-iterated exponential where h is the quantifier alternation depth.) We present some tools (based on common work in progress with I. Durand, see [5]) that help to overcome this difficulty, at least in some cases. First we use automata whose states are described in an appropriate syntax (and not listed) and whose transitions are computed only when needed (and not compiled in unmanageable tables). In particular, automata are not systematically determinized. They can take as input terms denoting graphs having a tree-width or clique-width that is not *a priori* bounded. Our second tool consists in attaching to each position of the input term a contextual information (computed by one or two preliminary top-down and/or bottom-up passes), that helps to reduce the size of the automata.

The automata approach to monadic second-order model-checking is flexible in that it is not problem specific. Another similar flexible one, based on games, is developed by Kneiss et al. in [11].

The *parsing problem* consisting in checking that the tree-width or clique-width of a given graph is at most some given integer and in constructing a witnessing decomposition is also difficult ([1], [2], [9]), but we do not discuss it in this communication.

References

1. Bodlaender, H., Koster, A.: Treewidth computations I. Upper bounds. Information and Computation 208, 259–275 (2010)
2. Bodlaender, H., Koster, A.: Treewidth computations II. Lower bounds. Information and Computation (in Press 2011)

G. Delzanno and I. Potapov (Eds.): RP 2011, LNCS 6945, pp. 26–27, 2011.

3. Courcelle, B.: Special tree-width and the verification of monadic second-order graph properties. In: Lodaya, K., Mahajan, M. (eds.) Foundations of Software Technology and Theoretical Computer Science, Chennai, India, vol. 8, pp. 13–29. LIPICs (2010)
4. Courcelle, B.: On the model-checking of monadic second-order formulas with edge set quantifications. Discrete Applied Mathematics (May 2010), http://hal.archives-ouvertes.fr/hal-00481735/fr/ (to appear)
5. Courcelle, B., Durand, I.: Automata for the verification of monadic second-order graph properties (in preparation 2011)
6. Courcelle, B., Engelfriet, J.: Graph structure and monadic second-order logic, a language theoretic approach. Cambridge University Press, Cambridge (2011), http://www.labri.fr/perso/courcell/Book/TheBook.pdf
7. Courcelle, B., Makowsky, J., Rotics, U.: On the fixed parameter complexity of graph enumeration problems definable in monadic second-order logic. Discrete Applied Mathematics 108, 23–52 (2001)
8. Downey, R., Fellows, M.: Parameterized complexity. Springer, Heidelberg (1999)
9. Fellows, M., Rosamond, F., Rotics, U., Szeider, S.: Clique-width is NP-complete. SIAM J. Discrete Math. 23, 909–939 (2009)
10. Flum, J., Grohe, M.: Parametrized complexity theory. Springer, Heidelberg (2006)
11. Kneiss, J., Langer, A., Rossmanith, P.: Courcelle's Theorem - A Game-Theoretic Approach, ArXiv, CoRR abs/1104.3905 (2011)

Reachability Problems for Hybrid Automata

Jean-François Raskin

Département d'Informatique – Université Libre de Bruxelles (U.L.B.), Belgium

Abstract. The reachability problem for hybrid automata is undecidable, even for linear hybrid automata. This negative result has triggered several research lines, leading among others to:

- the definition of subclasses of hybrid automata with a decidable reachability problem;
- the definition of semi-algorithms that are useful in practice to attack the reachability problem;
- the definition of variants of the reachability problem that are decidable for larger classes of hybrid automata.

In this talk, we summarize classical and more recent results about those three research lines.

Hybrid Automata. The formalism of hybrid automata [1] is a well-established model for hybrid systems. Important examples of hybrid systems are digital controllers embedded within physical environments. The state of a hybrid system changes both through discrete transitions (of the controller), and continuous evolutions (of the environment). The discrete state of a hybrid system is encoded by a *location* ℓ taken in the finite set Loc of *locations* of the hybrid automaton, and the continuous state is encoded by a valuation v of the *real-valued variables* X of the automaton. Discrete transitions are modeled by edges between locations of the automaton while continuous evolutions are modeled by dynamical laws constraining the first derivative $\dot{X}$ of the variables in each discrete location. Hybrid automata have proven useful in many applications, and their analysis is supported by several tools [11,10].

A central problem in hybrid-system verification is the *reachability problem* which is to decide if there exists an execution from a given initial location ℓ to a given goal location ℓ'.

Classes of HA with Decidable Reachability. While the reachability problem is undecidable for simple classes of hybrid automata (such as linear hybrid automata [1]), the decidability frontier of this problem is now sharply understood [13,14]. For example, the reachability problem is decidable for the class of *initialized rectangular* automata where (i) the flow constraints, guards, invariants and discrete updates are defined by rectangular constraints of the form $a \leq \dot{x} \leq b$ or $c \leq x \leq d$ (where a, b, c, d are rational constants), and (ii) whenever the flow constraint of a variable x changes between two locations ℓ and ℓ', then x is reset along the transition from ℓ to ℓ'. Of particular interest is the class of timed automata [2] which is a special class of initialized rectangular automata.

G. Delzanno and I. Potapov (Eds.): RP 2011, LNCS 6945, pp. 28–30, 2011.

Semi-algorithms. While the reachability problem is undecidable for the class of linear hybrid automata, there is a natural symbolic semi-algorithm to construct a symbolic representation of the reachable states for that class. This symbolic semi-algorithm relies on the following property: the set of one-step successors and the set of time successors of a set of states defined by a pair (ℓ, R), where ℓ is a location of the linear hybrid automaton and R is a polyhedra of valuations for its continuous variables, is definable by a finite set of such pairs [1]. This symbolic semi-algorithm has been implemented in tools like HyTech [11] and PhaVer [10] and used to analyze hybrid systems of practical interest, see for example [15,5]. On the other hand, if the evolution of continuous variables are subject to more complicated flow constraints, for example affine dynamics like $\dot{x} = 3x - y$, computing the flow successor is more difficult and only approximate methods are known. There is a rich literature on the problem of approximating the set of reachable states of complex hybrid automata, see for example [12,3,8,7]. Here we concentrate on [8], where we show how to efficiently compute over-approximations for the class of affine hybrid automata, that is, hybrid automata whose continuous dynamics are defined by systems of linear differential equations. More precisely, it is shown there (i) how to compute automatically rectangular approximations for affine hybrid automata, (ii) how to refine automatically and in an optimal way rectangular approximations that fail to establish a given safety property (the dual of a reachability property), (iii) how to target refinement only to relevant parts of the state space. Such techniques have been implemented with success within PhaVer [10].

The Bounded-Time Reachability Problem for HA. In recent years, it has been observed that new decidability results can be obtained in the setting of time-bounded verification of real-time systems [16,17]. Given a time bound $\mathbf{T} \in \mathbb{N}$, the time-bounded verification problems consider only traces with duration at most $\mathbf{T}$. Note that due to the density of time, the number of discrete transitions may still be unbounded. Several verification problems for timed automata and real-time temporal logics turn out to be decidable in the time-bounded framework (such as the language-inclusion problem for timed automata [16]), or to be of lower complexity (such as the model-checking problem for MTL [17]). The theory of time-bounded verification is therefore expected to be more robust and better-behaved in the case of hybrid automata as well. In [4], we revisit the reachability problem for hybrid automata with time-bounded traces. The *time-bounded reachability problem* for hybrid automata is to decide, given a time bound $\mathbf{T} \in \mathbb{N}$, if there exists an execution of duration less than $\mathbf{T}$ from a given initial location ℓ to a given goal location ℓ'. We study the frontier between decidability and undecidability for this problem and show how bounding time alters matters with respect to the classical reachability problem: the time-bounded reachability problem is *decidable* for *non-initialized rectangular automata* when *only positive rates* are allowed[1]. The time-bounded reachability problem can be reduced to the satisfiability of a formula in the first-order theory of real addition, decidable in EXPSPACE [9]. To characterize

---

[1] This class is interesting from a practical point of view as it includes, among others, the class of stopwatch automata [6], for which unbounded reachability is undecidable.

the boundary of decidability, we also show that simple relaxations of this class of hybrid automata leads to undecidability.

References

1. Alur, R., Courcoubetis, C., Halbwachs, N., Henzinger, T.A., Ho, P.-H., Nicollin, X., Olivero, A., Sifakis, J., Yovine, S.: The algorithmic analysis of hybrid systems. Theor. Comput. Sci. 138(1), 3–34 (1995)
2. Alur, R., Dill, D.L.: A theory of timed automata. Th. Comp. Sci. 126(2), 183–235 (1994)
3. Asarin, E., Dang, T., Maler, O., Bournez, O.: Approximate reachability analysis of piecewise-linear dynamical systems. In: Lynch, N.A., Krogh, B.H. (eds.) HSCC 2000. LNCS, vol. 1790, pp. 20–31. Springer, Heidelberg (2000)
4. Brihaye, T., Doyen, L., Geeraerts, G., Ouaknine, J., Raskin, J.-F., Worrell, J.: On reachability for hybrid automata over bounded time. In: Aceto, L., Henzinger, M., Sgall, J. (eds.) ICALP 2011, Part II. LNCS, vol. 6756, pp. 416–427. Springer, Heidelberg (To apper 2011) CoRR, abs/1104.5335
5. Cassez, F., Jessen, J.J., Larsen, K.G., Raskin, J.-F., Reynier, P.-A.: Automatic synthesis of robust and optimal controllers - an industrial case study. In: Majumdar, R., Tabuada, P. (eds.) HSCC 2009. LNCS, vol. 5469, pp. 90–104. Springer, Heidelberg (2009)
6. Cassez, F., Larsen, K.G.: The impressive power of stopwatches. In: Palamidessi, C. (ed.) CONCUR 2000. LNCS, vol. 1877, pp. 138–152. Springer, Heidelberg (2000)
7. Dang, T., Maler, O., Testylier, R.: Accurate hybridization of nonlinear systems. In: HSCC. ACM, New York (2010)
8. Doyen, L., Henzinger, T.A., Raskin, J.-F.: Automatic rectangular refinement of affine hybrid systems. In: Pettersson, P., Yi, W. (eds.) FORMATS 2005. LNCS, vol. 3829, pp. 144–161. Springer, Heidelberg (2005)
9. Ferrante, J., Rackoff, C.: A decision procedure for the first order theory of real addition with order. SIAM J. Comput. 4(1), 69–76 (1975)
10. Frehse, G.: Phaver: algorithmic verification of hybrid systems past hytech. Int. J. Softw. Tools Technol. Transf. 10, 263–279 (2008)
11. Henzinger, T.A., Ho, P.-H., Wong-Toi, H.: Hytech: A model checker for hybrid systems. In: Grumberg, O. (ed.) CAV 1997. LNCS, vol. 1254, pp. 460–463. Springer, Heidelberg (1997)
12. Henzinger, T.A., Horowitz, B., Majumdar, R., Wong-Toi, H.: Beyond hytech: Hybrid systems analysis using interval numerical methods. In: Lynch, N.A., Krogh, B.H. (eds.) HSCC 2000. LNCS, vol. 1790, p. 130. Springer, Heidelberg (2000)
13. Henzinger, T.A., Kopke, P.W., Puri, A., Varaiya, P.: What's decidable about hybrid automata? J. Comput. Syst. Sci. 57(1), 94–124 (1998)
14. Henzinger, T.A., Raskin, J.-F.: Robust undecidability of timed and hybrid systems. In: Lynch, N.A., Krogh, B.H. (eds.) HSCC 2000. LNCS, vol. 1790, pp. 145–159. Springer, Heidelberg (2000)
15. Henzinger, T.A., Wong-Toi, H.: Using hytech to synthesize control parameters for a steam boiler. In: Abrial, J.-R., Börger, E., Langmaack, H. (eds.) Dagstuhl Seminar 1995. LNCS, vol. 1165, pp. 265–282. Springer, Heidelberg (1996)
16. Ouaknine, J., Rabinovich, A., Worrell, J.: Time-bounded verification. In: Bravetti, M., Zavattaro, G. (eds.) CONCUR 2009. LNCS, vol. 5710, pp. 496–510. Springer, Heidelberg (2009)
17. Ouaknine, J., Worrell, J.: Towards a theory of time-bounded verification. In: Abramsky, S., Gavoille, C., Kirchner, C., Meyer auf der Heide, F., Spirakis, P.G. (eds.) ICALP 2010, Part II. LNCS, vol. 6199, pp. 22–37. Springer, Heidelberg (2010)

Synthesis of Timing Parameters Satisfying Safety Properties

Étienne André and Romain Soulat

LSV, ENS Cachan & CNRS

Abstract. Safety properties are crucial when verifying real-time concurrent systems. When reasoning parametrically, i.e., with unknown constants, it is of high interest to infer a set of parameter valuations consistent with such safety properties. We present here algorithms based on the inverse method for parametric timed automata: given a reference parameter valuation, it infers a constraint such that, for any valuation satisfying this constraint, the discrete behavior of the system is the same as under the reference valuation in terms of traces, i.e., alternating sequences of locations and actions. These algorithms do not guarantee the equality of the trace sets, but are significantly quicker, synthesize larger sets of parameter valuations than the original method, and still preserve various properties including safety (i.e., non-reachability) properties. Those algorithms have been implemented in IMITATOR II and applied to various examples of asynchronous circuits and communication protocols.

Keywords: Real-Time Systems, Timed Automata, Verification, IMITATOR.

1   Introduction

Timed Automata are finite-state automata augmented with clocks, i.e., real-valued variables increasing uniformly, that are compared within guards and transitions with timing delays [AD94]. Although techniques can be used in order to verify the correctness of a timed automaton for a given set of timing delays, these techniques become inefficient when verifying the system for a large number of sets of timing delays, and don't apply anymore when one wants to verify *dense* intervals of values, or optimize some of these delays. It is therefore interesting to reason parametrically, by assuming that those timing delays are unknown constants, or *parameters*, which give *Parametric Timed Automata (PTAs)* [AHV93].

We consider here the *good parameters problem* [FJK08]: "given a PTA $\mathcal{A}$ and a rectangular domain V bounding the value of each parameter, find all the parameter valuations within V such that $\mathcal{A}$ has a good behavior". Such good behaviors can refer to any kind of properties. We will in particular focus here on *safety properties*, i.e., the non-reachability of a given set of "bad" locations.

Parameters Synthesis for PTAs. The problem of parameter synthesis is known to be undecidable for PTAs, although semi-algorithms exist [AHV93] (i.e., if

G. Delzanno and I. Potapov (Eds.): RP 2011, LNCS 6945, pp. 31–44, 2011.
© Springer-Verlag Berlin Heidelberg 2011

the algorithm terminates, the result is correct). The synthesis of constraints has been implemented in the context of PTAs or hybrid systems, e.g., in [AAB00] using TREX [CS01], or in [HRSV02] using an extension of UPPAAL [LPY97] for linear parametric model checking. In [HRSV02], decidability results are given for a subclass of PTAs, viz., "L/U automata".

The problem of parameter synthesis for timed automata has been applied in particular to communication protocols (e.g., the Bounded Retransmission Protocol [DKRT97] using UPPAAL and Spin [Hol03], and the Root Contention Protocol in [CS01] using TREX) and asynchronous circuits (see, e.g., [YKM02, CC07]).

The authors of [KP10] synthesize a set of parameter valuations under which a given property specified in the existential part of CTL without the next operator ($ECTL_{-X}$) holds in a system modeled by a network of PTAs. This is done by using bounded model checking techniques applied to PTAs.

In the framework of Linear Hybrid Automata, techniques based on *counterexample guided abstraction refinement* (CEGAR) [CGJ$^+$00] have been proposed. In [JKWC07], a method of iterative relaxation abstraction is proposed, combining CEGAR and linear programming. In [FJK08], when finding a counterexample, the system obtains constraints that *make* the counterexample infeasible. When all the counterexamples have been eliminated, the resulting constraints describe a set of parameters for which the system is safe. Also note that an approach similar to the inverse method is proposed in [AKRS08], in order to synthesize initial values for the variables of a linear hybrid system.

Contribution. We introduced in [ACEF09] the inverse method *IM* for PTAs. Different from CEGAR-based methods, this original semi-algorithm for parameter synthesis is based on a "good" parameter valuation π_0 instead of a set of "bad" states. *IM* synthesizes a constraint K_0 on the parameters such that, for all parameter valuation π satisfying K_0, the trace set, i.e., the discrete behavior, of $\mathcal{A}$ under π is the same as for $\mathcal{A}$ under π_0. This preserves in particular linear time properties. However, this equality of trace sets may be seen as a too strong property in practice. Indeed, one is rarely interested in a strict ordering of the events, but rather in the partial match with the original trace set, or more generally in the non-reachability of a given set of bad locations.

We present here several algorithms based on *IM*, which do not preserve the equality of trace sets, but preserve various properties. In particular, they all preserve non-reachability: if a location is not reachable in $\mathcal{A}$ under π_0, it will not be reachable in $\mathcal{A}$ under π, for π satisfying K_0. The main advantage is that these algorithms synthesize weaker constraints, i.e., larger sets of parameters. Beside, termination is improved when compared to the original *IM* and the computation time is reduced, as shown in practice in the implementation IMITATOR II.

Plan of the Paper. We briefly recall *IM* in Section 2. We introduce in Section 3 algorithms based on *IM* synthesizing weaker constraints for safety properties, and show their interest compared to *IM*. We extend in Section 4 these algorithms in order to perform a behavioral cartography of the system. We show in

Section 5 the interest in practice by applying these algorithms to models of the literature. We also introduce algorithmic optimizations for two variants allowing to considerably reduce the state space. We conclude in Section 6.

2   The Inverse Method

Preliminaries. [1] Given a set X of clocks and a set P of parameters, a constraint C over X and P is a conjunction of linear inequalities on X and P. Given a parameter valuation (or point) π, we write $\pi \models C$ when the constraint where all parameters within C have been replaced by their value as in π is satisfied by a non-empty set of clock valuations. We denote by $\exists X : C$ the constraint over P obtained from C after elimination of the clocks in X.

Definition 1. *A PTA $\mathcal{A}$ is $(\Sigma, Q, q_0, X, P, K, I, \rightarrow)$ with Σ a finite set of actions, Q a finite set of locations, $q_0 \in Q$ the initial location, X a set of clocks, P a set of parameters, K a constraint over P, I the invariant assigning to every $q \in Q$ a constraint over X and P, and $\rightarrow$ a step relation consisting of elements (q, g, a, ρ, q'), where $q, q' \in Q$, $a \in \Sigma$, $\rho \subseteq X$ is the set of clocks to be reset, and the guard g is a constraint over X and P.*

The semantics of a PTA $\mathcal{A}$ is defined in terms of states, i.e., couples (q, C) where $q \in Q$ and C is a constraint over X and P. Given a point π, we say that a state (q, C) is π-compatible if $\pi \models C$. Runs are alternating sequences of states and actions, and traces are time-abstract runs, i.e., alternating sequences of *locations* and actions. The trace set of $\mathcal{A}$ corresponds to the traces associated with all the runs of $\mathcal{A}$. Given $\mathcal{A}$ and π, we denote by $\mathcal{A}[\pi]$ the (non-parametric) timed automaton where each occurrence of a parameter has been replaced by its constant value as in π. Given two states $s_1 = (q_1, C_1)$ and $s_1 = (q_2, C_2)$, we say that s_1 is *included* into s_2 if $q_1 = q_2$ and $C_1 \subseteq C_2$, where $\subseteq$ denotes the inclusion of constraints. One defines $Post^i_{\mathcal{A}(K)}(S)$ as the set of states reachable from a set S of states in exactly i steps under K, and $Post^*_{\mathcal{A}(K)}(S) = \bigcup_{i \geq 0} Post^i_{\mathcal{A}(K)}(S)$.

Description. Given a PTA $\mathcal{A}$ and a reference parameter valuation π_0, the inverse method *IM* synthesizes a constraint K_0 on the parameters such that, for all $\pi \models K_0$, $\mathcal{A}[\pi_0]$ and $\mathcal{A}[\pi]$ have the same trace sets [ACEF09]. We recall *IM* in Algorithm 1, which consists in two major steps.

1. The iterative removal of the π_0-incompatible states, i.e., states whose constraint onto the parameters is not satisfied by π_0, prevents for any $\pi \models K_0$ the behavior different from π_0 (by negating a π_0-incompatible inequality J).
2. The final intersection of the projection onto the parameters of the constraints associated with all the reachable states guarantees that all the behaviors under π_0 are allowed for all $\pi \models K_0$.

---

[1] Fully detailed definitions are available in [AS11].

Algorithm 1. Inverse method algorithm $IM(\mathcal{A}, \pi_0)$

---

    input : PTA $\mathcal{A}$ of initial state s_0, parameter valuation π_0
    output: Constraint K_0 on the parameters

1 $i \leftarrow 0$; $K \leftarrow$ **true**; $S \leftarrow \{s_0\}$
2 **while true do**
3   **while** *there are π_0-incompatible states in S* **do**
4   Select a π_0-incompatible state (q, C) of S (i.e., s.t. $\pi_0 \not\models C$) ;
5   Select a π_0-incompatible J in $(\exists X : C)$ (i.e., s.t. $\pi_0 \not\models J$) ;
6          $K \leftarrow K \wedge \neg J$; $S \leftarrow \bigcup_{j=0}^{i} Post_{\mathcal{A}(K)}^{j}(\{s_0\})$;
7   **if** $Post_{\mathcal{A}(K)}(S) \sqsubseteq S$ **then return** $\bigcap_{(q,C)\in S}(\exists X : C)$
8      $i \leftarrow i + 1$; $S \leftarrow S \cup Post_{\mathcal{A}(K)}(S)$;   // $S = \bigcup_{j=0}^{i} Post_{\mathcal{A}(K)}^{j}(\{s_0\})$

Item 1 is compulsory in order to prevent the system to enter "bad" (i.e., π_0-incompatible) states. However, item 2 can be lifted when one is only interested in safety properties. Indeed, in this case, it is acceptable that only part of the behavior of $\mathcal{A}[\pi_0]$ is available in $\mathcal{A}[\pi]$ (as long as the behavior absent from $\mathcal{A}[\pi_0]$ is also absent from $\mathcal{A}[\pi]$).

Properties. The main property of *IM* is the *preservation of trace sets*. As a consequence, linear-time properties are preserved. This is the case of properties expressed using the Linear Time Logics (LTL) [Pnu77], but also using the SE-LTL logics [CCO+04], constituted by both atomic state propositions and events.

It has been shown that *IM* is *non-confluent*, i.e., several applications of *IM* can lead to different results [And10b]. This comes from the non-deterministic selection of a π_0-incompatible inequality J (line 5 in Algorithm 1). *IM* behaves deterministically when such a situation of choice is non encountered. The non-confluence of *IM* leads to the *non-maximality* of the output constraint. In other words, given $\mathcal{A}$ and π_0, there may exist points $\pi \not\models IM(\mathcal{A}, \pi_0)$ such that $\mathcal{A}[\pi]$ and $\mathcal{A}[\pi_0]$ have the same trace sets. However, it can be shown that, when *IM* is deterministic, the output constraint is maximal.

Reachability analysis is known to be undecidable for PTAs [AHV93]. Hence, although we showed sufficient conditions, *IM* does not terminate in general.

3   Optimized Algorithms Based on the Inverse Method

A drawback of *IM* is that the notion of equality of trace sets may be seen as too strict in some cases. If one is interested in the non-reachability of a certain set of bad states, then there may exist different trace sets avoiding this set of bad states. We introduce here several algorithms derived from *IM*: none of them guarantee the strict equality of trace sets, but all synthesize weaker constraints than *IM* and still feature interesting properties. They all preserve in particular safety properties, i.e., non-reachability of a given location. In other words, if a given "bad" location is not reached in $\mathcal{A}[\pi_0]$, it will also not be reached by $\mathcal{A}[\pi]$, for π satisfying the constraint output by the algorithm. The corollary is that the set of locations reachable in $\mathcal{A}[\pi]$ is included into the set reachable in $\mathcal{A}[\pi_0]$.

We introduce algorithms derived from IM, namely $IM_\subseteq$, $IM^\cup$, and IM^K. We then introduce combinations between these algorithms. For each algorithm, we briefly state that the constraint is weaker than IM (when applicable), study the termination, and formally state the properties guaranteed by the output constraint. (We do not recall the preservation of non-reachability.) The fully detailed algorithms and all formal properties with proofs can be found in [AS11].

3.1   Algorithm with State Inclusion in the Fixpoint

The algorithm $IM_\subseteq$ is obtained from IM by terminating the algorithm, not when all new states are *equal* to a state computed previously, but when all new states are *included* into a previous state.

The constraint output by $IM_\subseteq$ is weaker than the one output by IM, and $IM_\subseteq$ entails an earlier termination than IM for the same input, and hence a smaller memory usage because states are merged as soon as one is included into another one. $IM_\subseteq$ preserves the equality of traces up to length n, where n is the number of iterations of $IM_\subseteq$ (i.e., the depth of the state space exploration).

Proposition 1. *Suppose that $IM_\subseteq(\mathcal{A}, \pi_0)$ terminates with output K_0 after n iterations of the outer **do** loop. Then, we have:*

1. *$\pi_0 \models K_0$,*
2. *for all $\pi \models K_0$, for each trace T_0 of $\mathcal{A}[\pi_0]$, there exists a trace T of $\mathcal{A}[\pi]$ such that the prefix of length n of T_0 and the prefix of length n of T are equal,*
3. *for all $\pi \models K_0$, for each trace T of $\mathcal{A}[\pi]$, there exists a trace T_0 of $\mathcal{A}[\pi_0]$ such that the prefix of length n of T_0 and the prefix of length n of T are equal.*

Proposition 2. *Suppose that $IM_\subseteq(\mathcal{A}, \pi_0)$ terminates with output K_0. Then, for all $\pi \models K_0$, the sets of reachable locations of $\mathcal{A}[\pi]$ and $\mathcal{A}[\pi_0]$ are the same.*

3.2   Algorithm with Union of the Constraints

The algorithm $IM^\cup$ is obtained from IM by returning, not the intersection of the constraints associated with *all* the reachable states, but the *union* of the constraints associated with the *last* state of each run. This notion of last state is easy to understand for finite runs. When considering infinite (and necessarily[2] cyclic) runs, it refers to the second occurrence of a same state within a run, i.e., the first time that a state is equal to a previous state of the same run.

The constraint output by $IM^\cup$ is weaker than the one output by IM. Note that the constraints output by $IM_\subseteq$ and $IM^\cup$ are incomparable (see example in Section 3.6 for which two incomparable constraints are synthesized). The termination is the same as for IM.

Although the equality of trace sets is no longer guaranteed for $\pi \models IM^\cup(\mathcal{A}, \pi_0)$, we have the guarantee that, for all $\pi \models K_0$, the trace set of $\mathcal{A}[\pi]$ is

---

[2] If the runs are infinite but not cyclic, the algorithm does not terminate.

a subset of the trace set of $\mathcal{A}[\pi_0]$. Furthermore, each trace of $\mathcal{A}[\pi_0]$ is reachable for *at least* one valuation $\pi \models K_0$.

Proposition 3. *Let* $K_0 = IM^{\cup}(\mathcal{A}, \pi_0)$. *Then:*

1. $\pi_0 \models K_0$;
2. *For all* $\pi \models K_0$, *every trace of* $\mathcal{A}[\pi]$ *is equal to a trace of* $\mathcal{A}[\pi_0]$;
3. *For all trace* T *of* $\mathcal{A}[\pi_0]$, *there exists* $\pi \models K_0$ *such that the trace set of* $\mathcal{A}[\pi]$ *contains* T.

Finally note that, due to the disjunctive form of the returned constraint, the synthesized constraint is not necessarily convex.

3.3   Algorithm with Direct Return

The algorithm IM^K is obtained from IM by returning only the constraint K computed during the algorithm instead of the intersection of the constraints associated to all the reachable states.

The constraint output by IM^K is weaker than the one output by IM. Also note that the constraints output by $IM_{\subseteq}$ and IM^K are incomparable (see example in Section 3.6). Termination is the same for IM^K and IM.

Proposition 4. *Let* $K_0 = IM^K(\mathcal{A}, \pi_0)$. *Then, for all* $\pi \models K_0$, *every trace of* $\mathcal{A}[\pi]$ *is equal to a trace of* $\mathcal{A}[\pi_0]$.

This algorithm only prevents π_0-incompatible states to be reached but, contrarily to IM and $IM^{\cup}$, does not guarantee that any "good" state will be reached. Hence, this algorithm only preserves the non-reachability of locations.

3.4   Combination: Inclusion in Fixpoint and Union

One combine the variant of the fixpoint (viz., $IM_{\subseteq}$) with the first variant of the constraint output (viz., $IM^{\cup}$), thus leading to $IM_{\subseteq}^{\cup}$. The constraint output by $IM^{\cup}$ is weaker than the ones output by both $IM_{\subseteq}$ and $IM^{\cup}$. Note that the constraints output by $IM_{\subseteq}^{\cup}$ and IM^K are incomparable (see example in Section 3.6 for which two incomparable constraints are synthesized). The termination is the same as for $IM_{\subseteq}$. This algorithm combines the properties of $IM_{\subseteq}$ and $IM^{\cup}$. Although not of high interest in practice, this result implies preservation of non-reachability. Finally note that, due to the disjunctive form of the returned constraint, the output constraint is not necessarily convex.

3.5   Combination: Inclusion in Fixpoint and Direct Return

One can also combine the variant of the fixpoint (viz., $IM_{\subseteq}$) with the second variant of the constraint output (viz., IM^K), thus leading to $IM_{\subseteq}^K$. The constraint output by $IM_{\subseteq}^K$ is weaker than the ones output by both IM^K and $IM_{\subseteq}^{\cup}$. Termination is the same as for $IM_{\subseteq}$. This algorithm only preserves the non-reachability of locations.

3.6   Summary of the Algorithms

We summarize in Table 1 the properties of each algorithm.

Table 1. Comparison of the properties of the variants of *IM*

Property	IM	$IM_\subseteq$	$IM^\cup$	IM^K	$IM_\subseteq^\cup$	$IM_\subseteq^K$
Equality of trace sets	√	×	×	×	×	×
Equality of trace sets up to n	√	√	×	×	×	×
Inclusion into the trace set of $\mathcal{A}[\pi_0]$	√	×	√	√	×	×
Preservation of at least one trace	√	×	√	×	×	×
Equality of location sets	√	√	×	×	×	×
Convex output	√	√	×	√	×	√
Preservation of non-reachability	√	√	√	√	√	√

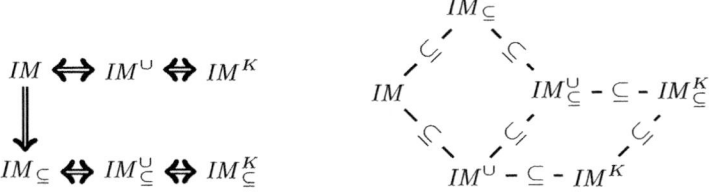

Fig. 1. Comparison of termination (left) and constraint output (right)

We give in Figure 1 (left) the relation between terminations: an oriented edge from A to B means that, for the same input, termination of variant A implies termination of B. We give in Figure 1 (right) the relations between the constraints synthesized by each variant: for example, given $\mathcal{A}$ and π_0, we have that $IM(\mathcal{A}, \pi_0) \subseteq IM_\subseteq(\mathcal{A}, \pi_0)$. Obviously, the weakest constraint is the one synthesized by $IM_\subseteq^K$. This variant should be thus used when one is interested only in safety properties; however, when one is interested in stronger properties (e.g., preservation of at least one trace of $\mathcal{A}[\pi_0]$), one may want to use another variant according to their respective properties. We believe that the most interesting algorithms are IM, for the equality of trace sets, $IM^\cup$, for the preservation of at least one maximal trace, and $IM_\subseteq^K$, for the sole preservation of non-reachability.

Non-maximality. Actually, none of these algorithms synthesize the maximal constraint corresponding to the property they are characterized with. This is due to their non-confluence, itself due to the random selection of a π_0-incompatible inequality. However, it can be shown that the constraint is maximal when the algorithm runs in a fully deterministic way. We address the issue of synthesizing a maximal constraint in Section 4. Also note that the comparison between the constraints (see Figure 1 (right)) holds only for deterministic analyses.

Comparison Using an Example of PTA. Let us consider the PTA $\mathcal{A}_{var}$ depicted below. We consider the following π_0: $p_1 = 1 \wedge p_2 = 4$. In $\mathcal{A}[\pi_0]$, location q_4 is not reachable, and can be considered as a "bad" location.

Let us suppose that a bad behavior of $\mathcal{A}_{var}$ corresponds to the fact that a trace goes into location q_4. Under π_0, the system has a good behavior. As a

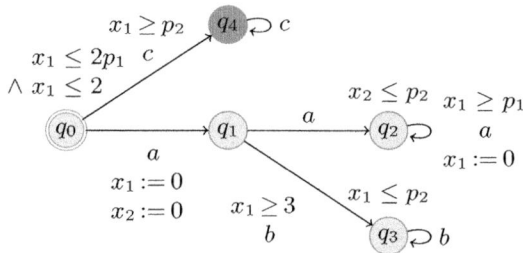

Fig. 2. A PTA $\mathcal{A}_{var}$ for comparing the variants of IM

consequence, by the property of non-reachability of a location met by all algorithms, the constraint synthesized by any algorithm also prevents the traces to enter q_4. One can show that the parameter valuations allowing the system to enter the bad location q_4 are comprised in the domain $2 * p_1 \leq p_2 \wedge p_2 \leq 2$. As a consequence, the (non-convex) maximal set of parameters avoiding the bad location q_4 is $2 * p_1 > p_2 \vee p_2 > 2$.

We give below the six constraints synthesized by the six versions of the inverse method. For each graphics, we depict in dark gray the parameter domain covered by the constraint, and in light gray the parameter domain corresponding to a bad behavior (the constraint itself is given in [AS11]). The "good" zone not covered by the constraint is depicted in very light gray. The dot represents π_0.

This example illustrates well the relationship between the different constraints. In particular, the constraint synthesized by $IM_{\subseteq}^{K}$ dramatically improves the set of parameters synthesized by IM. Also note that we chose on purpose an example such that none of the methods synthesizes a maximal constraint (observe that even $IM_{\subseteq}^{K}$ does not cover the whole "good" zone). This will be addressed in Section 4.

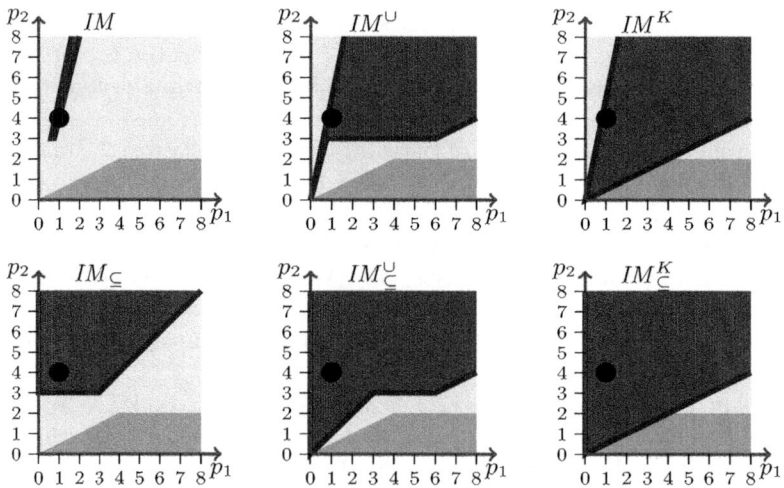

Fig. 3. Comparison of the constraints synthesized for $\mathcal{A}_{var}$

Experimental Validation. The example above shows clearly the gain of the algorithms w.r.t. *IM*. However, for real case studies, although checking the gain of these algorithms in terms of computation time is possible, measuring the gain in terms of the "size" of the constraints synthesized requires measures of polyhedra, which is not trivial when they are non-convex. Hence, we postpone this study to the framework of the cartography (see Section 5).

4   Behavioral Cartography

Although *IM* has been shown of interest for a large panel of case studies, its main shortcoming is the non-maximality of the output constraint. Moreover, the good parameters problem relates to the synthesis of parameter valuations corresponding to *any* good behavior, not to a single one.

The behavioral cartography algorithm *BC* relies on the idea of covering the parameter space within a rectangular real-valued parameter domain V_0 [AF10]. By iterating *IM* over all the *integer* points of V_0 (of which there are a finite number), one is able to decompose V_0 into a list *Tiling* of *tiles*, i.e., dense sets of parameters in which the trace sets are the same.

Then, given a property φ on trace sets (viz., a linear time property), one can partition the parameter space between good tiles (for which all points satisfy φ) and bad tiles. This can be done by checking the property for one point in each tile (using, e.g., UPPAAL, applied to the PTA instantiated with the considered point). Then the set of parameters satisfying φ corresponds to the union of the good tiles. Note that *BC* is independent from φ; only the partition between good and bad tiles involves φ.

In practice, not only the integer valuations of V_0 are covered by *Tiling*, but also most of the real-valued space of V_0. Furthermore, the space covered by *Tiling* often largely exceeds the limits of V_0. However, there may exist a finite number of "small holes" within V_0 (containing no integer point) that are not covered by any tile of *Tiling*. A refinement of *BC* is to consider a tighter grid, i.e., not only integer points, but rational points multiple of a smaller step than 1. We showed that, for a rectangle V_0 large enough and a grid tight enough, the full coverage of the whole real-valued parameter space (inside and outside V_0) is ensured for some classes of PTAs, in particular for acyclic systems (see [And10b] for details).

Combination with the Variants. By replacing within *BC* the call to *IM* by a call to one of the algorithms introduced in Section 3, one changes the properties of the tiles: for each tile, the corresponding trace set inherits the properties of the considered variant, and does not necessarily preserve the equality of trace sets. However, as shown in Section 3, they all preserve (at least) the non-reachability.

The main advantage of the combination of *BC* with one of the algorithms, say *IM'*, of Section 3 is that the coverage of V_0 needs *a smaller number of tiles*, i.e., of calls to *IM'*. Indeed, due to the weaker constraint synthesized by *IM'*, and hence larger sets of parameters, one needs less calls to *IM'* in order to cover V_0. Furthermore, due to the quicker termination of *IM'* when compared to *IM*, the computation time decreases considerably. Finally, due to an earlier termination *IM′* (i.e., less states computed) and the lower number of calls to *IM′* (hence, less trace sets to remember), the memory consumption also decreases.

5   Implementation and Experiments

All these algorithms, as well as the original *IM*, have been implemented in IMITATOR II [And10a]. We give in Table 2 the summary of various experiments of parameter synthesis applied to case studies from the literature as well as industrial case studies. For each case study, we apply each version of *BC* to a given V_0. Then, we split the tiles between good and bad w.r.t. a property. Finally, we synthesize a constraint corresponding to this property. For each case study, V_0 is either entirely covered, or "almost entirely covered". We give from left to right the name of the case study, the number of parameters varying in the cartography and the number of points within V_0. We then give the number of tiles and the computation time for each algorithm. We denote by $BC_\subseteq$ the variant of *BC* calling $IM_\subseteq$ instead of *IM* (and similarly for the other algorithms). All experiments were performed on an Intel Core 2 Duo 2,33 Ghz with 3,2 Go memory, using the no-random, no-dot and no-log options of IMITATOR II.

Table 2. Comparison of the algorithms for the behavioral cartography

Example			Tiles						Time (s)					
Name	$\lvert P \rvert$	$\lvert V_0 \rvert$	BC	BC^U	BC^K	$BC_\subseteq$	$BC^U_\subseteq$	$BC^K_\subseteq$	BC	BC^U	BC^K	$BC_\subseteq$	$BC^U_\subseteq$	$BC^K_\subseteq$
$\mathcal{A}_{var}$	2	72	14	10	10	7	5	5	0.101	0.079	0.073	0.036	0.028	0.026
Flip-flop	2	644	8	7	7	8	7	7	0.823	0.855	0.696	0.831	0.848	0.699
AND–OR	5	151 200	16	14	16	14	14	14	274	7154	105	199	551	68.4
Latch	4	73 062	5	3	3	5	3	3	16.2	25.2	9.2	15.9	25	9.1
CSMA/CD	3	2 000	139	57	57	139	57	57	112	276	76.0	46.7	88.0	22.6
SPSMALL	2	3 082	272	78	77	272	78	77	894	405	342	894	406	340

Since V_0 is always (at least "almost") entirely covered by *Tiling*, the number of tiles needed to cover V_0 gives a measure of the size of each tile in terms of parameter valuations: the lesser tiles needed, the larger the sets of parameter valuations are, the more efficient the corresponding algorithm is. Since the good property for all case studies is a property of (non-)reachability, the constraint computed is the same for all versions of *BC*. The latest version[3] of IMITATOR II as well as all the mentioned case studies can be found on IMITATOR II's Web page[4]. Details on case studies can be found in [AS11].

As expected from Section 2, all algorithms bring a significant gain in term of size of the constraint, because the number of tiles needed to cover V_0 is almost always smaller than for *IM*. Only $IM_\subseteq$ has a number of tiles often equal to *IM*; however, the computation is often much quicker for $IM_\subseteq$. As expected, the most

---

[3] Note that the software named IMITATOR 3 is an independent fork of IMITATOR II for hybrid systems [FK11]. The latest version of IMITATOR for PTAs is IMITATOR 2.3.

[4] http://www.lsv.ens-cachan.fr/Software/imitator/imitator2.3/

efficient algorithm is $IM_{\subseteq}^K$: both the number of tiles and the computation times decrease significantly. When one is only interested in reachability properties, one should then use this algorithm.

The only surprising result is the fact that $IM^{\cup}$ is sometimes slower than IM, although the number of tiles is smaller. This is due to the way it is implemented in IMITATOR II. Handling non-convex constraints is a difficult problem; hence, we compute a list of constraints associated with the last state of each trace. Unfortunately, many of these constraints are actually equal to each other. For systems with thousands of traces and hundreds of tiles, we manipulate hundreds of thousands of constraints; every time a new point is picked up, one should check whether it belongs to this huge set before calling (or not) IM on this point. This also explains the relatively disappointing speed performance of $IM_{\subseteq}^{\cup}$. Improving this implementation is the subject of ongoing work. A possible option would be to remove the constraints equal to each other in this constraint set; this would dramatically decrease the size of the set, but would induce additional costs for checking constraint equality.

On-the-fly Computation of K. We finally introduce here another modification of some of the algorithms in order to avoid the non-necessary duplication of some reachable states, leading to a dramatic diminution of the state space. Indeed, we met cases where two states (q, C) and (q, C') are not equal at the time they are computed, but are equal with the final intersection K of the constraints, i.e. $(q, C \wedge K) = (q, C' \wedge K)$. Such a situation is depicted in the trace sets of Figure 4, where identical states under $IM(\mathcal{A}, \pi_0)$ are unmerged on the left part of the figure and merged on the right part. We can solve this problem by performing *dynamically* the intersection of the constraints, i.e., adding $\exists X : C$ to all the states previously computed, every time a new state (q, C) is computed. This has the effect of merging such states, and hence often considerably decreasing the state space. With this modification, the algorithm only needs to return K at the end of the computation, since the intersection is performed on the fly. We give this algorithm IM_{otf} in Algorithm 2.

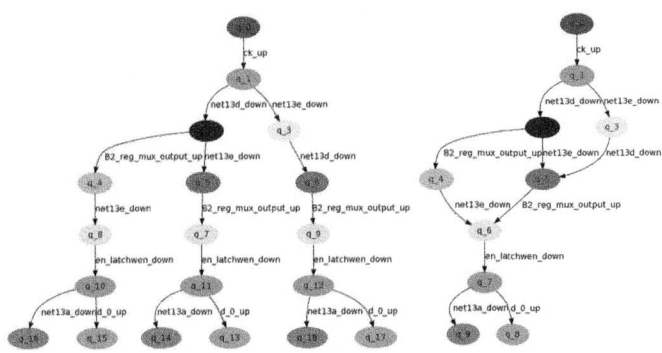

Fig. 4. Example of state space explosion due to unmerged states

---

Algorithm 2. $IM_{otf}(\mathcal{A}, \pi_0)$

---

input : PTA $\mathcal{A}$ of initial state s_0, parameter valuation π_0
output: Constraint K_0 on the parameters

1 $i \leftarrow 0$; $K \leftarrow$ **true** ; $S \leftarrow \{s_0\}$
2 **while true do**
3   **foreach** $s = (q, C) \in S$ **do**
4   **if** s is π_0-incompatible **then**
5   Select a π_0-incompatible J in $(\exists X : C)$
6              $K \leftarrow K \wedge \neg J$;
7   **foreach** $(q', C') \in S$ **do**
8                  $C' \leftarrow C' \wedge \neg J$

9   **else**
10             $K \leftarrow K \wedge \exists X : C$;
11   **foreach** $(q', C') \in S$ **do**
12                 $C' \leftarrow C' \wedge \exists X : C$

13   **if** $Post_{\mathcal{A}(K)}(S) \sqsubseteq S$ **then return** K
14     $i \leftarrow i + 1$; $S \leftarrow S \cup Post_{\mathcal{A}(K)}(S)$; $// \ S = \bigcup_{j=0}^{i} Post_{\mathcal{A}(K)}^{j}(\{s_0\})$

This modification can be extended to $IM_{\subseteq}$ in a straightforward manner, by applying to IM_{otf} the fixpoint modification as described in Section 3.1. However, applying it to other algorithms would modify their correctness, since the final intersection of the constraints is not performed in the other algorithms.

Using this modification, we successfully computed a set of parameters for the SPSMALL memory designed by ST-Microelectronics. We analyzed a much larger version of the "small" model considered above (see differences between these models in [And10b]). The larger model of this memory contains 28 clocks and 62 parameters. The computation consists in 98 iterations of the outer **DO** loop of IM. Without this optimization, IM crashed from lack of memory at iteration 27 (on a 2 GB memory machine), but the size of the state space was exponential, so we believe that the full computation would have required a huge amount of memory, preventing more powerful machine to perform the computation. Using this optimization, we computed quickly a constraint, made of the conjunction of 49 linear constraints. Full details are available in [And10b].

6 Conclusion

We introduced here several algorithms based on the inverse method for PTAs. Given a PTA $\mathcal{A}$ and a reference parameter valuation π_0, these algorithms synthesize a constraint K_0 around π_0, all preserving non-reachability properties: if a location (in general "bad") is not reached for π_0, it is also not reachable for any $\pi \models K_0$. Furthermore, each algorithm preserves different properties: strict equality of trace sets, inclusion within the trace set of $\mathcal{A}[\pi_0]$, preservation of at least one trace of $\mathcal{A}[\pi_0]$, etc. The major advantage of these variants is the faster computation of K_0, and a larger set of parameter valuations defined by K_0. These algorithms have been implemented in IMITATOR II and show significant gains of time and size of the constraint when compared to the original IM. When used in the behavioral cartography for synthesizing a constraint w.r.t. a given property, they cover both using less tiles and in general faster the parameter space.

Also recall that, although the algorithms preserve properties based on traces, i.e., *untimed* behaviors, it is possible to synthesize constraints guaranteeing *timed* properties by making use of an observer; this is the case in particular for the SPSMALL memory.

As a future work, the inverse method and the cartography algorithm, as well as the variants introduced here, could be extended in a rather straightforward way to the backward case.

Furthermore, we presented in [AFS09] an extension of the inverse method to probabilistic systems: given a parametric probabilistic timed automaton $\mathcal{A}$ and a reference valuation π_0, we synthesize a constraint K_0 by applying IM to a non-probabilistic version of $\mathcal{A}$ and π_0. Then, we guarantee that, for all $\pi \models K_0$, the values of the minimum (resp. maximum) probabilities of reachability properties are the same in $\mathcal{A}[\pi]$. Studying what properties each of the algorithms presented here preserves in the probabilistic framework is the subject of ongoing work.

It would also be of interest to consider the combination of these algorithms with the extension of the inverse method to linear hybrid automata [FK11].

References

[AAB00]   Annichini, A., Asarin, E., Bouajjani, A.: Symbolic techniques for parametric reasoning about counter and clock systems. In: Emerson, E.A., Sistla, A.P. (eds.) CAV 2000. LNCS, vol. 1855, pp. 419–434. Springer, Heidelberg (2000)

[ACEF09] André, É., Chatain, T., Encrenaz, E., Fribourg, L.: An inverse method for parametric timed automata. International Journal of Foundations of Computer Science 20(5), 819–836 (2009)

[AD94]   Alur, R., Dill, D.L.: A theory of timed automata. TCS 126(2), 183–235 (1994)

[AF10]   André, É., Fribourg, L.: Behavioral cartography of timed automata. In: Kučera, A., Potapov, I. (eds.) RP 2010. LNCS, vol. 6227, pp. 76–90. Springer, Heidelberg (2010)

[AFS09]   André, É., Fribourg, L., Sproston, J.: An extension of the inverse method to probabilistic timed automata. In: AVoCS 2009. Electronic Communications of the EASST, vol. 23 (2009)

[AHV93]   Alur, R., Henzinger, T.A., Vardi, M.Y.: Parametric real-time reasoning. In: STOC 1993, pp. 592–601. ACM, New York (1993)

[AKRS08] Alur, R., Kanade, A., Ramesh, S., Shashidhar, K.C.: Symbolic analysis for improving simulation coverage of simulink/stateflow models. In: EMSOFT 2008, pp. 89–98. ACM, New York (2008)

[And10a] André, É.: IMITATOR II: A tool for solving the good parameters problem in timed automata. In: INFINITY 2010. EPTCS, vol. 39, pp. 91–99 (2010)

[And10b]   André, É.: An Inverse Method for the Synthesis of Timing Parameters in Concurrent Systems. Ph.d. thesis, Laboratoire Spécification et Vérification, ENS Cachan, France (2010)

[AS11]   André, É., Soulat, R.: Synthesis of timing parameters satisfying safety properties (full version). Research report, Laboratoire Spécification et Vérification, ENS Cachan, France (2011), http://www.lsv.ens-cachan.fr/Publis/RAPPORTS_LSV/PDF/rr-lsv-2011-13.pdf

[CC07]   Clarisó, R., Cortadella, J.: The octahedron abstract domain. Sci. Comput. Program. 64(1), 115–139 (2007)

[CCO+04]   Chaki, S., Clarke, E.M., Ouaknine, J., Sharygina, N., Sinha, N.: State/event-based software model checking. In: Boiten, E.A., Derrick, J., Smith, G.P. (eds.) IFM 2004. LNCS, vol. 2999, pp. 128–147. Springer, Heidelberg (2004)

[CGJ+00]   Clarke, E.M., Grumberg, O., Jha, S., Lu, Y., Veith, H.: Counterexample-guided abstraction refinement. In: Emerson, E.A., Sistla, A.P. (eds.) CAV 2000. LNCS, vol. 1855, pp. 154–169. Springer, Heidelberg (2000)

[CS01]   Collomb–Annichini, A., Sighireanu, M.: Parameterized reachability analysis of the IEEE 1394 Root Contention Protocol using TReX. In: RT-TOOLS 2001 (2001)

[DKRT97]   D'Argenio, P.R., Katoen, J.P., Ruys, T.C., Tretmans, G.J.: The bounded retransmission protocol must be on time! In: Brinksma, E. (ed.) TACAS 1997. LNCS, vol. 1217, Springer, Heidelberg (1997)

[FJK08]   Frehse, G., Jha, S.K., Krogh, B.H.: A counterexample-guided approach to parameter synthesis for linear hybrid automata. In: Egerstedt, M., Mishra, B. (eds.) HSCC 2008. LNCS, vol. 4981, pp. 187–200. Springer, Heidelberg (2008)

[FK11]   Fribourg, L., Kühne, U.: Parametric verification and test coverage for hybrid automata using the inverse method. In: Delzanno, G., Potapov, I. (eds.) RP 2011. LNCS, vol. 6945, pp. 191–204. Springer, Heidelberg (2011)

[Hol03]   Holzmann, G.: The Spin model checker: primer and reference manual. Addison-Wesley Professional, Reading (2003)

[HRSV02]   Hune, T.S., Romijn, J.M.T., Stoelinga, M.I.A., Vaandrager, F.W.: Linear parametric model checking of timed automata. Journal of Logic and Algebraic Programming (2002)

[JKWC07]   Jha, S.K., Krogh, B.H., Weimer, J.E., Clarke, E.M.: Reachability for linear hybrid automata using iterative relaxation abstraction. In: Bemporad, A., Bicchi, A., Buttazzo, G. (eds.) HSCC 2007. LNCS, vol. 4416, pp. 287–300. Springer, Heidelberg (2007)

[KP10]   Knapik, M., Penczek, W.: Bounded model checking for parametric time automata. In: SUMo 2010 (2010)

[LPY97]   Larsen, K.G., Pettersson, P., Yi, W.: UPPAAL in a nutshell. International Journal on Software Tools for Technology Transfer 1(1-2), 134–152 (1997)

[Pnu77]   Pnueli, A.: The temporal logic of programs. In: SFCS 1977, pp. 46–57. IEEE Computer Society, Los Alamitos (1977)

[YKM02]   Yoneda, T., Kitai, T., Myers, C.J.: Automatic derivation of timing constraints by failure analysis. In: Brinksma, E., Larsen, K.G. (eds.) CAV 2002. LNCS, vol. 2404, pp. 195–208. Springer, Heidelberg (2002)

Formal Language Constrained Reachability and Model Checking Propositional Dynamic Logics[*]

Roland Axelsson[1] and Martin Lange[2]

[1] Dept. of Computer Science, University of Munich, Germany
[2] School of Electr. Eng. and Computer Science, University of Kassel, Germany

Abstract. We show interreducibility under (Turing) reductions of low polynomial degree between three families of problems parametrised by classes of formal languages: the problem of reachability in a directed graph constrained by a formal language, the problem of deciding whether or not the intersection of a language of some class with a regular language is empty, and the model checking problem for Propositional Dynamic Logic over some class of formal languages. This allows several decidability and complexity results to be transferred, mainly from the area of formal languages to the areas of modal logics and formal language constrained reachability.

1 Introduction

This paper investigates three families of decision problems from the domains of formal language theory, digraph reachability, and model checking. Each family is parametrised by a class $\mathcal{L}$ of formal languages which can be any class but we are mainly concerned with known and natural classes like the regular, context-free, context-sensitive languages, and also some equally natural but lesser known classes. We will always assume that there is a finite representation of any member of that class, for instance a finite-state automaton for a regular language or a context-free grammar for a context-free language, etc. The three families of problems are the following.

i) REG-Intersection for $\mathcal{L}$: determine for a given language $L \in \mathcal{L}$ and a regular language R, whether or not $L \cap R$ is empty.
ii) $\mathcal{L}$-Reachability: decide for a given directed graph with edge labels and node predicates whether or not there is a path from a designated source node to a designated target area s.t. the path is described by a given language $L \in \mathcal{L}$.
iii) Model checking PDL[$\mathcal{L}$]: decide for a given state of a Kripke structure and a given formula of Propositional Dynamic Logic over $\mathcal{L}$ whether or not the state satisfies the formula.

[*] The European Research Council has provided financial support under the European Community's Seventh Framework Programme (FP7/2007-2013) / ERC grant agreement no 259267.

G. Delzanno and I. Potapov (Eds.): RP 2011, LNCS 6945, pp. 45–57, 2011.

These problems have been considered each on its own so far, and the state of the art in knowing about decidability and complexity of these problems is the following.

(i) Closure under intersection with a regular language and decidability of the emptiness problem are two of the most important features usually investigated with any class of formal languages. Note that a class that is closed under intersections with regular languages and has a decidable emptiness problem also has a decidable REG-intersection problem. The converse need not necessarily be the case but to the best of our knowledge there is no natural class which would witness this. It can safely be said that the REG-intersection problem is well-understood in the domain of formal languages.

(ii) Digraph reachability is of course one of the most fundamental problems in computer science and related disciplines. The use of constraints restricting the paths under which certain vertices should be reachable has been outlined for multi-modal path planning for instance [5]. Complexity and decidability issues have been investigated with respect to a class of formal languages used to constrain the paths. It has been found that using context-free languages as opposed to no languages or regular ones increases the complexity from NLOGSPACE to PTIME, and using context-sensitive languages, this problem becomes undecidable [5]. To the best of our knowledge, the space between context-free and context-sensitive languages has not been looked at from the perspective of formal language constrained reachability problems.

(iii) Propositional Dynamic Logic (PDL) has been introduced by Fischer and Ladner [8] as a modal logic for reasoning about programs. Various applications of PDL have been identified, for instance in program verification because of its similarity to the branching-time temporal logic CTL and its apparent relation to Hoare logic; in knowledge representation and artificial intelligence because the test-free fragment for example turns out to equal the description logic $\mathcal{ALC}_{reg}$ [10], because it can be used to reason about knowledge [6] or about actions [25], etc. While much attention is being paid to its satisfiability problem, its model checking problem also has applications. It mainly occurs in automatic program verification, and certain inference problems in description logics for example can be reduced to model checking problems and then be tackled using database technology [3].

The original PDL is in fact *PDL over regular programs*—here written as PDL[REG]—since the programs which are interpreted as binary relations on program states, are built from atomic ones using the constructors union, composition and iteration. Such programs are denoted syntactically using regular expressions, and it is imminent that other formalisms for describing formal languages can be used instead, too.

Variants of PDL over richer classes of formal languages have been studied with a focus on their satisfiability problems [12,13,14,11]. This is undecidable for PDL[CFL] – *PDL over context-free languages* – already [12]. On the other hand, model checking for PDL[CFL] is PTIME-complete [18] as it is for PDL[REG]

[8], and larger classes as parameters have not been considered yet under this aspect.

In this paper we show that these problems are (Turing-)interreducible to each other in polynomial time: one reduction is a genuine Turing reduction of quadratic time, the others are many-one reductions of linear time. The constructions are simple, and so are their proofs of correctness. However, these simple constructions pave the way for a number of new decidability and complexity results on formal language constrained reachability analysis as well as on model checking extensions of Propositional Dynamic Logic. As a consequence, the gap between decidability and undecidability in terms of the class of formal languages used as the parameter has been narrowed down significantly: with the results obtained here we now know that it lies between a large subclass of CSL known as the multi-stack visibly pushdown languages (MVPL) and CSL itself.

The paper is organised as follows. Sect. 2 introduces the three problems formally. Sect. 3 motivates the study of the interconnection between these three problems and their respective areas by presenting exemplary applications of these problems. It gives a brief insight into the fact that these three problems, despite being very much related as decision problems, have been studied independently in different domains. Sect. 4 proves the interreducibility. New decidability and complexity results about model checking and reachability problems are derived in Sect. 5 from corresponding language-theoretic problems using this interreducibility. Finally, Sect. 6 provides a summary of the complexity and decidability results known in these areas.

2   Preliminaries

Classes of formal languages and their representations. Let Σ be a finite alphabet. As usual, a formal language is a $L \subseteq \Sigma^*$, and a class of formal languages is a $\mathcal{L} \subseteq 2^{\Sigma^*}$. We do not want to advertise nor restrict the use of a particular specification formalism for formal languages like automata, grammars, algebraic expressions, systems of equations, etc. We therefore identify a class $\mathcal{L}$ of formal languages with a class of its *acceptors* and restrict our attention to classes which can be represented by such acceptors. This means that we can assume a size measure $\|L\|$ which is a finite value for any L, even though it may contain infinitely many words. For instance, for $\mathcal{L} = $ REG this may be the size of a smallest nondeterministic finite automaton recognising L.

We make another very reasonable assumption on each $\mathcal{L}$: given an $L \in \mathcal{L}$, its alphabet must be computable in time $\mathcal{O}(\|L\|)$.[1] We write $\Sigma(L)$ to denote the alphabet that is underlying L.

The REG-intersection problem for classes of formal languages. Remember that the *non-emptiness problem* for a class $\mathcal{L}$ of languages is the following: given a

---

[1] This is true of virtually all known specification formalisms for formal languages and only precludes strange acceptors like encrypted strings representing automata, etc.

suitably represented $L \in \mathcal{L}$, decide whether or not $L \neq \emptyset$. Furthermore, a class $\mathcal{L}$ is *closed under intersections with regular languages* if for every $L \in \mathcal{L}$ and every regular language R we have $L \cap R \in \mathcal{L}$. These are combined into a decision problem which is of particular interest here. We assume familiarity with the theory of regular languages. Note that $L(\mathcal{A})$ denotes the language recognised by the automaton $\mathcal{A}$.

Definition 1. The problem of *non-emptiness of intersection with a regular language – REG-intersection problem* for short – for $\mathcal{L}$ is the following: given a suitably represented $L \in \mathcal{L}$ and a non-deterministic finite automaton (NFA) $\mathcal{A}$ over Σ, decide whether or not $L \cap L(\mathcal{A}) \neq \emptyset$.

Clearly, if a class of languages is closed under intersections with regular languages and has a decidable non-emptiness problem, then its REG-intersection problem is decidable, too. The converse may not be true in general.[2] Furthermore, if a class of languages is closed under intersections with regular languages but has an undecidable non-emptiness problem (like CSL for instance) then its REG-intersection problem is necessarily also undecidable.

Kripke structures, labeled digraphs, and words. Let Σ be a finite set of symbols and $\mathcal{P}$ be a countably infinite set of propositional constants. A *Kripke structure* is a triple $\mathcal{T} = (\mathcal{S}, \rightarrow, \ell)$, where $\mathcal{S}$ is a set of states, $\rightarrow \subseteq \mathcal{S} \times \Sigma \times \mathcal{S}$ is a transition relation and $\ell : \mathcal{S} \rightarrow 2^{\mathcal{P}}$ labels each state with a set of propositions that are true in that state. We write $s \xrightarrow{a} t$ instead of $(s, a, t) \in \rightarrow$. We will restrict ourselves to finite Kripke structures, i.e. those for which $|\mathcal{S}|$ is finite.

The accessibility relation $\rightarrow$ is inductively extended to words over Σ as follows.

$$s \xrightarrow{\epsilon} t \quad \text{iff} \quad s = t$$
$$s \xrightarrow{aw} t \quad \text{iff} \quad \exists u \in \mathcal{S} \text{ with } s \xrightarrow{a} u \text{ and } u \xrightarrow{w} t$$

An *edge-labeled directed graph* is a Kripke structure $\mathcal{T}$ as above such that $\ell(s) = \emptyset$ for all $s \in \mathcal{S}$. In the following, when we speak of a graph it is implicitly to be understood as an edge-labeled directed graph. We will denote such a structure like a Kripke structure but leaving out the labeling function, i.e. as $\mathcal{T} = (\mathcal{S}, \rightarrow)$.

The $\mathcal{L}$-reachability problem for a class of formal languages $\mathcal{L}$.

Definition 2. Let $\mathcal{L}$ be a class of languages over Σ. The *$\mathcal{L}$-reachability problem* is the following: given a graph $\mathcal{T} = (\mathcal{S}, \rightarrow)$, a state $s \in \mathcal{S}$, a set of target states $T \subseteq \mathcal{S}$ and a suitably represented $L \in \mathcal{L}$, decide whether or not there is a $w \in L$ and a $t \in T$ s.t. $s \xrightarrow{w} t$.

We also say that T is L-reachable from s in $\mathcal{T}$ if these form a positive instance of the $\mathcal{L}$-reachability problem.

---

[2] However, we are unaware of any (necessarily strange) class of languages that witnesses the failure of the converse direction. It also does not matter for our purposes here.

Propositional Dynamic Logic over a class of formal languages. Formulas of Propositional Dynamic Logic (with tests) over a class $\mathcal{L}$ of formal languages over some finite alphabet Σ — PDL[$\mathcal{L}$] — are defined recursively as the least set Form satisfying the following.

(i) $\mathcal{P} \subseteq$ Form
(ii) If $\varphi \in$ Form and $\psi \in$ Form then $\neg\varphi \in$ Form and $\varphi \vee \psi \in$ Form.
(iii) If L is a language over the alphabet $\Sigma \cup \{\psi? \mid \psi \in$ Form$\}$ s.t. $|\Sigma(L)| < \infty$
     and $\varphi \in$ Form then $\langle L \rangle \varphi \in$ Form.

We use the usual abbreviations: $\mathbf{tt} := q \vee \neg q$ for some $q \in \mathcal{P}$, $\mathbf{ff} := \neg\mathbf{tt}$, $\varphi \wedge \psi := \neg(\neg\varphi \vee \neg\psi)$, $\varphi \to \psi := \neg\varphi \vee \psi$, $[L]\varphi := \neg\langle L \rangle \neg\varphi$. We define $|\varphi|$ as the number of different subformulas of φ plus the sum over all $\|L\|$ s.t. $\langle L \rangle \psi$ is a subformula of φ for some ψ.

Suppose L is a language over $\Sigma \cup \{\psi? \mid \psi \in \Phi\}$ for some finite $\Phi \subset$ Form. We then write $Tests(L)$ for this set Φ. By the assumption made above about L being represented reasonably, we can also assume that $Tests(L)$ can be computed in time $\mathcal{O}(\|L\|)$.

Formulas of PDL[$\mathcal{L}$] are interpreted in states s of Σ-labeled Kripke structures $\mathcal{T} = (\mathcal{S}, \to, \ell)$ as follows.

$$\mathcal{T}, s \models q \quad \text{iff} \quad q \in \ell(s)$$
$$\mathcal{T}, s \models \neg\varphi \quad \text{iff} \quad \mathcal{T}, s \not\models \varphi$$
$$\mathcal{T}, s \models \varphi \vee \psi \quad \text{iff} \quad \mathcal{T}, s \models \varphi \text{ or } \mathcal{T}, s \models \psi$$
$$\mathcal{T}, s \models \langle L \rangle \varphi \quad \text{iff} \quad \text{there are } w \in L \text{ and } t \in \mathcal{S} \text{ s.t. } s \xrightarrow{w}' t \text{ and } \mathcal{T}, t \models \varphi \text{ where}$$
$$\to' := \to \cup \{(u, \psi?, u) \mid u \in \mathcal{S}, \psi \in Tests(L), \text{ and } \mathcal{T}, u \models \psi\}$$

Definition 3. The *model checking problem* for PDL[$\mathcal{L}$] is the following: given a Kripke structure $\mathcal{T} = (\mathcal{S}, \to, \ell)$, a state $s \in \mathcal{S}$ and a formula $\varphi \in$ PDL[$\mathcal{L}$], decide whether or not $\mathcal{T}, s \models \varphi$ holds.

3   Applications

We will briefly give some examples of the use of the three problems introduced in the previous section showing that each of them has attracted interest independent of the others.

Verification of Programs with Stack Inspection. In order to detect access violations in safety critical routines, inspection of the call stack may become necessary, e.g. in case of nested calls, where the initial call came from a method without the required permission. This has been implemented for instance in the runtime access control mechanism of JDK 1.2. In [22], such programs are modeled as the set of possible sequences of the call stack w.r.t. the program flow, called traces. The set of possible traces L_{tr} is an indexed language. The class IL of indexed languages [1] forms a subclass of the context-sensitive languages which properly includes the context-free languages and possesses some nice closure properties and decidability results.

One considers a regular language L_{safe} representing the set of safe traces. The verification itself is then performed by checking $L_{\mathsf{tr}} \subseteq L_{\mathsf{safe}}$, i.e. $L_{\mathsf{tr}} \cap \overline{L_{\mathsf{safe}}} = \emptyset$. and therefore is an instance of the REG-intersection problem for IL. Note that REG is closed under complement.

CFL- and REG-Reachability. The REG-reachability problem is at the core of several applications in network routing and intermodal route planning for instance [5]. It is known that the reachability problem in directed graphs when constrained with a regular language is not more difficult than the plain digraph reachability problem, i.e. NLOGSPACE-complete. However, it becomes PTIME-complete when the constraints are formed by context-free languages [5]. Such reachability problems, in particular for context-free languages have important applications in static analysis [26]. CFL-reachability for instance is used in type-based polymorphic flow analysis [7], field-sensitive points-to analysis or interprocedural dataflow analysis [27].

It is worth investigating decidability and complexity issues for classes beyond CFL which may allow more refined program analyses. Only little is known in this area so far, namely it is known that CSL-reachability is undecidable [5] which is very easily seen to be the case.

Model Checking PDL[CFL] in Abstract Interpretation. Consider the following system of mutually recursive functions where "+" denotes nondeterministic choice, ";" denotes sequential composition, and "term" denotes an anonymous terminating function.

$$
\begin{aligned}
f_0 &:= f_2; f_3 + f_2; f_1 \\
f_1 &:= f_3; f_1 + f_2; f_3 + f_1; f_3 \\
f_2 &:= f_1; f_2 + f_2; f_3 + \mathsf{term} \\
f_3 &:= f_1; f_1 + \mathsf{term}
\end{aligned}
$$

The function f_0 is the entry point of the system. Suppose we were interested in detecting whether on all possible system executions the call of f_3 is preceded by a successful return of f_1 (security check). Note that the stack behaviour, i.e. the sequences of function calls and returns is non-regular in general (for a non-fixed number of functions). We state the property we wish to verify as the regular expression $L_{\mathsf{safe}} = \Sigma^* c_1 \Sigma^* r_1 \Sigma^* c_3 \Sigma^*$, where a call of function f_i is indicated by c_i, a return by r_i respectively. It is possible to use abstract interpretation and overapproximate the system of recursive function into a one-state transition system with looping transitions for all elements in Σ. In order to restrict this overapproximation to non-spurious runs one can consider the context-free grammar

$$
\begin{aligned}
F_0 &\to c_0 F_2 F_3 r_0 \mid c_0 F_2 F_1 r_0 \\
F_1 &\to c_1 F_3 F_1 r_1 \mid c_1 F_2 F_3 r_1 \mid c_1 F_1 F_3 r_1 \\
F_2 &\to c_2 F_1 F_2 r_2 \mid c_2 F_2 F_3 r_2 \mid c_2 r_2 \\
F_3 &\to c_3 F_1 F_1 r_3 \mid c_3 r_3
\end{aligned}
$$

which is straight-forwardly derived from the recursive functions. Safety of the system is then established by checking the PDL[CFL] property $\varphi_{\mathsf{safe}} = \neg \langle L(G) \cap \overline{L_{\mathsf{safe}}} \rangle \mathsf{tt}$. It is easy to see that the only state s does not satisfy φ_{safe}: $F_0 \Rightarrow c_0 F_2 F_1 r_0 \Rightarrow^3 c_0 c_2 c_2 r_2 c_3 r_3 r_2 F_1 r_0$. Every derivation continuing from this point will end in a violation of L_{safe}, because every derivation from F_1 will be prefixed by c_1.

4   The Connection between the Three Problems

We first show that the three problems defined in Sect. 2 are interreducible onto each other. This is done as follows.

$$
\begin{array}{ccc}
\mathcal{L} - \text{reachability} & \xleftarrow{\;\;\mathcal{O}(n^2)\;\;} & \text{model checking PDL}[\mathcal{L}] \\
& \xrightarrow{\;\;\mathcal{O}(n)\;\;} & \\
\mathcal{O}(n) \Big\updownarrow \mathcal{O}(n) & & \\
\text{REG-intersection for } \mathcal{L} & &
\end{array}
$$

A single line from X to Y denotes a many-one reduction from X into Y transferring lower bounds along the arrow and upper bounds in the opposite direction. A double line denotes a Turing reduction transferring only an upper bound down the arrow but not a lower bound up the arrow.

We will begin with the forth and back between $\mathcal{L}$-reachability and model checking PDL[$\mathcal{L}$] (Lemmas 1 and 2), and then show the linear-time equivalence of $\mathcal{L}$-reachability and REG-intersection for $\mathcal{L}$ (Lemmas 3 and 4). Note that a circular series of reductions would not save any effort since the reduction from REG-intersection to model checking is very easily obtained as the composition of the two arrows via reachability. Moreover, a reduction from model checking to REG-intersection would also only be a Turing reduction, and it is not conceptually simpler than the composition of the two respective arrows via reachability.

4.1   Forth and Back between Graphs and Formulas

Lemma 1. *Let $\mathcal{L}$ be a class of languages. The $\mathcal{L}$-reachability problem reduces in linear time to the model checking problem for PDL[$\mathcal{L}$].*

Proof. Let $\mathcal{T} = (\mathcal{S}, \rightarrow)$ be a graph, $s \in \mathcal{S}$ and $T \subseteq \mathcal{S}$. Now take a proposition q_T and let $\mathcal{P} := \{q_T\}$. Define $\mathcal{T}' = (\mathcal{S}, \rightarrow, \ell')$ s.t. for all $u \in \mathcal{S}$: $q_T \in \ell'(u)$ iff $u \in T$. It is not hard to see that, for any $L \in \mathcal{L}$, there is a $w \in L$ and a $t \in T$ with $s \xrightarrow{w} t$ iff $\mathcal{T}', s \models \langle L \rangle q_T$. Furthermore, both $\mathcal{T}'$ and $\langle L \rangle q_T$ can be constructed in time $\mathcal{O}(|\mathcal{T}| + \|L\|)$.

The converse direction is not necessarily true. There does not seem to be a generic many-to-one reduction from the model checking problem for PDL[$\mathcal{L}$] to a single instance of the $\mathcal{L}$-reachability problem. However, a Turing reduction is possible. The following algorithm solves the model checking problem for PDL[$\mathcal{L}$]

given a procedure *Reach* which takes as arguments a graph $\mathcal{T}$, an $\mathcal{L}$-language, and a set U of target states and returns the set of all states which have an L-successor in U in this graph.

$$
\begin{aligned}
&\text{MC-PDL}(\mathcal{T}, \varphi) = \\
&\quad \texttt{let } (\mathcal{S}, \longrightarrow, \ell) = \mathcal{T} \texttt{ in} \\
&\quad \texttt{case } \varphi \texttt{ of} \\
&\qquad q \qquad\quad : \{s \in \mathcal{S} \mid q \in \ell(s)\} \\
&\qquad \neg\psi \qquad\ : \mathcal{S} \setminus \text{MC-PDL}(\mathcal{T}, \psi) \\
&\qquad \psi_1 \vee \psi_2 : \text{MC-PDL}(\mathcal{T}, \psi_1) \cup \text{MC-PDL}(\mathcal{T}, \psi_2) \\
&\qquad \langle L \rangle \psi \quad\ : \texttt{let } \longrightarrow' = \longrightarrow \cup \{(u, \psi?, u) \mid \psi \in \mathit{Tests}(L), \\
&\qquad\qquad\qquad\qquad\qquad\qquad\qquad\qquad u \in \text{MC-PDL}(\mathcal{T}, \psi)\ \} \\
&\qquad\qquad\qquad \texttt{in } \mathit{Reach}((\mathcal{S}, \longrightarrow'), L, \text{MC-PDL}(\mathcal{T}, \psi))
\end{aligned}
$$

Lemma 2. *Let $\mathcal{L}$ be a class of languages. The model checking problem for PDL[$\mathcal{L}$] Turing-reduces to the $\mathcal{L}$-reachability problem in quadratic time.*

Proof. It is not hard to see that algorithm MC-PDL can be made to run in time $\mathcal{O}(|\mathcal{T}| \cdot |\varphi|)$ when regarding *Reach* as an orcale. Using a dynamic programming approach one can restrict the numbers of recursive calls to one per subformula or test occurring in the input formula. Also, set operations and updates of the labeling function can be made to run in time $\mathcal{O}(|\mathcal{T}|)$. By assumption, $\mathit{Tests}(L)$ can be computed in time $\mathcal{O}(\|L\|)$ for every L occurring in φ.

Correctness of MC-PDL is straight-forward to show by induction on φ: for all states s of $\mathcal{T}$ we have: $s \in \text{MC-PDL}(\mathcal{T}, \varphi)$ iff $\mathcal{T}, s \models \varphi$.

4.2 Forth and Back between Graphs and Formal Languages

Lemmas 1 and 2 provide a connection between $\mathcal{L}$-reachability and model checking PDL[$\mathcal{L}$]. This allows to transfer lower complexity bounds from the graph-theoretic side to the logical side, and upper complexity bounds vice-versa. We will provide a further link from the formal-language side, allowing to transfer complexity results from that side — which are usually easier to achieve than for the more specialised graph-theoretic or logical problems. This can also be viewed as the aim to find a sufficient and necessary condition on the class $\mathcal{L}$ of languages that guarantees the model checking problem for PDL[$\mathcal{L}$] to be decidable.

Lemma 3. *Let $\mathcal{L}$ be a class of languages. The REG-intersection problem for $\mathcal{L}$ reduces in linear time to the $\mathcal{L}$-reachability problem.*

Proof. Let $\mathcal{A}$ be an NFA $(Q, \Sigma, \delta, q_{\text{init}}, F)$ and $L \in \mathcal{L}$. Define a graph $\mathcal{T_A} := (Q, \rightarrow)$ with $s \xrightarrow{a} t$ iff $t \in \delta(s, a)$ for any $s, t \in Q$. Now note that $L \cap L(\mathcal{A}) \neq \emptyset$ iff there is a $f \in F$ s.t. $q_0 \xrightarrow{w} f$ for some $w \in L$ iff F is L-reachable from q_0 in $\mathcal{T_A}$. Clearly, $|\mathcal{T_A}| + \|L\| = \mathcal{O}(|\mathcal{A}| + \|L\|)$.

Lemma 4. *Let $\mathcal{L}$ be a class of languages. The $\mathcal{L}$-reachability problem reduces in linear time to the REG-intersection problem for $\mathcal{L}$.*

Proof. Let $\mathcal{T} = (\mathcal{S}, \rightarrow)$ be a graph, $s \in \mathcal{S}$, $T \subseteq \mathcal{S}$, and $L \in \mathcal{L}$. Define an NFA $\mathcal{A}_{\mathcal{T},s,T} := (\mathcal{S}, \Sigma, \delta, s, T)$ s.t. for all $t \in \mathcal{S}$ and all $a \in \Sigma$: $\delta(t, a) := \{u \mid t \xrightarrow{a} u\}$. Now there is a $w \in L$ and a $t \in T$ with $s \xrightarrow{w} t$ iff there is a path in $\mathcal{T}$ from s to some $t \in T$ s.t. the transition labels along that path form the word w. This is the case iff $w \in L(\mathcal{A}_{\mathcal{T},s,T}) \cap L$. Hence, there is such a w iff $L \cap L(\mathcal{A}_{\mathcal{T},s,T}) \neq \emptyset$. Clearly, $|\mathcal{A}_{\mathcal{T},s,T}| + \|L\| = \mathcal{O}(|\mathcal{T}| + \|L\|)$.

Theorem 1. *The model checking problem for PDL[$\mathcal{L}$] is equivalent under quadratic-time Turing reductions to the REG-intersection problem for $\mathcal{L}$.*

Proof. Immediately from Lemmas 1–4.

5   New Decidability and Complexity Results on Model Checking and Formal Language Constrained Reachability

Thm. 1 allows many known results from the theory of formal languages to be transfered to the model checking theory of PDL[$\mathcal{L}$]. For example, regular languages are closed under intersections and have a decidable non-emptiness problem. Hence, their REG-intersection problem is decidable, too. In fact, it is decidable in linear time which then yields quadratic-time decidability of the model checking problem for PDL[REG]. This has of course been known for a while [8].[3]

It is also known that CFL is closed under intersections with regular languages and has a non-emptiness problem that is decidable in polynomial time. Hence, Thm. 1 reproves that model checking for PDL[CFL] is also P-complete [18].

Note that a Turing reduction, i.e. an algorithm using an oracle an arbitrary number of times, is only needed in the embedding of the model checking problem into the REG-intersection problem. The other direction is realised as an ordinary reduction. Hence, complexity-theoretic hardness results can be transferred in this direction, too. For example, the class CSL of context-sensitive languages is closed under intersections with regular languages but its non-emptiness problem is undecidable. Hence, its REG-intersection problem is undecidable, too. The same holds for the classes ACFL of *alternating context-free languages* [21] and CL of *conjunctive languages* generated by *conjunctive grammars* [23,17]. Both extend the class of context-free languages by introducing conjunctions into context-free grammars. It then also holds for their extension *boolean grammars* which gives rise to *boolean languages* BL [24].

Corollary 1. *Let $\mathcal{L} \in \{CSL, ACFL, CL, BL\}$. Then the model checking problems for PDL[$\mathcal{L}$] as well as the $\mathcal{L}$-reachability problem are undecidable.*

---

[3] Note that PDL[REG] is often said to be model checkable in linear time. However, standard algorithms are only linear in the formula and in the structure but not in both.

Note that the non-emptiness problem for context-sensitive languages is r.e. because the word problem is decidable. However, since the reduction in Lemma 2 is only a Turing-reduction, recursive enumerability does not extend to the model checking problem. This would also contradict undecidability because model checking problems for logics like PDL are closed under complement. Thus, if it was r.e. it would also be co-r.e. and therefore decidable.

The limits of decidability lie somewhere between the context-free and the context-sensitive ones. One class of languages that is known to contain CFL and be contained in CSL itself is the class IL of *indexed languages* [1]. It is known that indexed languages are closed under intersections with regular languages (with polynomial blow-ups only) and that their non-emptiness problem is EXPTIME-complete [1,28]. Hence, so is their REG-intersection problem. Applying Thm. 1 again yields positive results for model checking and graph reachability.

Corollary 2. *The model checking problem for PDL[IL] and the IL-reachability problem are EXPTIME-complete.*

There are other classes which contain CFL, have a decidable non-emptiness problem and are closed under intersections with regular languages. For example, there are *mildly context-sensitive* formalisms like the class LIL of *linear indexed languages* [9,30]. Again, they are closed under intersections with regular languages and their non-emptiness problem is decidable — even in polynomial time. Since the blow-up in the construction of intersecting a linear-indexed grammar with a regular language is polynomial, their REG-intersection problem is PTIME-complete as well. Thm. 1 then transfers the upper bound to the corresponding model checking as well as graph reachability problem. A matching lower bound follows trivially from the model checking problem for PDL[REG] or PDL[CFL] for instance. In [30] it is shown that linear indexed grammars are equivalent under polynomial-time reductions to several other at first glance different formalisms, namely head grammars, combinatory categorical grammars and tree adjoining grammars. We denote by HL, CCL and TAL the language classes generated by those formalisms respectively.

Corollary 3. *Let $\mathcal{L} \in \{\text{LIL}, \text{HL}, \text{CCL}, \text{TAL}\}$. Then the model checking problem for PDL[$\mathcal{L}$] and the $\mathcal{L}$-reachability problem are PTIME-complete.*

Finally, another class of context-sensitive languages with nice algorithmic properties has recently been discovered: MVPL is the class of languages recognised by *multi-stack visibly pushdown languages* [29]. Since it is closed under intersections in general and subsumes REG it is closed under intersections with regular languages in particular. Furthermore, its emptiness problem is decidable in double exponential time. A lower bound is currently not known. Thm. 1 then transfers this upper bound to the model checking problem of PDL[MVPL].

Corollary 4. *The model checking problem for PDL[MVPL] and MVPL-reachability are decidable in 2EXPTIME.*

language class $\mathcal{L}$	REG-inters. for $\mathcal{L}$	$\mathcal{L}$-reachability	mod. check. PDL[$\mathcal{L}$]
ACFL, CL, BL, CSL	undec. [16]	undec. [5]	undec. (here)
MVPL	2EXPTIME [29]	2EXPTIME (here)	
IL	EXPTIME-c [1],[28]	EXPTIME-c (here)	
LIL, HL, CCL, TAL	PTIME-c [9],↓	PTIME-c (here)	
DCFL, CFL	PTIME-c [4],↓	PTIME-c [5],↓	PTIME-c [18],↓
SML, SSML, VPL	PTIME-c ↑,↓		
REG	NLOGSPACE-c [15]		PTIME-c [8], folkl.

Fig. 1. Decidability and complexity results for REG-intersection, reachability and model checking

6 Summary

Fig. 1 summarises the decidability and complexity results about the REG-intersection problem for class $\mathcal{L}$, the $\mathcal{L}$-reachability problem, and the model checking problem for PDL[$\mathcal{L}$] with regards to some of the most popular classes $\mathcal{L}$ between REG and CSL. For a complexity class $\mathcal{C}$ we write $\mathcal{C}$-c to denote completeness for this class under the usual reductions. The table contains references to the location where the results have been shown first. In case of completeness, if two references are given then the first one concerns the upper, the second one the lower bound. An arrow down states that the lower bound follows from the line below, an arrow up states that the upper bound follows from the line above. All the results of the two rightmost columns — apart from PTIME-hardness in the case of $\mathcal{L}$ = REG — can be derived from Thm. 1 and the REG-intersection column. The complexities in the last row of $\mathcal{L}$ = REG do not coincide as opposed to the other rows because NLOGSPACE is presumably not closed under quadratic-time reductions.

References

1. Aho, A.V.: Indexed grammars - an extension of context-free grammars. J. ACM 15(4), 647–671 (1968)
2. Alur, R., Madhusudan, P.: Visibly pushdown languages. In: Proc. 36th Ann. ACM Symp. on Theory of Computing (STOC 2004), pp. 202–211. ACM Press, New York (2004)
3. Baader, F., Lutz, C., Turhan, A.-Y.: Small is again beautiful in description logics. KI – Künstliche Intelligenz (2010) (to appear)
4. Bar-Hillel, Y., Perles, M., Shamir, E.: On formal properties of simple phrase structure grammars. Zeitschrift für Phonologie, Sprachwissenschaft und Kommunikationsforschung 14, 113–124 (1961)
5. Barrett, C., Jacob, R., Marathe, M.: Formal-language-constrained path problems. SIAM Journal on Computing 30(3), 809–837 (2000)

6. Fagin, R., Halpern, J.Y., Moses, Y., Vardi, M.: Reasoning about Knowledge. MIT Press, Cambridge (1995)
7. Fähndrich, M., Rehof, J.: Type-based flow analysis and context-free language reachability. Mathematical Structures in Computer Science 18(5), 823–894 (2008)
8. Fischer, M.J., Ladner, R.E.: Propositional dynamic logic of regular programs. Journal of Computer and System Sciences 18, 194–211 (1979)
9. Gazdar, G.: Applicability of indexed grammars to natural languages. In: Reyle, U., Rohrer, C. (eds.) Natural Language Parsing and Linguistic Theories, pp. 69–94. Reidel, Dordrecht (1988)
10. De Giacomo, G., Lenzerini, M.: Boosting the correspondence between description logics and propositional dynamic logics. In: Proc. of the 12th National Conference on Artificial Intelligence (AAAI 1994), pp. 205–212. AAAI-Press/The MIT-Press (1994)
11. Harel, D., Kaminsky, M.: Strengthened results on nonregular PDL. Technical Report MCS99-13, Weizmann Institute of Science, Faculty of Mathematics and Computer Science (1999)
12. Harel, D., Pnueli, A., Stavi, J.: Propositional dynamic logic of nonregular programs. Journal of Computer and System Sciences 26(2), 222–243 (1983)
13. Harel, D., Raz, D.: Deciding properties of nonregular programs. SIAM J. Comput. 22(4), 857–874 (1993)
14. Harel, D., Singerman, E.: More on nonregular PDL: Finite models and Fibonacci-like programs. Information and Computation 128(2), 109–118 (1996)
15. Hunt, H.B.: On the time and tape complexity of languages I. In: ACM (ed.) Conf. Rec. of 5th Annual ACM Symp. on Theory of Computing (STOC 1973), pp. 10–19. ACM Press, New York (1973)
16. Landweber, P.S.: Three theorems on phrase structure grammars of type 1. Inform. and Control 6, 131–136 (1963)
17. Lange, M.: Alternating context-free languages and linear time μ-calculus with sequential composition. In: Proc. 9th Workshop on Expressiveness in Concurrency (EXPRESS 2002). ENTCS, vol. 68.2, pp. 71–87. Elsevier, Amsterdam (2002)
18. Lange, M.: Model checking propositional dynamic logic with all extras. Journal of Applied Logic 4(1), 39–49 (2005)
19. Löding, C., Lutz, C., Serre, O.: Propositional dynamic logic with recursive programs. J. Log. Algebr. Program 73(1-2), 51–69 (2007)
20. Mehlhorn, K.: Pebbling mountain ranges and its application to DCFL-recognition. In: de Bakker, J.W., van Leeuwen, J. (eds.) ICALP 1980. LNCS, vol. 85, pp. 422–435. Springer, Heidelberg (1980)
21. Moriya, E.: A grammatical characterization of alternating pushdown automata. TCS 67(1), 75–85 (1989)
22. Nitta, N., Seki, H., Takata, Y.: Security verification of programs with stack inspection. In: SACMAT, pp. 31–40 (2001)
23. Okhotin, A.: Conjunctive grammars. Journal of Automata, Languages and Combinatorics 6(4), 519–535 (2001)
24. Okhotin, A.: Boolean grammars. Information and Computation 194(1), 19–48 (2004)
25. Prendinger, H., Schurz, G.: Reasoning about action and change. A dynamic logic approach. Journal of Logic, Language and Information 5(2), 209–245 (1996)
26. Reps, T.: Shape analysis as a generalized path problem. In: Proc. ACM SIGPLAN Symp. on Partial Evaluation and Semantics-Based Program Manipulation, pp. 1–11 (1995)

27. Reps, T.W.: Program analysis via graph reachability. Information & Software Technology 40(11-12), 701–726 (1998)
28. Tanaka, S., Kasai, T.: The emptiness problem for indexed language is exponential-time complete. Systems and Computers in Japan 17(9), 29–37 (2007)
29. La Torre, S., Madhusudan, P., Parlato, G.: A robust class of context-sensitive languages. In: Proc. 22nd Conf. on Logic in Computer Science (LICS 2007), pp. 161–170. IEEE, Los Alamitos (2007)
30. Vijay-Shanker, K., Weir, D.J.: The equivalence of four extensions of context-free grammars. Mathematical Systems Theory 27, 27–511 (1994)

Completeness of the Bounded Satisfiability Problem for Constraint LTL[*]

Marcello M. Bersani, Achille Frigeri, Matteo Rossi, and Pierluigi San Pietro

Politecnico di Milano
{bersani,frigeri,rossi,sanpietr}@elet.polimi.it

Abstract. We show that the satisfiability problem for LTL (with past operators) over arithmetic constraints (Constraint LTL) can be answered by solving a finite amount of instances of *bounded* satisfiability problems when atomic formulae belong to certain suitable fragments of Presburger arithmetic. A formula is boundedly satisfiable when it admits an ultimately periodic model of the form $\delta\pi^\omega$, where δ and π are finite sequences of *symbolic valuations*. Therefore, for every formula there exists a *completeness bound* c, such that, if there is no ultimately periodic model with $|\delta\pi| \leq c$, then the formula is unsatisfiable.

1 Introduction

Given a formula ϕ expressed in a logical formalism, the *satisfiability* problem for ϕ is to determine whether there exists a model σ for ϕ (written $\sigma \models \phi$), i.e., an assignment of values to all of its atomic elements satisfying the formula. The logical formalism adopted here is $\mathrm{CLTLB}(\mathcal{D})$, which has already been studied in [7], though it was introduced in [4]. It is an extension of LTL in which atomic formulae are arithmetic constraints of a *constraint system* $\mathcal{D}$; e.g., $(\mathrm{X}x < y)\mathbf{U}(x \equiv_2 \mathrm{Y}z)$ is a legal formula when $\mathcal{D}$ is the structure $(\mathbb{N}, <, \equiv_d)$, x, y, z are variables over $\mathbb{N}$, $\mathbf{U}$ is the *until* temporal operator of LTL and X and Y are temporal operators on variables meaning the *next* value and the *previous* value, respectively. The satisfiability problem for $\mathrm{CLTLB}(\mathcal{D})$ was analyzed in depth when $\mathcal{D}$ is $(D, <, \equiv_d)$ for $D \in \{\mathbb{N}, \mathbb{Z}, \mathbb{Q}, \mathbb{R}\}$ in [5], and when $\mathcal{D}$ is the language of Integer Periodic Constraints in [7]. The decidability of the satisfiability problem for the above cases is shown by using an automata-based approach similar to the standard case for LTL. Given a $\mathrm{CLTLB}(\mathcal{D})$ formula ϕ, it is possible to define an automaton $\mathcal{A}_\phi$ such that ϕ is satisfiable iff $\mathscr{L}(\mathcal{A}_\phi)$ is not empty. Since the emptiness of $\mathscr{L}(\mathcal{A}_\phi)$ is decidable in PSPACE (polynomial space in the dimension of ϕ) [7], then the satisfiability problem is also decidable with the same complexity.

In this paper, we solve the satisfiability problem by following a different approach. We reduce the satisfiability of a $\mathrm{CLTLB}(\mathcal{D})$ formula ϕ to a finite number of instances of the *bounded* satisfiability problem (BSP) for ϕ. BSP is defined, given a formula ϕ and a constant $k \in \mathbb{N}$, as the problem of deciding whether

---

[*] This research was partially supported by Programme IDEAS-ERC and Project 227977-SMScom.

G. Delzanno and I. Potapov (Eds.): RP 2011, LNCS 6945, pp. 58–71, 2011.

ϕ admits an ultimately periodic model of length k, i.e., of the form $\delta\pi^\omega$ with $|\delta\pi| = k$. Each instance of BSP is then reduced to the satisfiability of Quantifier-Free Linear Integer (Real) Arithmetic with Uninterpreted Function QF-UFLIA (QF-UFLRA), by encoding the semantics of CLTLB($\mathcal{D}$) into a QF-UFLIA (QF-UFLRA) formula, following the approach introduced in [1]. Theory QF-UFLIA (QF-UFLRA) is the union of the theory of Equality with Uninterpreted functions and the quantifier-free fragment of Presburger (Real) arithmetic; its decidability can easily be proved by combining the decision procedures of the underlying theories, leading also to the efficient implementations of SMT-solvers such as Z3[1] or Yices[2]. Constant k is bounded by the length of the longest acyclic path of a Büchi automaton $\mathcal{A}_\phi$ representing the models of ϕ, which can effectively be computed. Since k is bounded, it is always possible, at least in theory, to decide the satisfiability of a CLTLB($\mathcal{D}$) formula. Hence, we say that our decision procedure for the satisfiability problem is *complete*. The value bounding the number of instances of BSP to be solved is called the *completeness bound*. Since the decision procedure is complete, the class of properties that can be automatically verified includes not only the reachability problems investigated in [1] ("the system will eventually be in configuration c"), but also more complex properties like liveness properties (e.g., $\mathbf{GF}(\mathbf{X}x > y)$).

To the best of our knowledge, the completeness bound for CLTLB($\mathcal{D}$) has never been investigated before. A related work about completeness for some classes of LTL formulae, but tailored to model checking over Kripke structures, is shown in [2], while further refinements can be found in [3], which gives a method to define the bound for full LTL.

The analysis presented herein is effective; the bounded satisfiability problem can be solved by using tools implementing the encoding of LTL with arithmetic constraints and temporal modalities over variables. Our version of the encoding has been already presented in [1] and implemented in the Zot[3] tool.

The main results of this paper are the following: (i) CLTLB($\mathcal{D}$) formulae using both the X and Y operators on variables are equisatisfiable w.r.t. CLTLB($\mathcal{D}$) formulae using only X. (ii) BSP is defined and it is shown to be complete with respect to the satisfiability problem of CLTLB($\mathcal{D}$); i.e., there is a finite bound on the number of BSP instances to be solved in order to answer the satisfiability problem. (iii) The satisfiability problem for CLTLB($\mathcal{D}$), when $\mathcal{D}$ is one of the constraint systems mentioned above, can be reduced to the satisfiability problem of a QF-UFLIA (QF-UFLRA) formula.

2   Languages

CLTL($\mathcal{D}$) is an extension of LTL enriched with temporal modalities over variables, where atomic formulae are defined by Boolean combinations of arithmetic formulae belonging to a constraint system $\mathcal{D}$. The logic CLTLB($\mathcal{D}$) extends

---

[1] http://research.microsoft.com/en-us/um/redmond/projects/z3/
[2] http://yices.csl.sri.com/
[3] http://home.dei.polimi.it/pradella/Zot/

CLTL($\mathcal{D}$) allowing the use of past operators. This section recalls the definitions of constraint system and of CLTLB($\mathcal{D}$).

Constraint Language. Let V be a finite set of variables; a *constraint system* is a pair $\mathcal{D} = \langle D, \Pi \rangle$ where D is a specific domain of interpretation for variables and constants and Π is a family of relations on D that is closed under complement. An *atomic $\mathcal{D}$-constraint* is a term of the form $R(x_1, \ldots, x_n)$, where R is an n-ary relation on D and $x_1, \ldots, x_n$ are variables. A $\mathcal{D}$-valuation is a mapping $v : V \to D$, i.e., an assignment of a value in D to each variable. A constraint is *satisfied* by a $\mathcal{D}$-valuation v, written $v \models_{\mathcal{D}} R(x_1, \ldots, x_n)$, if $(v(x_1), \ldots, v(x_n)) \in R$.

In this paper we consider $\mathcal{D}$ to be one of the following constraint systems: Integer Periodic Constraints (IPC*) or fragments (e.g., $(\mathbb{Z}, <, =)$ or $(\mathbb{N}, <, =)$) and $(D, <, =)$ when $<$ is a dense order without endpoints, e.g., $D = \mathbb{R}, \mathbb{Q}$. The language IPC* is defined by the following grammar, where ξ is the axiom:

$$\xi := \theta \mid x < y \mid \xi \wedge \xi \mid \neg \xi$$
$$\theta := x \equiv_c d \mid x \equiv_c y + d \mid x = y \mid x < d \mid x = d \mid \theta \wedge \theta \mid \neg \theta$$

where $x, y \in V$, $c \in \mathbb{N}^+$ and $d \in \mathbb{Z}$. The first definition of IPC* can be found in [6]; it is different from ours since it allows the existentially quantified formulae (i.e., $\theta := \exists x \, \theta$) to be part of the language. However, since IPC* is a fragment of Presburger arithmetic, it has the same expressivity of the above quantifier-free version (but with an exponential blow-up to remove quantifiers). The restriction IPC^{++} is the language defined by considering θ, rather than ξ, as the axiom in the above grammar.

Given a valuation v, the satisfaction relation $\models_{\mathcal{D}}$ is defined:

- $v \models_{\mathcal{D}} x \sim y$ iff $v(x) \sim v(y)$;
- $v \models_{\mathcal{D}} x \sim d$ iff $v(x) \sim d$;
- $v \models_{\mathcal{D}} x \equiv_c d$ iff $v(x) - d = kc$ for some $k \in \mathbb{Z}$;
- $v \models_{\mathcal{D}} x \equiv_c y + d$ iff $v(x) - v(y) - d = kc$ for some $k \in \mathbb{Z}$;
- $v \models_{\mathcal{D}} \xi_1 \wedge \xi_2$ iff $v \models_{\mathcal{D}} \xi_1$ and $v \models_{\mathcal{D}} \xi_2$;
- $v \models_{\mathcal{D}} \neg \xi$ iff $v \not\models_{\mathcal{D}} \xi$.

A constraint is *satisfiable* if there is a valuation v such that $v \models_{\mathcal{D}} \xi$. Given a set of IPC* constraints C, we write $v \models_{\mathcal{D}} C$ when $v \models_{\mathcal{D}} \xi$ for every $\xi \in C$.

Temporal Language. CLTLB($\mathcal{D}$) includes Boolean connectives as well as the usual temporal modalities of LTL $\mathbf{X}$ (next) and $\mathbf{U}$ (until), together with their past counterparts $\mathbf{Y}$ (yesterday) and $\mathbf{S}$ (since). Let x be a variable in V, an *arithmetic temporal term* (a.t.t.) α is defined by the grammar:

$$\alpha := x \mid \mathbf{X}\alpha \mid \mathbf{Y}\alpha.$$

The *depth* $|\alpha|$ of an a.t.t. α is defined by induction as: $|x| = 0$, $|\mathbf{X}\alpha| = |\alpha| + 1$, $|\mathbf{Y}\alpha| = |\alpha| - 1$. The syntax of (well formed) formulae of CLTLB($\mathcal{D}$) is defined as follows:

$$\phi := \alpha_1 \sim \alpha_2 \mid \phi \wedge \phi \mid \neg \phi \mid \mathbf{X}\phi \mid \mathbf{Y}\phi \mid \phi \mathbf{U} \phi \mid \phi \mathbf{S} \phi.$$

Let ϕ be a CLTLB($\mathcal{D}$) formula. Let $at(\phi)$ be the set of all a.t.t's occurring in formula ϕ; moreover, if x is a variable, $at_x(\phi) \subseteq at(\phi)$ is the set of all a.t.t.'s in which x appears. We define the "look-forwards" $\lceil\phi\rceil_x$ and "look-backwards" $\lfloor\phi\rfloor_x$ of ϕ relatively to x as: $\lceil\phi\rceil_x = \max_{\alpha_i \in at_x(\phi)}\{0, |\alpha_i|\}$, $\lfloor\phi\rfloor_x = \min_{\alpha_i \in at_x(\phi)}\{0, |\alpha_i|\}$. The above definitions naturally extend to V by letting $\lceil\phi\rceil = \max_{x \in V}\{\lceil\phi\rceil_x\}$, $\lfloor\phi\rfloor = \min_{x \in V}\{\lfloor\phi\rfloor_x\}$. Hence, $\lceil\phi\rceil$ ($\lfloor\phi\rfloor$) is the largest (smallest) depth of all the a.t.t.'s of ϕ, representing the length of the future (past) segment needed to evaluate ϕ in the current instant.

The semantics of a formula ϕ of CLTLB($\mathcal{D}$) is defined w.r.t. a sequence of $\mathcal{D}$-valuations $\sigma : \mathbb{Z} \times V \to D$. The satisfaction relation $\models$ is defined for $i \geq 0$ as follows, for every formulae ϕ, ψ and for every a.t.t. α (where x_{α_i} is the variable that appears in α_i):

$$\sigma, i \models_{\mathcal{D}} \alpha_1 \sim \alpha_2 \Leftrightarrow \sigma(i + |\alpha_1|, x_{\alpha_1}) \sim \sigma(i + |\alpha_2|, x_{\alpha_2})$$
$$\sigma, i \models \neg\phi \Leftrightarrow \sigma, i \not\models \phi$$
$$\sigma, i \models \phi \wedge \psi \Leftrightarrow \sigma, i \models \phi \wedge \sigma, i \models \psi$$
$$\sigma, i \models \mathbf{X}\phi \Leftrightarrow \sigma, i + 1 \models \phi$$
$$\sigma, i \models \mathbf{Y}\phi \Leftrightarrow \sigma, i - 1 \models \phi \wedge i > 0$$
$$\sigma, i \models \phi\mathbf{U}\psi \Leftrightarrow \exists j \geq i : \sigma, j \models \psi \wedge \sigma, n \models \phi \,\forall i \leq n < j$$
$$\sigma, i \models \phi\mathbf{S}\psi \Leftrightarrow \exists 0 \leq j \leq i : \sigma, j \models \psi \wedge \sigma, n \models \phi \,\forall j < n \leq i$$

A formula $\phi \in$ CLTLB($\mathcal{D}$) is *satisfiable* if there exists a sequence σ such that $\sigma, 0 \models \phi$ (in which case σ is a *model* of ϕ). The CLTLB($\mathcal{D}$) language admits the use of the "previous" operator Y on a.t.t.'s, nevertheless, we shall only consider the future fragment of the language defining the a.t.t.'s as Y can be removed from formulae. Let X^i (resp. Y^i) represent the nesting of X (resp. Y) i times and let $p : \text{CLTL}(\mathcal{D}) \to \text{CLTL}(\mathcal{D})$ be the rewriting function defined recursively as:

$$p(X^i x) \stackrel{\text{def}}{=} X^{i-|\phi|}x \quad p(Y^i x) \stackrel{\text{def}}{=} Y^{i+|\phi|}x \quad p(\alpha_1 \sim \alpha_2) \stackrel{\text{def}}{=} p(\alpha_1) \sim p(\alpha_2)$$
$$p(\neg\phi) \stackrel{\text{def}}{=} \neg p(\phi) \quad p(\phi \wedge \psi) \stackrel{\text{def}}{=} p(\phi) \wedge p(\psi) \quad p(\mathbf{X}\phi) \stackrel{\text{def}}{=} \mathbf{X}p(\phi)$$
$$p(\mathbf{Y}\phi) \stackrel{\text{def}}{=} \mathbf{Y}p(\phi) \quad p(\phi\mathbf{U}\psi) \stackrel{\text{def}}{=} p(\phi)\mathbf{U}p(\psi) \quad p(\phi\mathbf{S}\psi) \stackrel{\text{def}}{=} p(\phi)\mathbf{S}p(\psi)$$

Given a CLTLB($\mathcal{D}$) formula ϕ it is easy to see that Y does not occur in $p(\phi)$ since Y^{-i} can be rewritten as X^i (e.g., $X^3 y = Y^{-3}y$). The equisatisfiability of formulae is guaranteed by moving the origin of ϕ by $-\lfloor\phi\rfloor$ instants in the past. Since only X occurs in $p(\phi)$, then models for CLTLB($\mathcal{D}$) formulae without Y are now sequences of $\mathcal{D}$-valuations $\sigma : \mathbb{N} \times V \to D$.

Proposition 1. *Let ϕ be a CLTLB($\mathcal{D}$) formula, then $\sigma, 0 \models \phi \Leftrightarrow \sigma, \lfloor\phi\rfloor \models p(\phi)$.*

Proof. Let $s = \lfloor\phi\rfloor$. We show that for all $i \geq 0$, $\sigma, i \models \phi \Leftrightarrow \sigma, i + s \models p(\phi)$ by induction on the structure of the formula ϕ.

The **base case** of the induction is given on the atomic formulae $\phi = \alpha_1 \sim \alpha_2$. Since $\sigma, i \models_{\mathcal{D}} \phi \Leftrightarrow \sigma(i + |\alpha_1|, x_{\alpha_1}) \sim \sigma(i + |\alpha_2|, x_{\alpha_2})$, by shifting the instant i of s the satisfaction relation is $\sigma, i \models_{\mathcal{D}} \phi \Leftrightarrow \sigma(i + s + |\alpha_1| - s, x_{\alpha_1}) \sim \sigma(i + s + |\alpha_2| - s, x_{\alpha_2})$. Then, we can equivalently write $\sigma, i \models_{\mathcal{D}} \phi \Leftrightarrow \sigma(i + s + |p(\alpha_1)|, x_{\alpha_1}) \sim$

$\sigma(i + s + |p(\alpha_2)|, x_{\alpha_2})$ that is $\sigma, i + s \models p(\alpha_1) \sim p(\alpha_2)$ and $\sigma, i + s \models p(\alpha_1 \sim \alpha_2)$. In fact, if $\alpha = X^i x$ then $p(\alpha) = X^{i-s} x$ and $|p(\alpha)| = |\alpha| - s$. If $\alpha = Y^i x$ then $p(\alpha) = Y^{i+s} x$ and $|p(\alpha)| = -(i + s) = |\alpha| - s$, since $|\alpha| = -i$.

Inductive Step

- If $\phi = \neg\psi$ then $\sigma, i \models \phi \Leftrightarrow \sigma, i \not\models \psi$. By inductive hypothesis, this is equivalent to $\sigma, i + s \not\models p(\psi)$, i.e. $\sigma, i + s \models p(\phi)$, as $p(\phi) = \neg p(\psi)$.
- If $\phi = \psi_1 \wedge \psi_2$ then $\sigma, i \models \phi \Leftrightarrow \sigma, i \models \psi_1$ and $\sigma, i \models \psi_2$. By inductive hypothesis, this is equivalent to $\sigma, i + s \models p(\psi_1)$ and $\sigma, i + s \models p(\psi_2)$, i.e. $\sigma, i + s \models p(\psi_1) \wedge p(\psi_2)$, and $\sigma, i + s \models p(\phi)$.
- If $\phi = X\psi$ then $\sigma, i \models \phi \Leftrightarrow \sigma, i + 1 \models \psi$. By inductive hypothesis, this is equivalent to $\sigma, i + 1 + s \models p(\psi)$, i.e., $\sigma, i + s \models Xp(\psi)$, which corresponds to $\sigma, i + s \models p(\phi)$.
- If $\phi = Y\psi$ then $\sigma, i \models \phi \Leftrightarrow \sigma, i - 1 \models \psi$. By inductive hypothesis, this is the same as $\sigma, i - 1 + s \models p(\psi)$, i.e., $\sigma, i + s \models Yp(\psi)$, and $\sigma, i + s \models p(\phi)$, as $p(\phi) = Yp(\psi)$.
- If $\phi = \psi_1 U\psi_2$ then $\sigma, i \models \phi$ iff there exists $j \geq i$ s.t. $\sigma, j \models \psi_2$ and $\sigma, n \models \psi_1$ forall $i \leq n < j$, that is, by inductive hypothesis, $\sigma, j+s \models p(\psi_2)$ and $\sigma, n \models p(\psi_1)$ forall $i + s \leq n < j + s$, which in turn is equivalent to $\sigma, i + s \models p(\psi_1)Up(\psi_2)$ and $\sigma, i + s \models p(\phi)$.
- If $\phi = \psi_1 S\psi_2$ then $\sigma, i \models \phi$ iff there exists $0 \leq j \leq i$ s.t. $\sigma, j \models \psi_2$ and $\sigma, n \models \psi_1$ forall $j < n \leq i$, that is, by inductive hypothesis $\sigma, j+s \models p(\psi_2)$ and $\sigma, n \models p(\psi_1)$ forall $j + s < n \leq i + s$, which is equivalent to $\sigma, i + s \models p(\psi_1)Sp(\psi_2)$ and $\sigma, i + s \models p(\phi)$.

Finally, $\sigma, 0 \models \phi \Leftrightarrow \sigma, s \models p(\phi)$ by taking $i = 0$.   □

3   Symbolic Valuations

In order to represent exactly models of a CLTL($\mathcal{D}$) formula ϕ by means of automata, we need to represent symbolically all sequences σ such that $\sigma \models \phi$. The same representation can be adopted also for CLTLB($\mathcal{D}$) formulae (without occurrences of Y). In this section, we briefly recall some useful notions. At the end, we give a slightly different construction than that of [5] of the automaton recognizing models of CLTLB($\mathcal{D}$) formulae, one that is tailored to results given in Section 5.1.

Let ϕ be a CLTLB($\mathcal{D}$) formula, $terms(\phi)$ be the set of arithmetic terms of the form $X^i x$ for all $0 \leq i \leq \lceil \phi \rceil$ and for all $x \in V$ and $c(\phi)$ be the set of constants occurring in ϕ. A set of $\mathcal{D}$-constraints over $terms(\phi)$ is *maximally consistent* if, for every $\mathcal{D}$-constraint θ over $terms(\phi)$ and $c(\phi)$, either θ or $\neg\theta$ is in the set. A *symbolic valuation sv* for ϕ is a maximally consistent set of $\mathcal{D}$-constraints over $terms(\phi)$ and $c(\phi)$; the set of all symbolic valuations for ϕ is denoted by $SV(\phi)$.

A valuation $v : V \rightarrow D$ naturally extends to a valuation $v' : terms(\phi) \rightarrow D$, such that $v' \models_{\mathcal{D}} \alpha_1 \sim \alpha_2$ iff $v'(\alpha_1) \sim v'(\alpha_2)$. Then, a symbolic valuation sv for ϕ is *satisfiable* if there exists a $\mathcal{D}$-valuation $v' : terms(\phi) \rightarrow D$ such that $v' \models_{\mathcal{D}} \xi$, for all ξ belonging to sv. We write $v' \models_{\mathcal{D}} sv$ when sv is satisfied by v'.

Given a symbolic valuation sv and a $\mathcal{D}$-constraint ξ over a.t.t.'s, we write $sv \models_s \xi$ if for every $\mathcal{D}$-valuation v' such that $v' \models_{\mathcal{D}} sv$ then $v' \models \xi$. Observe that in the considered constraint systems, the problem of checking whether $sv \models_s \xi$ is decidable. All symbolic valuations may be defined by means of a syntactic construction on formula ϕ by using a procedure similar to the one in [5].

A pair of symbolic valuations (sv_1, sv_2) is *locally consistent* if, for all $\sim$ in $\mathcal{D}$:

$$X^{i_1} x_1 \sim X^{i_2} x_2 \in sv_1 \text{ implies } X^{i_1-1} x_1 \sim X^{i_2-1} x_2 \in sv_2.$$

A sequence of symbolic valuations $sv_0, sv_1, \ldots$ is *locally consistent* if all pairs (sv_i, sv_{i+1}), $i \geq 0$, are locally consistent. A locally consistent infinite sequence $\rho : \mathbb{N} \to SV(\phi)$ of symbolic valuations *admits a model*, written $\sigma \models \rho$, if there exists a model σ of ϕ such that for every $i \geq 0$, $\sigma, i \models \rho(i)$. In this case, ρ is called a *symbolic model* for ϕ.

The satisfaction relation $\models_s$ can also be extended to sequences of symbolic valuations; it is the same as $\models$ for all temporal operators except for atomic formulae:

$$\rho, i \models_s \xi \Leftrightarrow \rho(i) \models_s \xi.$$

The following fundamental proposition draws a link between the satisfiability by sequences of symbolic valuations and by sequences of $\mathcal{D}$-valuations.

Proposition 2 ([5]). *A CLTL($\mathcal{D}$) formula ϕ is satisfiable iff there exists a symbolic model for ϕ.*

Given a CLTL($\mathcal{D}$) formula ϕ, it is possible [5] to define an automaton $\mathcal{A}_\phi$ recognizing symbolic models of ϕ, which reduces satisfiability of CLTL($\mathcal{D}$) to emptiness of Buchi automata. The idea is that automaton $\mathcal{A}_\phi$ should accept the intersection of the following languages, which defines exactly the language of symbolic models of ϕ:

(i) the language of LTL models ρ;
(ii) the language of sequences of locally consistent symbolic valuations;
(iii) the language of sequences of symbolic valuations for ϕ which admit an arithmetic model.

Languages (i) and (ii) can be accepted by Büchi automata, called respectively $\mathcal{A}_s$ and $\mathcal{A}_\ell$. In general, however, the language (iii) may *not* be ω-regular. Nonetheless, automaton $\mathcal{A}_\phi$ can be defined to accept a superset of the language of the sequences of locally consistent symbolic valuations that are models for ϕ, such that the *ultimately periodic* models of $\mathcal{A}_\phi$ are all *ultimately periodic* models of ϕ. Then, from Lemma (1) below, it follows that ϕ is satisfiable iff $\mathcal{A}_\phi$ recognizes an *ultimately periodic* word.

$\mathcal{A}_\phi$ is defined as the product of automata $\mathcal{A}_\ell$, $\mathcal{A}_s$, and $\mathcal{A}_C$, where $\mathcal{A}_C$ defines a condition C guaranteeing the existence of a sequence σ such that $\sigma \models \rho$. In particular, for constraint systems IPC* and $(\mathbb{N}, <, =)$, $(\mathbb{Z}, <, =)$, $\mathcal{A}_C$ can effectively be built. Condition C is given by considering the graph representation G_ρ of sequences of symbolic valuations ρ. It enforces the absence of infinite $<$-strict paths in graph G_ρ, i.e., that between any two nodes of G_ρ there are no paths of infinite length in which relation $<$ occurs (see details in [5]). When the condition C is sufficient and necessary for the existence of models σ such that $\sigma \models \rho$, then automaton $\mathcal{A}_\phi$ represents all the sequences of symbolic valuations which admit a model. A fundamental lemma, on which Proposition 3 below relies on, draws a sufficient and necessary condition for the existence of models of sequences of symbolic valuations.

Lemma 1 ([5]). *Let ρ be an ω-periodic sequence of symbolic valuations of the form $\rho = \alpha(\beta)^\omega$ that is locally consistent. Then ρ admits a model σ iff ρ satisfies C.*

Proposition 3 ([5]). *A CLTL($\mathcal{D}$) formula is satisfiable iff the language $\mathscr{L}(\mathcal{A}_\phi)$ is not empty.*

3.1   Automaton Construction

It is worth noticing that the definition of $\mathcal{A}_\phi$ is given by considering as alphabet the set $SV(\phi)$ of all symbolic valuations of ϕ. This construction can be slightly modified in the definition of $\mathcal{A}_s$ since this automaton can be built using a restricted alphabet instead of $SV(\phi)$. This allows us to use the fixpoint representation of the semantics of formula ϕ, as we will actually do in Section 5.1. In this section we define the synchronization of the three automata.

Let ϕ be a CLTL($\mathcal{D}$) formula, let $A \subseteq \mathcal{D}$ be the closure under negation of the set of arithmetic constraints occurring in ϕ, and let $\mathcal{A}_s = (\Sigma, Q', Q_0, \eta, F)$ be the symbolic Büchi automaton of ϕ. Alphabet Σ is the subset $valid(A) \subseteq \wp(A)$ such that for every atomic formula ξ of ϕ, $\beta \in valid(A)$ iff either ξ or $\neg\xi$ belongs to β. The closure of ϕ, denoted $cl(\phi)$, is the smallest set containing all subformulae of ϕ that is also closed under negation. An *atom* $\Gamma \subseteq cl(\phi)$ is a subset of formulae of $cl(\phi)$ that is maximally consistent, i.e., such that, for each formula ξ in ϕ, either $\xi \in \Gamma$ or $\neg\xi \in \Gamma$. It is worth noticing that an atom so defined might be unsatisfiable, i.e., there does not exist a valuation v' over a.t.t.'s such that $v' \models_\mathcal{D} \xi$, for all ξ in Γ, since $cl(\phi)$ is closed under negation. A pair (Γ_1, Γ_2) of atoms is *one-step temporally consistent* when Γ_1 and Γ_2 agree on the structure of temporal operators, that is:

- for every $\mathbf{X}\psi \in cl(\phi)$, then $\mathbf{X}\psi \in \Gamma_1 \Leftrightarrow \psi \in \Gamma_2$,
- for every $\mathbf{Y}\psi \in cl(\phi)$, then $\mathbf{Y}\psi \in \Gamma_2 \Leftrightarrow \psi \in \Gamma_1$,
- if $\psi_1\mathbf{U}\psi_2 \in \Gamma_1$, then $\psi_2 \in \Gamma_1$ or ($\psi_1 \in \Gamma_1$ and $\psi_1\mathbf{U}\psi_2 \in \Gamma_2$),
- if $\psi_1\mathbf{S}\psi_2 \in \Gamma_2$, then $\psi_2 \in \Gamma_2$ or ($\psi_1 \in \Gamma_2$ and $\psi_1\mathbf{S}\psi_2 \in \Gamma_1$).

The automaton $\mathcal{A}_s = (\Sigma, Q, Q_0, \eta, F)$ is then defined as follows:

- Q is the set of atoms;
- $Q_0 = \{\Gamma \in Q : \phi \in \Gamma, \mathbf{Y}\psi \notin \Gamma$ for all $\psi \in cl(\phi), \psi_1\mathbf{S}\psi_2 \in \Gamma$ iff $\psi_2 \in \Gamma\}$;
- $\Gamma_1 \xrightarrow{\beta} \Gamma_2 \in \eta$ iff

- $\beta = \Gamma_1 \cap A$,
- (Γ_1, Γ_2) is one-step consistent;

- $F = \{F_1, \ldots, F_m\}$, where $F_i = \{\Gamma \in Q \mid \phi_i \mathbf{U} \psi_i \notin \Gamma$ or $\psi_i \in \Gamma\}$ and $\{\phi_1 \mathbf{U} \psi_1, \ldots, \phi_m \mathbf{U} \psi_m\}$ is the set of Until formulae occurring in $cl(\phi)$.

The automaton $\mathcal{A}_s$ is a generalized Büchi automaton. In order to provide the automaton $\mathcal{A}_\phi$, we shall translate $\mathcal{A}_s$ into a classical Büchi automaton, still preserving the language of accepted $\omega-$words. For ease of writing, we also denote this automaton with $\mathcal{A}_s$.

Let $\mathcal{A} = (SV(\phi), Q', Q'_0, \delta', F')$ be the automaton over the alphabet of symbolic valuations given by the intersection of automata $\mathcal{A}_\ell$ and $\mathcal{A}_{\neg C}$, as shown in [5]. Automaton $\mathcal{A}_\phi = (SV(\phi), Q'', Q''_0, \eta, F'')$ is defined as the product of $\mathcal{A}_s$ and $\mathcal{A}$, according to the standard intersection of Büchi automata but adapted in the definition of η:

- $Q'' = Q \times Q' \times \{0, 1, 2\}$;
- $Q''_0 = \{(\Gamma, q', 0) : \Gamma \in Q_0$ and $q' \in Q'_0\}$;
- $(\Gamma_1, q', i) \xrightarrow{sv} (\Gamma_2, p', j) \in \eta$ iff $\Gamma_1 \xrightarrow{\beta} \Gamma_2 \in \delta$, $q' \xrightarrow{sv} p' \in \delta'$ and $sv \models_s \xi$, for all $\xi \in \Gamma_1$, and:
  - if $i = 0$ then $j = 1$;
  - if $i = 1$ and $\Gamma_1 \in F$, then $j = 2$;
  - if $i = 2$ and $q' \in F'$, then $j = 0$;
  - otherwise $i = j$;
- $F'' = Q \times Q' \times \{0\}$.

Proposition 3 holds also for $\mathcal{A}_\phi$. The proof follows the line of Lemma 6.3 of [5].

Proof. Suppose that ϕ is satisfiable. From Proposition 2, there exists a sequence ρ such that $\rho \models_s \phi$ and ρ admits a model σ. The sequence of symbolic valuations is such that, at each step, a set of $\mathcal{D}$-constraints $\beta \in A$ is satisfied, $\rho(i) \models_s \beta$. Then sequence $\rho \in \mathcal{L}(\mathcal{A}_s)$, and ρ is locally consistent; hence, $\rho \in \mathcal{L}(\mathcal{A}_\ell)$. Moreover, by Lemma 1, $\rho \in \mathcal{L}(\mathcal{A}_C)$, and so also $\rho \in \mathcal{A}_\phi$.

Conversely, suppose $\mathcal{A}_\phi$ accepts a word ρ. By the nature of its acceptance condition, $\mathcal{A}_\phi$ must also accept some ultimately periodic word, say ρ'. As $\rho' \in \mathcal{L}(\mathcal{A}_s)$, it follows that $\rho' \models_s \phi$, with ρ' locally consistent and satisfying C. By Lemma 1, ρ' admits a sequence of $\mathcal{D}$-valuations σ such that $\sigma \models \rho'$. Moreover, since $\rho' \models_s \phi$ and $\sigma \models \rho'$ then, by Proposition 2, $\sigma \models \phi$. Hence, ϕ is satisfiable.

4   Extensions

This section is based on concepts given in [5] in order to simplify and extend the results presented in Section 3. In particular, the construction of $\mathcal{A}_\phi$ can be made simpler when $\mathcal{D}$ benefits of a property of *completion*, that, when only order relations are consider, reduces to require that it is a dense and open ordered set, as $(\mathbb{R}, <, =)$ and $(\mathbb{Q}, <, =)$. Moreover, we prove that some fragments of IPC* can be enriched by constants allowing the use of constraints like $x < d$ with $d \in D$ without affecting the construction of the automaton $\mathcal{A}_\phi$.

Completion Property. As explained before, each automaton involved in the definition of $\mathcal{A}_\phi$ has the function of "filtering" sequences of symbolic valuations so that 1) they are locally consistent, 2) they satisfy an LTL property and 3) they admit a (arithmetic) model. For some constraint systems, admitting a model is a consequence of local consistency. A constraint systems $\mathcal{D}$ has the *completion* property if, given:

(i) a symbolic valuation sv over a finite set of variables $H \subseteq V$,
(ii) a subset $H' \subseteq H$,
(iii) a $\mathcal{D}$-valuation v' over H' such that $v' \models sv'$, where sv' is the subset of $\mathcal{D}$-constraints in sv which uses only variables in H'

then there exists a $\mathcal{D}$-valuation v over V extending v' such that $v \models sv$. An example of such a constraint system is $(\mathbb{R}, <, =)$.

Lemma 2 ([5]). *Let $\mathcal{D}$ be a constraint system of the form $(D, <, =)$, where D is infinite and $<$ is a total order. Then, $\mathcal{D}$ satisfies the completion property iff D is dense and open.*

The following result relies on the fact that every locally consistent sequence of symbolic valuations with respect to the constraint system $\mathcal{D}$ admits a model.

Proposition 4. *Let $\mathcal{D}$ be a constraint system satisfying the completion property and ϕ be a CLTL($\mathcal{D}$) formula. Then, the language of sequences of symbolic valuations which admit a model is ω-regular.*

In this case, automaton $\mathcal{A}_\phi$ recognizing the sequence of symbolic valuations may be defined by $\mathcal{A}_\phi = \mathcal{A}_s \cap \mathcal{A}_\ell$.

Adding Constants. Languages CLTL($D, <, =$) can be extended by allowing the use of constants. As shown in [5], both the satisfiability and the model-checking problems are still decidable. This follows by introducing new fresh variables to replace the occurrences of constants, with some CLTL constraints.

If the constraint system satisfies the completion property, let $c_1, \ldots, c_n$ be the constants occurring in a formula ϕ, with $c_1 < c_2 < \cdots < c_n$, and let ϕ' be the formula

$$\phi[c_1 \leftarrow y_1, \ldots, c_n \leftarrow y_n] \wedge \bigwedge_{i=1}^{n-1} (y_i < y_{i+1}) \wedge \mathbf{G}(\bigwedge_{i=1}^{n} y_i = Xy_i)$$

where $y_1, \ldots, y_n$ are new variables not occurring in ϕ and $\phi[c_1 \leftarrow y_1, \ldots, c_n \leftarrow y_n]$ is the formula obtained from ϕ by the replacing all occurrences of c_i with y_i, $1 \leq i \leq n$. It is easy to see that ϕ and ϕ' are equisatisfiable.

Otherwise, if the constraint system $(D, <, =)$ does not have the completion property (e.g., $D \in \{\mathbb{N}, \mathbb{Z}\}$); the satisfiability of a CLTL formula involving constants can still be reduced to the satisfiability of a formula in CLTL($D, <, =$) without constants. Indeed, let ϕ be a CLTL($D, <, =$) formula using constants, with m and M being, respectively, the minimum and the maximum value of such constants. Let $n = |M - m|$ and let $c_i = m + i - 1$, $1 \le i \le n$, then, by introducing n new variables y_i not occurring in ϕ, the formula

$$\phi'' = \phi' \wedge \bigwedge_{x \in var(\phi)} \mathbf{G} \left((x < y_1) \vee \bigvee_{i=1}^{n} (x = y_i) \vee (x > y_n) \right),$$

where $var(\phi)$ is the set of variables of V occurring in ϕ, is equisatisfiable with ϕ.

5   Bounded Satisfiability Problem

The Bounded Satisfiability Problem is defined by considering bounded symbolic models of CLTLB($\mathcal{D}$) formulae. A bounded symbolic model is, informally, a finite representation of infinite CLTLB($\mathcal{D}$) models over the alphabet of symbolic valuations $SV(\phi)$. We restrict the analysis to ultimately periodic models, i.e., sequences of symbolic valuations of the form $\alpha(\beta)^\omega$, where α, β are finite words. BSP is defined with respect to a partial model $\sigma_k : \{0, \dots, k + \lceil \phi \rceil\} \times V \to D$, a finite sequence ρ', $|\rho'| = k$, of symbolic valuations and a partial satisfaction relation $\models_k$ defined as follows:

$$\sigma_k \models_k \rho' \text{ iff } \sigma_k, i \models_s \rho'(i) \text{ for all } 0 \le i \le k.$$

The *k-bounded satisfiability problem* of CLTLB($\mathcal{D}$) is defined as follows:

Input: a CLTLB($\mathcal{D}$) formula ϕ, a constant $k \in \mathbb{N}$;
Problem: is there an ultimately periodic sequence of symbolic valuations $\rho = \delta(\pi)^\omega$ such that $k = |\delta\pi|$ and $\rho, 0 \models_s \phi$, and which admits a partial model σ_k such that $\sigma_k \models_k \rho'$ with $\rho' = \delta\pi$?

Since the length k is fixed, the satisfiability of CLTLB($\mathcal{D}$) formulae over bounded models is not complete: even if automaton $\mathcal{A}_\phi$ has no accepting runs of length k, it might have one of length $k' > k$. The next section explains how to make BSP complete. The completeness property is defined as follows:

Definition 1. *A CLTLB(D) formula ϕ has the completeness property if there is a constant $K \in \mathbb{N}$, depending on ϕ, such that ϕ is satisfiable if, and only if, ϕ is K-bounded satisfiable.*

Hence, if ϕ has the completeness property for a value K and there is no finite model σ_K of ϕ, then ϕ is unsatisfiable.

5.1   Completeness Bound for CLTLB($\mathcal{D}$)

Completeness has been studied in depth for Bounded Model Checking. Given a state-transition system M and a temporal logic property ϕ, BMC looks for a witness of bounded length k to prove $\neg\phi$. If the model does not admit a bounded witness then length k is increased. The process terminates when a witness is found or when k reaches a value, the *completeness threshold*, which guarantees that if no counterexample has been found so far, then no counterexample disproving property ϕ exists for M. For LTL it is shown that a completeness threshold always exists; [3] shows a procedure to estimate an over-approximation of the value, by satisfying a formula representing the existence of an accepting run of the product automaton $M \times B_{\neg\phi}$, with $B_{\neg\phi}$ the Büchi automaton for $\neg\phi$.

In this section, we study the existence of a completeness threshold for the satisfiability problem of CLTLB($\mathcal{D}$) formulae. Since model checking and satisfiability problems are reducible to each other when $\mathcal{D}$-automata or $\mathcal{D}$-Kripke structures are considered, then a completeness threshold for satisfiability may be used to derive one also for the model checking problem, by using the traditional transformation proposed in [9].

Informally, the idea for finding a completeness threshold for a CLTLB($\mathcal{D}$) formula is based on the fact that ultimately periodic symbolic models ρ of CLTLB($\mathcal{D}$) formulae admit an arithmetic model σ if condition C holds (see Proposition 3). Also, if a CLTLB($\mathcal{D}$) formula ϕ is satisfiable, then all ultimately periodic symbolic models ρ, such that $\rho \models_s \phi$, admit a model σ such that $\sigma \models \rho$. Completeness is a consequence of the existence of a finite value c for which all initialized runs of $\mathcal{A}_\phi$, representing models for ϕ, of length greater than c visit at least one control state twice. Consequently, if a CLTLB($\mathcal{D}$) formula is not (boundedly) satisfiable by any ultimately periodic model of length less than or equal to the value $c+1$, then the formula is unsatisfiable. Let c be the length of the longest loop-free path of automaton $\mathcal{A}_\phi$. c is commonly known as *recurrence diameter* of $\mathcal{A}_\phi$, and, in general, it can be defined for every transition system with a finite set of states. This value can be computed by using a SAT-based procedure which builds a sequence of control states such that none of them is repeated along the path. A SAT-based procedure for the computation of the recurrence diameter is proposed in [2].

In order to define a procedure to decide the satisfiability for ϕ by reducing the problem to a finite amount of bounded satisfiability problem, we use the framework proposed in [1] to effectively represent the automata and the fixpoint representation of formula ϕ. Instead of defining the automaton $\mathcal{A}_\phi$ we define a CLTLB($\mathcal{D}$) formula ϕ' such that if it is boundedly satisfiable for some $k \in \mathbb{N}$ then the formula ϕ is satisfiable. In particular, we firstly define formulae ϕ_ℓ and $\phi_{\mathcal{A}_C}$ for automata $\mathcal{A}_\ell$ and $\mathcal{A}_C$ whose models are exactly words of the language recognized by the automata. Finally, ϕ' is the conjunction of the two formulae above, ϕ_ℓ and $\phi_{\mathcal{A}_C}$, with ϕ which is, then, checked for bounded satisfiability. Both automata $\mathcal{A}_\ell$ and $\mathcal{A}_C$ involved in the construction of $\mathcal{A}_\phi$ do not depend on the LTL temporal modalities appearing in ϕ, but only on the constraint system $\mathcal{D}$, i.e., the set of variables V, the set of constants and, also, the length $\lceil\phi\rceil$ of symbolic valuations. Let $\mathcal{A}_\ell = (SV(\phi), Q_{\mathcal{A}_\ell}, Q_0, \delta_{\mathcal{A}_\ell}, F_{\mathcal{A}_\ell})$ be the Büchi automaton such that $Q = Q_0 = F = SV(\phi)$ and its transition relation is such that $sv \xrightarrow{sv'} sv' \in \delta$ iff (sv, sv') are locally consistent. Observe that sequences of locally consistent symbolic valuations recognized by automaton $\mathcal{A}_\ell$ are also models of the formula $\mathbf{G}(\bigvee_1^m sv_i)$. In fact, since (i) in the encoding introduced in [1] the representation of formulae is not contradictory, i.e. two consecutive symbolic valuations are satisfiable iff they are locally consistent, and (ii) the symbolic valuation sv satisfied in a position i is unique (because of the maximal consistency of symbolic valuations, Lemma 4 of [7]), then $\mathbf{G}(\bigvee_1^m sv_i)$ represents exactly the words of $\mathscr{L}(\mathcal{A}_\ell)$. According to [5], automaton $\mathcal{A}_C$ is not directly built from condition C. Instead, an automaton $\mathcal{A}_{\neg C}$, recognizing the complement language of $\mathscr{L}(\mathcal{A}_C)$, is built first. Then, automaton $\mathcal{A}_C$ is obtained through Safra's method [8] for complementing Büchi automata. In general, $\mathcal{A}_C$ is defined by $\mathcal{A}_C = (SV(\phi), Q_{\mathcal{A}_C}, Q_0, \delta_{\mathcal{A}_C}, F_{\mathcal{A}_C})$. We are going to use the reduction of the model-checking problem to the satisfiability problem, given in [9], to represent the automaton $\mathcal{A}_C$ by a CLTL formula. Let $\phi_{\mathcal{A}_c}$ be the formula representing $\mathcal{A}_c$:

$$\bigvee_{q_i \in Q_0} q_i \wedge \mathbf{GF}(\bigvee_{q_i \in F} q_i) \wedge \mathbf{G}\left(\bigvee_{i \in N}(q_i \wedge \bigwedge_{j \in N\setminus\{i\}} \neg q_j) \wedge \bigwedge_{i \in N}(q_i \Rightarrow \bigvee_{t \in \delta}(sv \wedge \mathbf{X}q_j))\right)$$

where $N = \{1, \dots, |Q|\}$, $t = (q_i, sv, q_f)$, $\mathbf{G}\psi = \neg(\top\mathbf{U}\neg\psi)$ and $\mathbf{F}\psi = \top\mathbf{U}\psi$.
We want to verify if the following formula is boundedly satisfiable with respect to $k \in \mathbb{N}$:

$$I(\mathbf{x}_0) \wedge \phi \wedge \phi_{\mathcal{A}_c} \wedge \phi_\ell \tag{1}$$

where $I(\mathbf{x}_0)$ is a general initialization of variables and $\phi_\ell = \mathbf{G}(\bigvee_1^m sv_i)$ with $m = |SV(\phi)|$. If the formula (1) is boundedly unsatisfiable for all $k \in [1, c+1]$ then there does not exist an ultimately periodic symbolic model ρ such that $\rho \models_s \phi$ and such that there exists an arithmetic model σ with $\sigma \models \rho$. Hence, formula ϕ is unsatisfiable. Otherwise, there exists an ultimately periodic symbolic model ρ of length $k > 0$ which admits a model σ. From the bounded solution, we know exactly the model $\rho = \delta(\pi)^\omega$ and the bounded model σ_k. Then, the infinite model σ is defined from σ_k by iterating infinitely many times the sequence of symbolic valuations in π.

Lemma 3. *Formula (1) is satisfiable, for some $k \in [1, c+1]$, if and only if there exists an ultimately periodic model which is accepted by automaton $\mathcal{A}_\phi$.*

The completeness bound for BSP of CLTLB($\mathcal{D}$) formulae is defined by the recurrence diameter of $\mathcal{A}_\phi$. We are ready to give the main result of the paper.

Proposition 5. *For constraint systems IPC*, $(\mathbb{N}, <, =)$, $(\mathbb{Z}, <, =)$, $(D, <, =)$, where D is dense and open, and their extensions with constants, there exists a finite completeness threshold for BSP.*

Proof. The statement is a consequence of Proposition 3. In particular, if $\mathcal{A}_\phi$ accepts a word ρ then it must accept also ultimately periodic words (by the nature of the acceptance condition of the automaton) which admit arithmetic models since they respect condition C. From Lemma 3, if there does not exist a value of k which makes the formula satisfiable, then language $\mathscr{L}(\mathcal{A}_\phi)$ is empty; otherwise, there exists a model σ and an ultimately periodic sequence of symbolic valuations ρ such that $\sigma \models \rho$.

In practice, when the domain of $\mathcal{D}$ is $\mathbb{N}$ or $\mathbb{Z}$, formula (1) can be simplified. In fact, if it is satisfiable, the sequence accepted by the automaton $\mathcal{A_C}$ is already locally consistent, as two consecutive symbolic valuations are satisfiable iff they are locally consistent, due to the consistency of the encoding. Then, formula ϕ_ℓ can be removed and formula (1) becomes $I(\mathbf{x_0}) \wedge \phi \wedge \phi_{\mathcal{A_c}}$. When the domain of $\mathcal{D}$ has the completion property, instead, formula (1) becomes $I(\mathbf{x_0}) \wedge \phi \wedge \phi_\ell$. In this case, formula ϕ_ℓ is necessary to define the sequence of locally consistent symbolic valuations, since the automaton $\mathcal{A_C}$ is not needed anymore. Moreover, we can estimate the value of the completeness bound without building automaton $\mathcal{A}_\phi$. Since the size of the set of control states of $\mathcal{A}_\phi$ is $\mathcal{O}(2^{|\phi|})$, we can consider a rough estimation for the completeness bound defined by the value $d \times |SV(\phi)| \times 2^{|\phi|}$, where d is the cardinality of the control state set of $\mathcal{A}_c$ and $|SV(\phi)|$ is representative of the dimension of ϕ_ℓ (which is again exponential in the size of the formula).

6 Conclusions

We provide a novel approach to solve the satisfiability problem for CLTLB($\mathcal{D}$) formulae by means of a reduction to the satisfiability problem over ultimately periodic models. Although the finite representation captures only models of the form $\delta\pi^\omega$, it is possible to solve the general satisfiability problem since the number of instances of BSP to be solved is bounded. This allows us to claim that the bounded satisfiability problem is complete w.r.t. the satisfiability problem. We know that this value is, in the worst case, the recurrence diameter of the automaton $\mathcal{A}_\phi$ accepting the models of ϕ. The decision procedure presented in Section 5.1 which we used to solve the satisfiability problem for a CLTLB($\mathcal{D}$) formula is effective. Given a CLTLB($\mathcal{D}$) formula ϕ and a natural k, our tool Zot reduces an instance of bounded satisfiability for ϕ into the satisfiability problem for a formula of the decidable theory QF-UFLIA or QF-UFLRA. Since decision procedures for these theories are supported by many SMT-solvers, like Z3 and Yices, we are able to effectively solve the satifiability problem for CLTLB($\mathcal{D}$).

Acknowledgments. We thank Stéphane Demri for fruitful conversations about this work.

References

1. Bersani, M.M., Frigeri, A., Morzenti, A., Pradella, M., Rossi, M., San Pietro, P.: Bounded reachability for temporal logic over constraint systems. In: Markey, N., Wijsen, J. (eds.) TIME, pp. 43–50. IEEE Computer Society, Los Alamitos (2010)
2. Biere, A., Cimatti, A., Clarke, E.M., Strichman, O., Zhu, Y.: Bounded model checking. Advances in Computers 58, 118–149 (2003)
3. Clarke, E.M., Kroening, D., Ouaknine, J., Strichman, O.: Completeness and complexity of bounded model checking. In: Steffen, B., Levi, G. (eds.) VMCAI 2004. LNCS, vol. 2937, pp. 85–96. Springer, Heidelberg (2004)

4. Comon, H., Cortier, V.: Flatness is not a weakness. In: Clote, P., Schwichtenberg, H. (eds.) CSL 2000. LNCS, vol. 1862, pp. 262–276. Springer, Heidelberg (2000)
5. Demri, S., D'Souza, D.: An automata-theoretic approach to constraint LTL. Inf. Comput. 205(3), 380–415 (2007)
6. Demri, S., Gascon, R.: Verification of qualitative $\mathbb{Z}$ constraints. In: Abadi, M., de Alfaro, L. (eds.) CONCUR 2005. LNCS, vol. 3653, pp. 518–532. Springer, Heidelberg (2005)
7. Demri, S., Gascon, R.: The effects of bounding syntactic resources on Presburger LTL. In: TIME, pp. 94–104. IEEE Computer Society, Los Alamitos (2007)
8. Safra, S.: On the complexity of omega-automata. In: FOCS, pp. 319–327 (1988)
9. Sistla, A.P., Clarke, E.M.: The complexity of propositional linear temporal logics. J. ACM 32(3), 733–749 (1985)

Characterizing Conclusive Approximations by Logical Formulae

Yohan Boichut[1], Thi-Bich-Hanh Dao[1], and Valérie Murat[2]

[1] LIFO - Université Orléans, France
[2] IRISA - Université Rennes 1, France

Abstract. Considering an initial set of terms E, a rewriting relation $\mathcal{R}$ and a goal set of terms Bad, reachability analysis in term rewriting tries to answer to the following question: does there exists at least one term of Bad that can be reached from E using the rewriting relation $\mathcal{R}$? Some of the approaches try to show that there exists at least one term of Bad reachable from E using the rewriting relation $\mathcal{R}$ by computing the set of reachable terms. Some others tackle the unreachability problem i.e. no term of Bad is reachable by rewriting from E. For the latter, over-approximations are computed. A main obstacle is to be able to compute an over-approximation precise enough that does not intersect Bad i.e. a conclusive approximation. This notion of precision is often defined by a very technical parameter of techniques implementing this over-approximation approach. In this paper, we propose a new characterization of conclusive approximations by logical formulae generated from a new kind of automata called symbolic tree automata. Solving a such formula leads automatically to a conclusive approximation without extra technical parameters.

1 Introduction

In the rewriting theory, the reachability problem is the following: given a term rewriting system (TRS) $\mathcal{R}$ and two terms s and t, can we decide whether $s \rightarrow^*_{\mathcal{R}} t$ or not? This problem, which can easily be solved on strongly terminating TRSs (by rewriting s into all its possible reduced forms and compare them to t), is undecidable on non terminating TRSs. There exists several syntactic classes of TRSs for which this problem becomes decidable: some are surveyed in [8], more recent ones are [14,19]. In general, the decision procedures for those classes compute a finite tree automaton recognising the possibly infinite set of terms reachable from a set $E \subseteq \mathcal{T}(\mathcal{F})$ of initial terms, by $\mathcal{R}$, denoted by $\mathcal{R}^*(E)$. Then, provided that $s \in E$, those procedures check whether $t \in \mathcal{R}^*(E)$ or not. On the other hand, outside of those decidable classes, one can prove $s \not\rightarrow^*_{\mathcal{R}} t$ using over-approximations of $\mathcal{R}^*(E)$ [16,8] and proving that t does not belong to this approximation.

Recently, reachability analysis turned out to be a very efficient verification technique for proving properties on infinite systems modeled by TRSs. Some of the most successful experiments, using proofs of $s \not\rightarrow^*_{\mathcal{R}} t$, were done on cryptographic protocols [17,11,4] where protocols and intruders are described using

G. Delzanno and I. Potapov (Eds.): RP 2011, LNCS 6945, pp. 72–84, 2011.

a TRS $\mathcal{R}$, E represents the set of initial configurations of the protocol and t a possible flaw. Some other have been carried out on Java byte code programs [2] and in this context, $\mathcal{R}$ encodes the byte code instructions and the evolution of the Java Virtual Machine (JVM), E specifies the set of initial configurations of the JVM and t a possible flaw.

Then reachability analysis can prove the absence of flaws (if $\forall s \in E : s \not\rightarrow_{\mathcal{R}}^* t$). In [8], given a TRS $\mathcal{R}$, a set of terms E and an abstraction function γ, a sequence of sets of terms $App_0^\gamma, App_1^\gamma, \ldots, App_k^\gamma$ is built such that $App_0^\gamma = E$ and $\mathcal{R}(App_i^\gamma) \subseteq App_{i+1}^\gamma$. This technique is called *tree automata completion*. The role of the abstraction γ is to define equivalence classes of terms and to allot each term to an equivalence class. The computation stops when on the one hand, the number of equivalence classes introduced by the abstraction function is bounded, and on the other hand, each equivalence class is $\mathcal{R}$−closed, i.e. when there exists $N \in \mathbb{N}$ such that $\mathcal{R}(App_N^\gamma) = App_N^\gamma$. Then, App_N^γ represents an over-approximation of terms reachable by $\mathcal{R}$ from E. The abstraction function γ should be well designed in such a way that on one hand App_N^γ exists, and on the other hand $t \notin App_N^\gamma$. However, the main drawback of this technique based on tree automata, is that if $t \notin \mathcal{R}^*(E)$ then it is not trivial (when it is possible) to compute a such fix-point over-approximation App_N^γ. Indeed, a high-level expertise in this technique is required for defining a pertinent abstraction function. In a recent work [12], the approximation function is seen as a set of equations γ. Let C_1 and C_2 be two equivalence classes and t_1 and t_2 be respectively terms of classes C_1 and C_2. If $t_1 = t_2$ modulo the set of equations γ then C_1 and C_2 are merged together.

In [5], the authors propose a similar technique in which they use tree transducers instead of TRSs. The approximation functions they use can be seen as predicates. More precisely, if two different equivalence classes satisfy the same set of predicates then they are merged together. So, one has to define carefully the set of predicates. Once again, a high-level expertise in the technique itself is required for obtaining conclusive analysis. But contrary to the technique presented in [8], their technique is equipped with a refinement process acting when the approximation function leads to inconclusive analysis. Nevertheless, the refinement process may be expensive since it involves backward computations for detecting the point where the approximation has become too coarse.

So, to summarize, both of the techniques mentioned previously are instrumented either by equations or predicates for the computation of over-approximations. Both of them use tree automata to represent over-approximations. More precisely, set of terms are represented by tree automata languages. However, these parameters often require a highly specialized expertise for expecting a conclusive analysis.

In this paper, we characterize by a logical formula all the criteria of such a conclusive analysis performed with the technique proposed in [10,8,12]. The idea is that instead of reasoning with a tree automaton A, we generalize A to a symbolic tree automaton (STA) A_s, whose states are represented by variables. The rewriting relations and "bad" terms are represented by boolean

combinations of equalities and inequalities on these variables. An instantiation
of these variables by states gives a tree automaton, and each valid instantiation
of this formula ensures that, as soon as the STA is instantiated, the language
of the resulting tree automaton is a conclusive over-approximation of the set
of terms reachable from the language of A according to the rewriting relation.
With this formulation, finding a conclusive analysis becomes solving logical formulae. Thus, different solving and search techniques, for example in artificial
intelligence, can be applied.

The paper is organized as follows: Section 2 recalls background on terms,
rewriting and tree automata as well as the connection between rewriting and
tree automata. In this section we also describe the kind of formulae we manipulate and notion of instantiations. Section 3 introduces symbolic tree automata.
In this section, we point out the connection between an STA and traditional tree
automata. Section 4 describes the cornerstone of our contribution: the matching
algorithm for STA. In other words, given a term t, we characterize each solution
of this pattern as well as its existence condition by a formula. Section 5 presents
our main contribution: the characterization of a conclusive over-approximation
by a formula. Before concluding, in Section 6, given a TRS $\mathcal{R}$, a tree automaton
A and a set of goal terms Bad, we describe a semi-algorithm for computing automatically a conclusive approximation. For a lack of space, the proofs of this paper are available at http://www.univ-orleans.fr/lifo/prodsci/rapports/
RR/RR2011/RR-2011-04.pdf.

2   Background and Notations

In this section, we introduce some definitions and concepts that will be used
throughout the rest of the paper (see also [1,7,13]). Let $\mathcal{F}$ be a finite set of
symbols, each one is associated with an arity, and let $\mathcal{X}$ be a countable set of
variables. $\mathcal{T}(\mathcal{F}, \mathcal{X})$ denotes the set of *terms* and $\mathcal{T}(\mathcal{F})$ denotes the set of *ground
terms* (terms without variables). The set of variables of a term t is denoted by
$Var(t)$. A term t is said *linear* if there is no variable appearing more than once
in t. A *substitution* is a function σ from $\mathcal{X}$ into $\mathcal{T}(\mathcal{F}, \mathcal{X})$, which can be uniquely
extended to an endomorphism of $\mathcal{T}(\mathcal{F}, \mathcal{X})$. The substitution σ applied to the
term t (denoted $t\sigma$) is constructed such that $x\sigma = \sigma(x)$, where $x \in \mathcal{X}$, and
$f(t_1, ..., t_n)\sigma = f(t_1\sigma, ..., t_n\sigma)$. Let A, B and C be three sets of elements. Let σ
and μ be two substitutions such that $\sigma : A \mapsto B$ and $\mu : B \mapsto C$. We denote by
$\sigma \circ \mu$ the substitution such that $\sigma \circ \mu(x) = \mu(\sigma(x))$ where $x \in A$.

A *term rewriting system* (TRS) $\mathcal{R}$ is a set of *rewrite rules* $l \to r$, where
$l, r \in \mathcal{T}(\mathcal{F}, \mathcal{X})$, $l, r \notin \mathcal{X}^1$, and $Var(l) \supseteq Var(r)$. A rewrite rule $l \to r$ is *left-
linear* if l is linear. A TRS $\mathcal{R}$ is left-linear if every rewrite rule $l \to r$ of $\mathcal{R}$ is
left-linear. The TRS $\mathcal{R}$ induces a rewriting relation $\to_\mathcal{R}$ on terms as follows. Let
$s, t \in \mathcal{T}(\mathcal{F}, \mathcal{X})$ and $l \to r \in \mathcal{R}$, $s \to_\mathcal{R} t$ denotes that there exists a subterm u
of s and a substitution σ such that $u = l\sigma$ and t is obtained by substituting
u by $r\sigma$ in s. The reflexive transitive closure of $\to_\mathcal{R}$ is denoted by $\to_\mathcal{R}^*$. The

---

[1] A more general definition is that only l must not be a variable.

set of $\mathcal{R}$-descendants of a set of ground terms I is $\mathcal{R}^*(I) = \{t \in \mathcal{T}(\mathcal{F}) \mid \exists s \in I \text{ s.t. } s \rightarrow_{\mathcal{R}}^* t\}$. We now define tree automata that are used to recognize possibly infinite sets of terms. Let Q be a finite set of symbols with arity 0, called *states*, such that $Q \cap \mathcal{F} = \emptyset$. $\mathcal{T}(\mathcal{F} \cup Q)$ is called the set of *configurations*. A *transition* is a rewrite rule $c \rightarrow q$, where c is a configuration and q is a state. A transition is *normalized* when $c = f(q_1, \ldots, q_n)$, $f \in \mathcal{F}$ is of arity n, and $q_1, \ldots, q_n \in Q$.

Definition 1 (Bottom-up nondeterministic finite tree automaton). *A bottom-up nondeterministic finite tree automaton (tree automaton for short) over the alphabet $\mathcal{F}$ is a tuple $A = \langle Q, \mathcal{F}, Q_F, \Delta \rangle$, where $Q_F \subseteq Q$ is the set of final states, Δ is a set of normalized transitions.*

The transitive and reflexive *rewriting relation* on $\mathcal{T}(\mathcal{F} \cup Q)$ induced by all the transitions of A is denoted by $\rightarrow_A^*$. The tree language recognized by A in a state q is $\mathcal{L}(A, q) = \{t \in \mathcal{T}(\mathcal{F}) \mid t \rightarrow_A^* q\}$. We define $\mathcal{L}(A) = \bigcup_{q \in Q_F} \mathcal{L}(A, q)$.

Some of the techniques marry ([8,18,5]) tree automata and rewriting for computing the set of reachable terms from a given tree automata A i.e. $\mathcal{R}^*(\mathcal{L}(A))$. Unfortunately, enumerating reachable terms may never terminate. There is thus a need to "accelerate" the search through the term space in order to reach, in a finite amount of time, terms at unbounded depths.

Definition 2. *A tree automaton B is $\mathcal{R}$-closed if for each rule $l \rightarrow r \in \mathcal{R}$, for any substitution $\sigma : \mathcal{X} \mapsto Q$, $l\sigma$ is recognized by B into state q then so is $r\sigma$. The situation is represented with the following graph:*

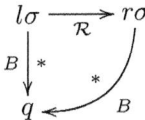

It is easy to see that if B is $\mathcal{R}$-closed and $\mathcal{L}(B) \supseteq \mathcal{L}(A)$, then $\mathcal{L}(B) \supseteq \mathcal{R}^*(\mathcal{L}(A))$[6].

In the following definitions, we introduce the logical formulae that we manipulate as well as notions of instantiation and satisfaction of a formula.

Definition 3 ($\mathcal{W}[\mathcal{X}_Q]$). *Let $\mathcal{X}_Q$ be a set of variables. We define $\mathcal{W}[\mathcal{X}_Q]$ to be the set of logical formulae on $\mathcal{X}_Q$ as following:*

- $\top, \bot \in \mathcal{W}[\mathcal{X}_Q]$;
- $X = Y$, $X \neq Y \in \mathcal{W}[\mathcal{X}_Q]$ *with* $X, Y \in \mathcal{X}_Q$;
- *if* $\alpha, \beta \in \mathcal{W}[\mathcal{X}_Q]$ *then* $\neg\alpha$, $\alpha \wedge \beta$, $\alpha \vee \beta$, $\alpha \Rightarrow \beta$ *are in* $\mathcal{W}[\mathcal{X}_Q]$.

Definition 4 (Instantiation/satisfaction). *Let D be a domain which is a non-empty set. An instantiation ι of variables of $\mathcal{X}_Q$ is a function $\iota : \mathcal{X}_Q \rightarrow D$. The instantiation ι satisfies a formula $\alpha \in \mathcal{W}[\mathcal{X}_Q]$, denoted by $\iota \models \alpha$, iff:*

- $\iota \models \top$;
- $\iota \models X = Y$ *iff* $\iota(X) = \iota(Y)$; $\iota \models X \neq Y$ *iff* $\iota(X) \neq \iota(Y)$;
- $\iota \models \neg\alpha$ *iff* $\iota \not\models \alpha$; $\iota \models \alpha \wedge \beta$ *iff* $\iota \models \alpha$ *and* $\iota \models \beta$;
  $\iota \models \alpha \vee \beta$ *iff* $\iota \models \alpha$ *or* $\iota \models \beta$; $\iota \models \alpha \Rightarrow \beta$ *iff* $\iota \not\models \alpha$ *or* $\iota \models \alpha \wedge \beta$.

Example 1. Let $\mathcal{X}_Q$ be the set of variables such that $\mathcal{X}_Q = \{X_1, X_2, X_3\}$. Thus, $(X_1 \neq X_2) \wedge ((X_1 = X_3) \vee (X_2 = X_3))$ is a formula in $\mathcal{W}[\mathcal{X}_Q]$. Let $D = \{1, 2\}$ and ι be the instantiation such that $\iota(X_1) = 2$, $\iota(X_2) = \iota(X_3) = 1$. We have $\iota \not\models X_1 = X_2$ and $\iota \models (X_1 = X_2) \vee (X_2 = X_3)$.

Note that instantiations will be also considered as substitutions in the remainder of the paper.

3   Symbolic Tree Automata

Let $\mathcal{X}_Q$ be a set of variables that we call symbolic states. Symbolic tree automata (STA) are tree automata where states are variables. An STA is composed of normalized symbolic transitions as defined below.

Definition 5 (Normalized symbolic transition). *Let $\mathcal{X}_Q$ be a set of symbolic states. A normalized symbolic transition is of the form $f(X_1, .., X_n) \to X$ where $f \in \mathcal{F}$ of arity n and $X, X_1, .., X_n \in \mathcal{X}_Q$.*

Definition 6 (STA). *An STA is a tuple $\langle \mathcal{X}_Q, \mathcal{F}, \mathcal{X}_Q^f, \Delta \rangle$ where $\mathcal{X}_Q$ is a set of symbolic states, $\mathcal{F}$ a set of functional symbols, $\mathcal{X}_Q^f \subseteq \mathcal{X}_Q$ a set of final symbolic states and Δ a set of normalized symbolic transitions.*

Example 2. Let $\mathcal{F}$ be a set of functional symbols such that $\mathcal{F} = \{a : 0, s : 1\}$. Let $\mathcal{X}_Q$ and $\mathcal{X}_Q^f$ be two sets of symbolic states such that $\mathcal{X}_Q = \{X_{q_0}, X_{q_1}\}$ and $\mathcal{X}_Q^f = \{X_{q_1}\}$. Let Δ be a set of symbolic transitions such that $\Delta = \{a \to X_{q_0}, a \to X_{q_1}, s(X_{q_0}) \to X_{q_1}\}$. Thus, considering $A_S = \langle \mathcal{X}_Q, \mathcal{F}, \mathcal{X}_Q^f, \Delta \rangle$, A_S is an STA.

The following definition gives details on how a tree automaton can be obtained from a STA and a given instantiation from $\mathcal{X}_Q$ to a domain Q.

Definition 7 (Instance of a STA). *Let Q be a non-empty set of states. Let A_S be an STA $\langle \mathcal{X}_Q, \mathcal{F}, \mathcal{X}_Q^f, \Delta \rangle$ and ι be an instantiation $\mathcal{X}_Q \to Q$. An instance of A_S by ι, denoted by A_S^ι, is a tree automaton $\langle Q^{A_S^\iota}, \mathcal{F}, Q_f^{A_S^\iota}, \Delta^{A_S^\iota} \rangle$ where:*

- $Q^{A_S^\iota} = \{\iota(X) \mid X \in \mathcal{X}_Q\}$; $Q_f^{A_S^\iota} = \{\iota(X) \mid X \in \mathcal{X}_Q^f\}$;
- $\Delta^{A_S^\iota} = \{f(\iota(X_1), \ldots, \iota(X_n)) \to \iota(X) \mid f(X_1, \ldots, X_n) \to X \in \Delta\}$.

Example 3. Let A_S be the STA defined in Example 2. Let ι_1 and ι_2 be two instantiations such that $\iota_1 = \{X_{q_0} \mapsto q, X_{q_1} \mapsto q\}$ and $\iota_2 = \{X_{q_0} \mapsto q', X_{q_1} \mapsto q\}$. Thus, $A_S^{\iota_1} = \langle \{q\}, \mathcal{F}, \{q\}, \{a \to q, s(q) \to q\} \rangle$ and $A_S^{\iota_2} = \langle \{q', q\}, \mathcal{F}, \{q\}, \{a \to q', a \to q, s(q') \to q\} \rangle$. Note that $\mathcal{L}(A_S^{\iota_1}) = \{s^n(a) \mid n \geq 0\}$ and $\mathcal{L}(A_S^{\iota_2}) = \{a, s(a)\}$.

For a term $t \in \mathcal{T}(\mathcal{F}, \mathcal{X}_Q)$, a formula $\alpha \in \mathcal{W}[\mathcal{X}_Q]$ and a symbolic state X, we define the relation $t \xrightarrow{\alpha}_{A_S} X$. In a couple of words, if an instantiation ι satisfies

α then the relation ensures that A_S^ι accepts the term t in the state $\iota(X)$. Note that if $t \xrightarrow{\alpha}_{A_S} X$ then α is a conjunction of equalities between symbolic states. This is involved by a straightforward reduction of the term t using transitions of A_S.

Definition 8 ($t \xrightarrow{\alpha}_{A_S} X$). *Let A_S be an STA $\langle \mathcal{X}_Q, \mathcal{F}, \mathcal{X}_Q^f, \Delta \rangle$. Let t be a term of $\mathcal{T}(\mathcal{F}, \mathcal{X}_Q)$ and X a symbolic state of $\mathcal{X}_Q$. One has:*

- $X \xrightarrow{\top}_{A_S} X$
- *If $t \to Y \in \Delta$ then $t \xrightarrow{X=Y}_{A_S} X$*
- *If $t = f(t_1, ..., t_n)$ and $t_1 \xrightarrow{\alpha_1}_{A_S} X_1$, ..., $t_n \xrightarrow{\alpha_n}_{A_S} X_n$ and $f(X_1, ..., X_n) \to Y \in \Delta$ then $t \xrightarrow{\alpha_1 \wedge \cdots \wedge \alpha_n \wedge X=Y}_{A_S} X$*

Example 4. Let A_S be the STA defined in Example 2. Let t be a term of $\mathcal{T}(\mathcal{F})$ such that $t = s(s(s(a)))$. According to Definition 8, one has $t \xrightarrow{\alpha}_{A_S} X_{q_1}$ with $\alpha = X_{q_0} = X_{q_1}$. Let ι_1 be the instantiation defined in Example 3. Note that, according to Definition 4, $\iota_1 \models \alpha$. Note also that $s(s(s(a))) \to^*_{A_S^{\iota_1}} \iota_1(X_{q_1})$.

Usually, given an STA A_S, a term t, a formula α, a symbolic state X and an instantiation ι, one cannot deduce that $t \not\to^*_{A_S^\iota} \iota(X)$ if $\iota \not\models \alpha$. Nevertheless, if for any formula α such that $t \xrightarrow{\alpha}_{A_S} X$ one has $\iota \not\models \alpha$ then one can conclude that $t \not\to^*_{A_S^\iota} \iota(X)$.

The following proposition presents the characterization by a formula of the acceptance of a term t by a given STA. Consequently, each instantiation satisfying this formula leads to an automaton recognizing t.

Proposition 1. *Let $A_S = \langle \mathcal{X}_Q, \mathcal{F}, \mathcal{X}_Q^f, \Delta \rangle$ be an STA and ι be an instantiation. Let $t \in \mathcal{T}(\mathcal{F}, \mathcal{X}_Q)$ and $X \in \mathcal{X}_Q$. Let $\mathtt{Reco}(t, X) = \bigvee_{\{t \xrightarrow{\alpha}_{A_S} X\}} \alpha$. Then, one has:*

$$\iota \models \mathtt{Reco}(t, X) \quad \textit{iff} \quad t\iota \to^*_{A_S^\iota} \iota(X).$$

4   Solutions for Patterns in STA

Let t be a term of $\mathcal{T}(\mathcal{F}, \mathcal{X})$. For a classical tree automaton A and a state q, the matching problem $t \trianglelefteq q$ has a solution if there exists a substitution $\sigma : \mathcal{X} \mapsto Q$ such that $t\sigma \to^*_A q$. Let us recall that this point is essential for testing whether an automaton is $\mathcal{R}$−closed or not (see Definition 2).

In this section, we propose to solve this problem in the context of STA. Thus, the matching problem is formalized on symbolic states instead of classical states i.e. $t \trianglelefteq X$ with $X \in \mathcal{X}_Q$. Actually, in this context and considering an STA A_S, solutions are represented as a set of pairs (α, σ) where σ is a substitution from $\mathcal{X}$ to $\mathcal{X}_Q$ and α a formula such that $t\sigma \xrightarrow{\alpha}_{A_S} X$. Suppose $\iota : \mathcal{X}_Q \mapsto Q$ is an instantiation. Semantically, a solution (α, σ) means that, as soon as $\iota \models \alpha$, the substitution $\sigma \circ \iota$ is a solution of the matching problem $t \trianglelefteq \iota(X)$ in the tree automaton A_S^ι.

Definition 9 (Matching Algorithm – S_X^t). *Let A_S be an STA $\langle \mathcal{X}_\mathcal{Q}, \mathcal{F},$ $\mathcal{X}_\mathcal{Q}^f, \Delta \rangle$. We denote by S_X^t the solution set of the matching problem $t \trianglelefteq X$ where t is a linear term. S_X^t is built recursively as follows:*

$$
S_X^t = \begin{cases}
\{(\top, \{t \mapsto X\})\} & \text{if } t \in \mathcal{X} & (Var) \\
\{(X = Y, \emptyset)\} & \text{if } t = Y \in \mathcal{X}_\mathcal{Q} \text{ or } t \to Y \in \Delta \ (SymbVar) \\
\bigotimes_{k=1\ldots n}^{X=Y}(S_{X_k}^{t_k}) & \text{if } t = f(t_1, \ldots, t_n) \text{ and} & (Delta) \\
& f(X_1, \ldots, X_n) \to Y \in \Delta
\end{cases}
$$

where $\bigotimes_{k=1\ldots n}^{\phi}(S_{X_k}^{t_k}) = \{(\phi, \emptyset) \oplus (\phi_1, \sigma_1) \oplus \cdots \oplus (\phi_n, \sigma_n) \mid (\phi_i, \sigma_i) \in S_{X_i}^{t_i}\}$, and $(\phi, \sigma) \oplus (\phi', \sigma') = (\phi \wedge \phi', \sigma \cup \sigma')$.

The following proposition shows that this algorithm is sound and complete.

Proposition 2. *Let A_S be an STA $\langle \mathcal{X}_\mathcal{Q}, \mathcal{F}, \mathcal{X}_\mathcal{Q}^f, \Delta \rangle$, X be a symbolic state in $\mathcal{X}_\mathcal{Q}$, t be a linear term in $\mathcal{T}(\mathcal{F}, \mathcal{X})$ and σ be a substitution from $Var(t)$ into $\mathcal{X}_\mathcal{Q}$. One has*

$$
\forall (\alpha, \sigma), \ t\sigma \xrightarrow{\alpha}_{A_S} X \text{ iff } (\alpha, \sigma) \in S_X^t.
$$

Example 5. Let A_S be an STA whose symbolic transition set $\Delta = \{a \to X_{q_0}, a \to X_{q_1}, s(X_{q_0}) \to X_{q_1}\}$. Using the rules we can find that $S_{X_{q_0}}^a = \{(\top, \emptyset), (X_{q_1} = X_{q_0}, \emptyset)\}$, $S_{X_{q_1}}^{s(a)} = \{(\top, \emptyset), (X_{q_1} = X_{q_0}, \emptyset)\}$ and $S_{X_{q_1}}^{s(s(a))} = \{(X_{q_0} = X_{q_1}, \emptyset), (X_{q_0} = X_{q_1}, \emptyset)\}$.

5   Finding a Conclusive Fix-Point Automaton

Let us recall that the *Graal* of the tree automata completion is to detect a conclusive fix-point automaton. Given a set of terms *Bad*, a TRS $\mathcal{R}$ and an initial tree automaton A, a conclusive fix-point automaton is a tree automaton $A^\star$ such that $A^\star$ is $\mathcal{R}$-closed with regard to A and $\mathcal{L}(A^\star) \cap Bad = \emptyset$. Note also that the tree automata completion is only sound for left linear TRSs. So, we only consider left linear TRSs .

In this section, given an STA A_S, a TA A, a TRS $\mathcal{R}$ and a set of bad terms *Bad*, we propose two formulae $\phi_{\mathcal{R}, A_S}^{FP}$ and $\phi_{A_S}^{Bad}$ such that any instantiation ι of A_S satisfying both formulae leads to a conclusive automaton. Moreover, we define a notion of compatibility between A and A_S ensuring that the automaton A_S^ι is a conclusive automaton with regard to A.

The constraint presented below depicts a condition, built from A_S, to satisfy for any instantiation ι in order to ensure that A_S^ι is $\mathcal{R}$-closed. In [8], a TA A is $\mathcal{R}$-closed (fix-point automaton) if $\forall l \to r \in \mathcal{R}$, $\forall \sigma : \mathcal{X} \mapsto Q$ and $\forall q$, if $l\sigma \to_A^* q$ then $r\sigma \to_A^* q$.

Definition 10 ($\phi_{\mathcal{R}, A_S}^{FP}$). *Let A_S be an STA $\langle \mathcal{X}_\mathcal{Q}, \mathcal{F}, \mathcal{X}_\mathcal{Q}^f, \Delta \rangle$ and let $\mathcal{R}$ be a left-linear TRS. We denote by $\phi_{\mathcal{R}, A_S}^{FP}$ the formula defined as follows:*

$$
\phi_{\mathcal{R}, A_S}^{FP} \overset{def}{=} \bigwedge_{l \to r \in \mathcal{R}} \bigwedge_{X \in \mathcal{X}_\mathcal{Q}} \bigwedge_{(\alpha, \sigma) \in S_X^l} \left(\alpha \Rightarrow \bigvee_{(\beta, \_) \in S_X^{r\sigma}} \beta \right)
$$

Example 6. Let A_S be the STA of the example 5 and let $\mathcal{R}$ be a TRS such that $\mathcal{R} = \{s(a) \rightarrow s(s(a))\}$. The formula $\phi^{FP}_{\mathcal{R},A_S}$ is then:

$$(\top \Rightarrow (X_{q_0} = X_{q_1} \vee X_{q_0} = X_{q_1})) \wedge (X_{q_0} = X_{q_1} \Rightarrow (X_{q_0} = X_{q_1} \vee X_{q_0} = X_{q_1}))$$

We state in the following proposition the use of $\phi^{FP}_{\mathcal{R},A_S}$.

Proposition 3. *Let A_S be an STA and $\mathcal{R}$ be a left-linear TRS. Let Q be a set of states and ι be an instantiation $\mathcal{X}_Q \rightarrow Q$. Thus, $\iota \models \phi^{FP}_{\mathcal{R},A_S}$ iff A^ι_S is $\mathcal{R}$-closed.*

At this point, for a given STA A_S, we are able to formalize a fix-point condition. However, a particular fix-point is needed. Suppose that there exists an instantiation ι such that $\iota \models \phi^{FP}_{\mathcal{R},A_S}$. We recall that our goal is to find a fix-point automaton $A^\star$ such that $\mathcal{L}(A^\star) \cap Bad = \emptyset$. The following Definition proposes a formula characterizing the no-recognition of the whole set Bad by any instance of A^ι_S as soon as ι also satisfies this formula.

Definition 11 ($\phi^{Bad}_{A_S}$). *Let A_S be an STA $\langle \mathcal{X}_Q, \mathcal{F}, \mathcal{X}^f_Q, \Delta \rangle$ and Bad be a finite set of ground terms. We denote by $\phi^{Bad}_{A_S}$ the formula defined as follows:*

$$\phi^{Bad}_{A_S} \overset{def}{=} \bigwedge_{t \in Bad} \bigwedge_{X \in \mathcal{X}^f_Q} \bigwedge_{(\alpha,\_) \in S^t_X} \neg\alpha.$$

Proposition 4. *Let A_S be a STA $\langle \mathcal{X}_Q, \mathcal{F}, \mathcal{X}^f_Q, \Delta \rangle$. Let Bad be a finite set of ground terms. Let Q be a set of states and ι be an instantiation $\mathcal{X}_Q \mapsto Q$. Thus, $\iota \models \phi^{Bad}_{A_S}$ iff $\mathcal{L}(A^\iota_S) \cap Bad = \emptyset$.*

We are close to the claimed goal. Indeed, given a STA A_S, a TRS $\mathcal{R}$ and a set of terms Bad, we can deduce that for any instantiation ι satisfying $\phi^{Bad}_{A_S} \wedge \phi^{FP}_{\mathcal{R},A_S}$, $\mathcal{R}(\mathcal{L}(A^\iota_S)) \subseteq \mathcal{L}(A^\iota_S)$ and $\mathcal{L}(A^\iota_S) \cap Bad = \emptyset$. Is it sufficient to ensure that this fix-point is interesting for our input data i.e. A, $\mathcal{R}$ and Bad? In other words, can we deduce that $\mathcal{R}^*(\mathcal{L}(A)) \cap Bad = \emptyset$ from $\iota \models \phi^{Bad}_{A_S} \wedge \phi^{FP}_{\mathcal{R},A_S}$? Trivially the answer is no since no relation is specified between A_S and A. So, we define a compatibility notion between A_S and A leading to our expected result.

Definition 12 (*A-compatibility*). *Let A_S be an STA $\langle \mathcal{X}_Q, \mathcal{F}, \mathcal{X}^f_Q, \Delta_S \rangle$ and A be a TA $\langle Q, \mathcal{F}, Q_F, \Delta \rangle$. The STA A_S is said to be A-compatible iff these three criteria are satisfied: (1) $\{X_q | q \in Q\} \subseteq \mathcal{X}_Q$; (2) $\{X_q | q \in Q_F\} \subseteq \mathcal{X}^f_Q$; and (3) $\{f(X_{q_1}, \ldots, X_{q_n}) \rightarrow X_q | f(q_1, \ldots, q_n) \rightarrow q \in \Delta\} \subseteq \Delta_S$.*

The notion of *A-compatibility* presented above ensures that each instantiation of a STA A_S contains the language $\mathcal{L}(A)$.

Proposition 5. *Let A_S be a STA and A be a TA such that A_S is A-compatible. For any $\iota : \mathcal{X}_Q \mapsto Q$, one has $\mathcal{L}(A) \subseteq \mathcal{L}(A^\iota_S)$.*

Consequently, our main result is that we are able to characterize a conclusive fix-point automaton, that can be found using a technique such as completion, by a single formula of $\mathcal{W}[\mathcal{X}_Q]$.

Theorem 1. *Let A_S be a STA and A be a TA such that A_S is A-compatible. Let $\mathcal{R}$ be left-linear TRS and Bad be a finite set of ground terms. Let ι be an instantiation from $\mathcal{X}_Q$ to Q. Thus,*

$$\iota \models \phi_{A_S}^{Bad} \wedge \phi_{\mathcal{R},A_S}^{FP} \quad \text{iff } A_S^{\iota} \text{ is } \mathcal{R}\text{-closed, } \mathcal{L}(A) \subseteq \mathcal{L}(A_S^{\iota}) \text{ and } \mathcal{L}(A_S^{\iota}) \cap Bad = \emptyset.$$

Another way to interpret this result is the following:

Theorem 2. *Let A_S be a STA and A be a TA such that A_S is A-compatible. Let $\mathcal{R}$ be left-linear TRS and Bad be a finite set of ground terms. Let ι be an instantiation from $\mathcal{X}_Q$ to Q. Thus,*

$$\iota \models \phi_{A_S}^{Bad} \wedge \phi_{\mathcal{R},A_S}^{FP} \quad \text{implies that } \mathcal{R}^*(\mathcal{L}(A)) \subseteq \mathcal{L}(A_S^{\iota}) \text{ and } \mathcal{R}^*(\mathcal{L}(A)) \cap Bad = \emptyset.$$

6   Reachability Analysis via Logical Formula Solving

In this section we synthesize our contribution in the semi-algorithm Algorithm 6.1. Given a TRS $\mathcal{R}$, a tree automaton A and a set of goal terms Bad, Algorithm 6.1 searches an STA for which there exists an instantiation leading to a conclusive fix-point. It is indeed a semi-algorithm since a such conclusive fix-point may not exist (see [3]). In this case, the computation will not terminate. In a couple of words, the algorithm starts with the STA immediately obtained from A. If the whole formula has no solution then the current STA is improved by adding new symbolic transitions (using Norm defined in Algorithm 6.1). The whole formula is computed for the new STA and its satisfiability is checked using *hasNoValidSolution*. The process is iterated until finding a solution. We have used Mona [15] for solving formulae (*hasNoValidSolution*). Mona is a tool handling monadic second-order logic. Given a formula, Mona computes an automaton recognizing all of its solutions.

Algorithm 6.1 *Given a left-linear TRS $\mathcal{R}$, a tree automaton $A = \langle Q, \mathcal{F}, Q_F, \Delta \rangle$ and a set of goal terms Bad, areTermsUnreachable?$(A, \mathcal{R}, Bad)$ is defined as follows*

Variables
(* Starting STA *)
$A_S := \langle \{X_q \mid q \in Q\}, \mathcal{F}, \{X_q \mid q \in Q_F\}, \{f(X_{q_1}, \ldots, X_{q_n}) \to X_q \mid f(q_1, \ldots, q_n) \in \Delta\} \rangle;$
(* Starting Formula *)
$\phi := \phi_{\mathcal{R},A_S}^{FP} \wedge \phi_{A_S}^{Bad};$
00 **Begin**
01   **While** *(hasNoValidSolution(ϕ))* **do**
02   **Foreach** $l \to r \in \mathcal{R}$ **do**
03 $\sigma := \{x_1 \mapsto X_1, \ldots, x_n \mapsto X_n\}$ *where $X_1, \ldots, X_n$ are new symbolic states*
04 $(\Delta', \mathcal{X}'_Q) := $ **Norm**$(r\sigma, X_{n+1})$ *where X_{n+1} is new symbolic state*
05 $A_S := \langle \mathcal{X}_Q \cup \mathcal{X}'_Q \cup \{X_1, \ldots, X_n\}, \mathcal{F}, \mathcal{X}_Q^f, \Delta' \cup \Delta \rangle;$
06   **done;**
07 $\phi := \phi_{\mathcal{R},A_S}^{FP} \wedge \phi_{A_S}^{Bad};$
08 **EndWhile**

09 return true;
10 **End**

The function `Norm` used at Line 04 is defined as follows:

$$\text{Norm}(t, X) = \begin{cases} (\emptyset, \emptyset), & \text{if } t \in \mathcal{X}_{\mathcal{Q}} \\ (\Delta', \mathcal{X}'_{\mathcal{Q}}) & \text{if } t = f(t_1, \ldots, t_n) \end{cases}$$

where $\Delta' = \{f(X_1, \ldots, X_n) \to X\} \cup \bigcup_{i=1}^{n}(\Delta^i)$, $\mathcal{X}'_{\mathcal{Q}} = \{X\} \bigcup_{i=1}^{n}(\mathcal{X}^i_{\mathcal{Q}})$, $\text{Norm}(t_i, X_i) = (\Delta^i, \mathcal{X}^i_{\mathcal{Q}})$ and X_i is either a new symbolic states or equal to t_i if $t_i \in \mathcal{X}_{\mathcal{Q}}$.

We present now a complete example. The idea is to show that all terms of the form $f(s^{(k)}(0)))$ reachable from $f(0)$ using the following TRS $\mathcal{R}_f = \{f(x) \to f(s(s(x)))\}$ are such that k is even. So, we define the parity test using three rules: $\mathcal{R}_{parity} = \{even(f(s(s(x)))) \to even(f(x)), even(f(0)) \to true, even(f(s(0))) \to false\}$. Thus, the given inputs are: $\mathcal{R} = \mathcal{R}_f \cup \mathcal{R}_{parity}$, $Bad = \{false\}$ and $A = \langle Q, \mathcal{F}, Q^f, \delta \rangle$ with $Q = \{q_0, q_1, q_2\}$, $\mathcal{F} = \{f : 1, s : 1, 0 : 0, even : 1, true : 0, false : 0\}$, $Q^f = \{q_2\}$ and $\delta = \{even(q_1) \to q_2, f(q_0) \to q_1, 0 \to q_0\}$. So, if a conclusive over-approximation can be found then it ensures that the set of terms reachable from $f(0)$ using the rule $f(x) \to f(s(s(x)))$ is necessarily of the form $f(s^k(0))$ with k an even integer.

So, the starting STA is such that $\Delta = \{even(X_{q_1}) \to X_{q_2}, 0 \to X_{q_0}, f(X_{q_0}) \to X_{q_1}\}$ and $\mathcal{X}_{\mathcal{Q}} = \{X_{q_0}, X_{q_1}, X_{q_2}\}$. Applying Definition 10, one obtains the following formula:

	Rule involved
$\phi^{FP}_{\mathcal{R}, A_S} = \bigwedge_{Y \in \{X_{q_1}, X_{q_2}, X_{q_3}\}}(\top \wedge X_{q_1} = Y \Rightarrow \bot) \wedge$	$f(x) \to f(s(s(x)))$
$\bigwedge_{X, Y \in \{X_{q_1}, X_{q_2}, X_{q_3}\}}(\bot \Rightarrow \top \wedge X = Y) \wedge$	$even(f(s(s(x)))) \to even(f(x))$
$\bigwedge_{Y \in \{X_{q_1}, X_{q_2}, X_{q_3}\}}(\top \wedge X_{q_2} = Y \Rightarrow \bot) \wedge$	$even(f(0)) \to true$
$\bot \Rightarrow \bot$	$even(f(s(0))) \to false$

Clearly, $\phi^{FP}_{\mathcal{R}, A_S}$ is unsatisfiable. Indeed, $\phi^{FP}_{\mathcal{R}, A_S} = \phi_1 \wedge (\top \wedge X_{q_1} = X_{q_1} \Rightarrow \bot) \wedge \phi_2$ with $\phi_1, \phi_2 \in \mathcal{W}[\mathcal{X}_{\mathcal{Q}}]$. By simplifying the formula, one obtains that $\phi^{FP}_{\mathcal{R}, A_S} = \phi_1 \wedge (\top \Rightarrow \bot) \wedge \phi_2 = \bot$. So, ϕ (at line 01 in Algorithm 6.1) is also unsatisfiable. Consequently, the STA A_S needs to be extended. In this example, four substitutions (one per rule following the order of the table above) $\sigma_1, \sigma_2, \sigma_3$ and σ_4 are created such that $\sigma_1 = \{x \mapsto X_{q_4}\}$, $\sigma_2 = \{x \mapsto X_{q_8}\}$ and $\sigma_3 = \sigma_4 = \emptyset$ where X_{q_4} and X_{q_8} are two new symbolic states.

Consequently, applying the substitutions $\sigma_1, \sigma_2, \sigma_3, \sigma_4$ respectively on terms $f(s(s(x)))$, $even(f(x))$, $true$ and $false$, one has to normalize the terms $f(s(s(X_{q_4})))$, $even(f(X_{q_8}))$, $true$ and $false$. Let X_{q_3}, X_{q_7}, $X_{q_{10}}$ and $X_{q_{11}}$ be four new symbolic states, the normalization steps at Line 04 – $\text{Norm}(f(s(s(X_{q_4}))), X_{q_3})$, $\text{Norm}(even(f(X_{q_8})), X_{q_7})$, $\text{Norm}(true, X_{q_{10}})$ and $\text{Norm}(false, X_{q_{11}})$ – may produce the following STA: $A_S = \langle \mathcal{X}_{\mathcal{Q}}, \mathcal{F}, \mathcal{X}^f_{\mathcal{Q}}, \Delta \rangle$ $\mathcal{X}_{\mathcal{Q}} = \{X_{q_0}, \ldots, X_{q_{11}}\}$, $\mathcal{X}^f_{\mathcal{Q}} = \{X_{q_2}\}$ and $\Delta = \{true \to X_{q_{10}}, false \to X_{q_{11}}, s(X_{q_5}) \to X_{q_6}, s(X_{q_4}) \to X_{q_5}, 0 \to X_{q_0}, even(X_{q_9}) \to X_{q_7}, even(X_{q_1}) \to X_{q_2}, f(X_{q_8}) \to X_{q_9}, f(X_{q_6}) \to X_{q_3}, f(X_{q_0}) \to X_{q_1}\}$. Note that A_S is still A-compatible.

Following Definition 11, one obtains $\phi_{A_S}^{Bad} = X_{q_2} \neq X_{q_{11}}$.

Let us construct the instantiation ι from the solution returned by Mona[2]. We obtain: $\iota = \{X_{q_0} \mapsto q_0, X_{q_1} \mapsto q_0, X_{q_2} \mapsto q_0, X_{q_3} \mapsto q_1, X_{q_4} \mapsto q_1, X_{q_5} \mapsto q_0, X_{q_6} \mapsto q_1, X_{q_7} \mapsto q_1, X_{q_8} \mapsto q_1, X_{q_9} \mapsto q_1, X_{q_{10}} \mapsto q_0, X_{q_{11}} \mapsto q_1\}$. Applying ι on A_S, the resulting TA is: $A_S^\iota = \langle \{q_0, q_1\}, \mathcal{F}, \{q_0\}, \Delta^\iota \rangle$ with $\Delta^\iota = \{false \to q_1, 0 \to q_0, s(q_0) \to q_1, s(q_1) \to q_0, even(q_1) \to q_1, even(q_0) \to q_0, true \to q_0, f(q_1) \to q_1, f(q_0) \to q_0\}$.

This tree automaton is actually $\mathcal{R}$-closed. Indeed, concerning the rule $f(x) \to f(s(s(x)))$, note that $f(q_0)$ and $f(s(s(q_0)))$ can both be reduced to q_0. Similarly, $f(q_1)$ and $f(s(s(q_1)))$ can be reduced on q_1. For the rule $even(f(s(s(x)))) \to even(f(x))$, one has $even(f(s(s(q_0)))) \to_{A_S^\iota}^* q_0$ and $even(q_0) \to_{A_S}^* q_0$. Similarly, one has $even(f(s(s(q_1)))) \to_{A_S^\iota}^* q_1$ and $even(q_1) \to_{A_S^\iota}^* q_1$. Finally, for the rule $even(f(0)) \to true$ one has $even(f(0)) \to_{A_S^\iota}^* q_0$ and $true \to_{A_S^\iota}^* q_0$. Moreover, the term $false$ is not in $\mathcal{L}(A_S^\iota)$. Thus, A_s^ι is a conclusive fix-point automaton.

7   Conclusion

To summarize, given an STA A_S, a set of forbidden terms Bad, a TA A and a TRS $\mathcal{R}$, we have characterized by a logical formula what a conclusive fix-point in terms of reachability analysis is. Each solution of such a formula is an instantiation that can be applied on A_S. The automatically obtained automaton is an automaton that could have been obtained using a technique as in [8]. Such a technique requires a technical parameter (a set of equations or an approximation function) influential on the quality of the approximation computed. This parameter requires a certain expertise of the technique itself. For instance in [12], one has to define a set of equations whose goal is to define a finite number of equivalence classes of terms. A finite number of equivalence classes ensures the computation terminates. But, the crucial point remains in finding a conclusive approximation. Thus, the set of equation has to be defined very carefully. In [5], they used a set of predicates for defining a finite set of equivalence classes of terms. Once again, a highly specialized expertise in the technique itself is needed. Concerning ours, we generate an STA from the initial TA A and we are looking for solutions. If no solution is found then we are sure that there is no conclusive $\mathcal{R}$-closed automaton for the given A_S. So, we increase the size of the starting STA and so on.

In [9], the authors encode the tree automata completion by logic programs (Horn clauses). Consequently, they use several results obtained on static analysis of logic programs in order to compute precise approximations in the sense of static analysis. In the field of the tree automata completion, an approximation is precised enough as soon as this latter allows us to show the unreachability of a term or a set of terms. In this context, our proposition allows us to find

---

[2] The Mona program and thus the formula $\phi_{\mathcal{R},A_S}^{FP}$, can be downloaded at http://www.univ-orleans.fr/lifo/Members/Yohan.Boichut/research/exampleMona.txt

only conclusive approximations, contrary to the ones obtained in [9]. Indeed, our approach is in some sense goal oriented while the one proposed in [9] check the reachability of a term only after having computed an approximation.

This work is a first step towards a verification technique based on formula solving. In the verification framework, it allows us to prove safety property. We claim that it is only a first step since specifications involving STA containing more than 20 variables or a bigger TRS are out of the Mona scope. Even if the formulas involved by such a specification present a certain regularity in their form, their size may be huge (in particular for $\phi_{\mathcal{R},A_S}^{FP}$ see Definition 10). We are also aware that the solving problem is not elementary, but we are working on dedicated solving techniques and search heuristics for handling huge formulae. We are also studying a symbolic technique *à la Mona*. First results of both techniques are very promising.

References

1. Baader, F., Nipkow, T.: Term Rewriting and All That. Cambridge University Press, Cambridge (1998)
2. Boichut, Y., Genet, T., Jensen, T., Leroux, L.: Rewriting Approximations for Fast Prototyping of Static Analyzers. In: Baader, F. (ed.) RTA 2007. LNCS, vol. 4533, pp. 48–62. Springer, Heidelberg (2007)
3. Boichut, Y., Héam, P.-C.: A theoretical limit for safety verification techniques with regular fix-point computations. Inf. Process. Lett. 108(1), 1–2 (2008)
4. Boichut, Y., Héam, P.-C., Kouchnarenko, O.: Approximation-based tree regular model-checking. Nord. J. Comput. 14(3), 216–241 (2008)
5. Bouajjani, A., Habermehl, P., Rogalewicz, A., Vojnar, T.: Abstract regular tree model checking. ENTCS 149(1), 37–48 (2006)
6. Boyer, B., Genet, T., Jensen, T.: Certifying a Tree Automata Completion Checker. In: Armando, A., Baumgartner, P., Dowek, G. (eds.) IJCAR 2008. LNCS (LNAI), vol. 5195, pp. 523–538. Springer, Heidelberg (2008)
7. Comon, H., Dauchet, M., Gilleron, R., Jacquemard, F., Lugiez, D., Löding, C., Tison, S., Tommasi, M.: Tree automata techniques and applications (2008)
8. Feuillade, G., Genet, T., Viet TriemTong, V.: Reachability Analysis over Term Rewriting Systems. Journal of Automated Reasonning 33(3-4), 341–383 (2004)
9. Gallagher, J., Rosendahl, M.: Approximating term rewriting systems: a horn clause specification and its implementation. In: Cervesato, I., Veith, H., Voronkov, A. (eds.) LPAR 2008. LNCS (LNAI), vol. 5330, pp. 682–696. Springer, Heidelberg (2008)
10. Genet, T.: Decidable approximations of sets of descendants and sets of normal forms. In: Nipkow, T. (ed.) RTA 1998. LNCS, vol. 1379, pp. 151–165. Springer, Heidelberg (1998)
11. Genet, T., Klay, F.: Rewriting for Cryptographic Protocol Verification. In: McAllester, D. (ed.) CADE 2000. LNCS (LNAI), vol. 1831, pp. 271–290. Springer, Heidelberg (2000)
12. Genet, T., Rusu, R.: Equational tree automata completion. JSC 45, 574–597 (2010)
13. Gilleron, R., Tison, S.: Regular tree languages and rewrite systems. Fundamenta Informaticae 24, 157–175 (1995)

14. Gyenizse, P., Vágvölgyi, S.: Linear Generalized Semi-Monadic Rewrite Systems Effectively Preserve Recognizability. TCS 194(1-2), 87–122 (1998)
15. Henriksen, J., Jensen, J., Jørgensen, M., Klarlund, N., Paige, B., Rauhe, T., Sandholm, A.: Mona: Monadic second-order logic in practice. In: Brinksma, E., Steffen, B., Cleaveland, W.R., Larsen, K.G., Margaria, T. (eds.) TACAS 1995. LNCS, vol. 1019, pp. 89–110. Springer, Heidelberg (1995)
16. Jacquemard, F.: Decidable approximations of term rewriting systems. In: Ganzinger, H. (ed.) RTA 1996. LNCS, vol. 1103, pp. 362–376. Springer, Heidelberg (1996)
17. Monniaux, D.: Abstracting Cryptographic Protocols with Tree Automata. In: Cortesi, A., Filé, G. (eds.) SAS 1999. LNCS, vol. 1694, pp. 149–163. Springer, Heidelberg (1999)
18. Takai, T.: A Verification Technique Using Term Rewriting Systems and Abstract Interpretation. In: van Oostrom, V. (ed.) RTA 2004. LNCS, vol. 3091, pp. 119–133. Springer, Heidelberg (2004)
19. Takai, T., Kaji, Y., Seki, H.: Right-linear finite-path overlapping term rewriting systems effectively preserve recognizability. In: Bachmair, L. (ed.) RTA 2000. LNCS, vol. 1833, pp. 246–260. Springer, Heidelberg (2000)

Decidability of LTL for Vector Addition Systems with One Zero-Test

Rémi Bonnet

LSV, CNRS, ENS Cachan

Abstract. We consider the class of Vector Addition Systems with one zero-test and we show that the model-checking problem for LTL is decidable thanks to a reduction to the computability of the cover and the decidability of reachability. Our proof uses the notion of increasing loop, that we refine to fit the non-standard monotony of our system.

1 Introduction

Petri Nets. Vector Addition Systems (VAS) are a well-known classes of counter systems, equivalent to Petri Nets. The reachability problem is known to be decidable [14,15,16,17] even if its complexity is still an open problem. As the equality of the reachability sets (the set of states that are reachable from an initial state) of two such systems is undecidable [13], one cannot compute a canonical finite representation of the reachability set. However, there is such an effective finite representation for the *cover*, the downward closure of the reachability set, which is connected to various verification problems, like the control state reachability problem.

If we add to VASS the ability to test at least two counters to zero, one obtains a model equivalent to Minsky machines, for which all nontrivial properties are undecidable. The study of VASS with *a single* zero-test transition began recently, and a reasonable number of results are now known. Reinhardt [18] has shown that the reachability problem is decidable. Abdulla and Mayr [2] have provided an algorithm based on the backward procedure of Well Structured Transition Systems [1,9] to decide coverability of a state. Termination and Boundedness were shown by Finkel and Sangnier [8], while an algorithm to compute the maximal elements of the cover has been found by Bonnet, Finkel, Leroux and Zeitoun [3].

LTL. Linear-time logic is a widely used logic in order to express safety and liveness properties of a system. Emerson [4] provided an algorithm based on a covering graph that worked on well structured transition systems, but that was not guaranteed to terminate. Esparza [5,6] showed that LTL on the actions of a VASS was decidable, but that CTL was not, and that LTL became undecidable when predicates regarding the states were added. Habermehl [12] completed this proof by showing EXPSPACE-completeness of LTL satisfiability.

G. Delzanno and I. Potapov (Eds.): RP 2011, LNCS 6945, pp. 85–95, 2011.
© Springer-Verlag Berlin Heidelberg 2011

Our contribution. We complete the works of [3] by showing decidability of LTL model checking. We start by the usual reduction of LTL model-checking to repeated control state reachability by defining the synchronized product of a $VASS_0$ and a Buchi automaton. Then, we show that repeated control state reachability can be decided by looking at the existence of a special kind of increasing loop. We first provide a reduction of this problem of existence of a loop to the reachability problem for $VASS_0$ when the starting point of there is a finite number of such subsets, and hence that if one is able to compute a finite representation of the cover, existence of an increasing loop can be decided by looking at all the subsets.

2 Preliminaries

2.1 Generalities

Sets and Vectors. The cartesian product of two sets X and Y is noted $X \times Y$ and the disjoint union $X \uplus Y$. For $d \geq 1$, we write any $x \in X^d$ as $x = (x[0], \ldots, x[d-1])$, with $x[i] \in X$. For $x_1 \in X^{d_1}$ and $x_2 \in X^{d_2}$, we let (x_1, x_2) be the vector of $X^{d_1+d_2}$ obtained by gluing x_1 and x_2. Addition of vectors is defined by $(x+y)[i] = x[i] + y[i]$ and substraction similarly.

We denote by $\mathbb{N}_\omega$ the set $\mathbb{N} \cup \{\omega\}$ where ω is an element strictly greater than all integers. We will use the notations 0^d to denote the vector composed of d 0's, ω^d for the vector composed of d ω's, and e_i^d be the vector of $\mathbb{N}^d$ such that $e_i^d[i] = 1$ and $e_i^d[j] = 0$ if $i \neq j$.

Orderings. An *ordering* $\preceq$ on a set X is a reflexive, transitive and antisymmetric binary relation on X. Given $x, y \in X$, we write $x \prec y$ for $x \preceq y$ and $x \neq y$. The *pointwise ordering* on X^d, still denoted $\preceq$, is defined by $x \preceq y$ if $x[i] \preceq y[i]$ for all i. Given $Y \subseteq X$, $\downarrow_\preceq Y = \{x \in X \mid \exists y \in Y, \ x \preceq y\}$ denotes the downward closure of Y with respect to $\preceq$. The set Y is *downward closed* if $Y = \downarrow_\preceq Y$. In $\mathbb{N}^d$, we shorten $\downarrow_\leq$ as $\downarrow$.

An ordering $\preceq$ on X is *well* if, given any sequence $(x_i)_{i \in \mathbb{N}}$ of elements of X, one can find $i < j$ such that $x_i \leq x_j$. The usual ordering on $\mathbb{N}^d$ is well.

Basis in $\mathbb{N}_\omega^d$. Given a downward-closed set $X \subseteq \mathbb{N}^d$, a *basis* of X is a finite subset B of $\mathbb{N}_\omega^d$ such that $\downarrow B \cap \mathbb{N}^d = X$. Any downward-closed set of $\mathbb{N}^d$ admits a basis [7] and one can show that the maximal elements of any basis B of X still form a basis which does not depend of B. It is minimal for inclusion among all basis, and is called the *minimal basis*.

Words. The set of finite words (shortly words) on A is denoted A^*. A word $u \in A^*$ is written $a_1 a_2 \ldots a_n$, $a_i \in A$, and we will also use the notation $u[i]$ to refer to the i-th letter of u. The concatenation of two words u and v is simply written uv and the empty word ε, with $\varepsilon a = a\varepsilon = a$. A^+ denotes the set of non-empty words. An infinite word on A is a sequence $(a_i)_{i \in \mathbb{N}}$. Given an infinite word u, we use the notation $u[k \ldots]$ to refer to the subsequence $(u[k+i])_{i \in \mathbb{N}}$. The set of infinite words on A is written A^ω and the union of finite and infinite words is written $A^{\leq \omega}$.

2.2   Transition Systems

Definition 1. *A* Labelled Transition System *(LTS) S is a tuple $\langle X, A, \rightarrow, s_{in} \rangle$ where X is the set of states, A is the set of transition labels, $\rightarrow \subseteq X \times (A \cup \{\varepsilon\}) \times X$ is the transition relation and s_{in} is the initial state.*

We will use the notations $States(S)$, $Actions(S)$ and $Init(S)$ to refer respectively to X, A, s_{in}. Moreover, we write $s \xrightarrow{a} s'$ if $(s, a, s') \in \rightarrow$ and we extend this notation to words by $s \xrightarrow{\varepsilon} s$ and $s \xrightarrow{uv} s'$ iff $\exists s''$, $s \xrightarrow{u} s'' \xrightarrow{v} s'$. Note that transitions may be labelled by ε and hence that $s \xrightarrow{a} s'$ where $a \in A$ doesn't mean that s' is reached from s by one transition, but by one transition labelled by a and any number of ε-transitions.

A run w of S is a sequence $(s_i, t_i) \in (States(S) \times Actions(S))^{\leq \omega}$ such that $s_0 = Init(S)$ and $\forall i, s_i \xrightarrow{t_i} s_{i+1}$. Given a run $(s_i, t_i)_i$, we define $actions(w)$ as $(t_i)_i$. The *reachability set* is defined as $Reach(S) = \{y \in States(S) \mid \exists u \in Actions(S)^* \mid Init(S) \xrightarrow{u} y\}$. If $States(S)$ is ordered by $\leq$, the *cover* is $Cover_{\leq}(S) = \downarrow_{\leq} Reach(S)$. The subscript $\leq$ will be omitted when it is clear from the context.

2.3   Vector Addition Systems

Definition 2. *A* Vector Addition System with States and one zero-test *(shortly VASS$_0$) of dimension d is a tuple $\mathcal{V} = \langle Q, A, a_Z, T, s_{in} \rangle$ where Q is a finite set of control states, A is a finite alphabet of actions, $a_Z \in A$ is called the* zero-test, *$T \subseteq Q \times \mathbb{Z}^d \times A \times Q$ is the finite set of transitions, and $s_{in} = (q_{in}, x_{in}) \in Q \times \mathbb{N}^d$ is the* initial state.

Intuitively, a VASS$_0$ works on d counters, one for each component, whose initial values are given by x_{in}. If $(q, v, a, q') \in T$, $a \in A$ when the VASS$_0$ is in control state q adds the vector v to the counters and moves the system in the control state q'. This action can be executed only if the resulting counters values are non-negative. Moreover, we have the restriction that a_Z can be fired only if the first counter is zero.

More formally, a VASS$_0$ $\langle Q, A, a_Z, T, s_{in} \rangle$ induces a transition system S by:

$$
\begin{aligned}
States(S) &= Q \times \mathbb{N}^d \\
Actions(S) &= A \\
Init(S) &= (q_{in}, x_{in}) \\
(q, x) \xrightarrow{a}_S (q', x') &\iff (q, x' - x, a, q') \in T \qquad \text{for } a \neq a_Z \\
(q, x) \xrightarrow{a_Z}_S (q', x') &\iff \begin{cases} (q, x' - x, a_Z, q') \in T \\ x[0] = 0 \end{cases}
\end{aligned}
$$

A finite automaton (FA) is a VASS$_0$ of dimension 0. We get back the usual definition of VASS (without zero-test) as $\langle Q, A, T, s_{in} \rangle$ whose semantics are the same as the VASS$_0$ $\langle Q, A \uplus \{a_Z\}, a_Z, T, s_{in} \rangle$ where a_Z doesn't appear in T.

We recall from previous works the following properties of VASS$_0$ that we will use in the sequel :

Theorem 1. *(Reachability [18], Coverability [2]) Let S be the transition system associated to a VASS$_0$. Membership in Reach(S) and Cover(S) is decidable.*

Regarding coverability, we can be even more precise. Actually, Cover(S) is not only recursive, but also has a finite representation.

Theorem 2. *(Cover [3]) Let S be the transition system associated to a VASS$_0$. One can compute the minimal basis of Cover(S).*

To simplify some proofs, we will only consider *normed* VASS$_0$, i.e. VASS$_0$ such that there exists a unique (q, q', δ) for which $(q, \delta, a_z, q') \in T$. We show in the appendix (proposition 3) that any VASS$_0$ can be rewritten in a normed VASS$_0$ satisfying the same LTL formulas.

3   The LTL Logic

3.1   Buchi Automata and LTL

Definition 3. *A Buchi automaton is a pair $\langle A, F \rangle$ where A is a finite automaton and $F \subseteq States(S)$.*

An infinite run $((q_i, x_i), t_i)_{i \in \mathbb{N}}$ of a Buchi Automata is accepted iff $\{i \in \mathbb{N} \mid q_i \in F\}$ is infinite.

Definition 4. *Given a set A, the set of LTL formulae is given by the following grammar, where a ranges over A :*

$$\varphi ::= true \mid a \mid \neg\varphi \mid \varphi_1 \wedge \varphi_2 \mid \mathcal{X}\varphi \mid \varphi_1 \mathcal{U} \varphi_2$$

Formulas are interpreted on infinite words over the alphabet A. We denote that w satisfies a formula φ by $w \models \varphi$. This relation is defined inductively on the structure of φ by:

$$
\begin{aligned}
&w \models true \\
&w \models a &&\iff w[0] = a \\
&w \models \neg\varphi &&\iff w \not\models \varphi \\
&w \models \varphi_1 \wedge \varphi_2 &&\iff w \models \varphi_1 \text{ and } w \models \varphi_2 \\
&w \models \mathcal{X}\varphi &&\iff w[1\ldots] \models \varphi \\
&w \models \varphi_1 \mathcal{U} \varphi_2 &&\iff \exists i, \forall 0 \leq j < i, \ w[j\ldots] \models \varphi_1 \wedge w[i\ldots] \models \varphi_2
\end{aligned}
$$

Given a LTL formula φ, one can build a Buchi automaton $\mathcal{B}_\varphi$ such that the set of infinite words satisfying φ is exactly the infinite words accepted by $\mathcal{B}_\varphi$. We refer to the abundant literature on this subject for the construction (Proposition 4.1 of [5], but also [11] and [10]).

3.2   Model Checking

We consider two problems on $VASS_0$. *LTL Model Checking* consists in, given a $VASS_0$ $\mathcal{V}$ inducing a transition system $\mathcal{S}$ and a LTL formula φ on $Actions(\mathcal{S})$, determining whether there exists an infinite run w of $\mathcal{S}$ such that $actions(w) \models \varphi$. *Repeated Control State Reachability* consists in, given a $VASS_0$ $\mathcal{S} = \langle Q, A, a_Z, T, s_{in} \rangle$ and a control state $q_f \in Q$, determining whether there exists an infinite run $w = (s_1, t_1) \dots (s_k, t_k) \dots$ of $\mathcal{S}$ such that $\{ j \in \mathbb{N} \mid \exists x_j,\ s_j = (q_f, x_j) \}$ is infinite.

We have the following usual reduction :

Proposition 1. *LTL Model Checking on $VASS_0$ reduces to Repeated Control State Reachability on $VASS_0$.*

Proof. Let $\mathcal{V} = \langle Q, A, a_Z, T, (q_{in}, x_{in}) \rangle$ be a $VASS_0$ and φ a LTL formula on A. Let $\mathcal{B} = \langle Q_\mathcal{B}, A, T_\mathcal{B}, q_{in\,\mathcal{B}}, F \rangle$ be a Buchi automaton representing φ. The synchronized product of $\mathcal{V}$ and $\mathcal{B}$ is defined as the $VASS_0$ $\mathcal{V}' = \langle Q \times Q_\mathcal{B}, A, a_Z, T', ((q_{in}, q_{in\,\mathcal{B}}), x_{in}) \rangle$ with :

$$T' = \{ ((q_1, q_2), \delta, a, (q_1', q_2')) \mid (q_1, \delta, a, q_1') \in T \wedge (q_2, a, q_2') \in T_\mathcal{B} \}$$

This $VASS_0$ induces a transition system $\mathcal{S}'$, and it is easy to check that a sequence $((q_i^1, q_i^2, x_i), a_i)_i$ is a run of $\mathcal{S}'$ if and only if $((q_i^1, x_i), a_i)_i$ is a run of $\mathcal{S}$ and $(q_i^2, a_i)_i$ is a run of $\mathcal{B}$. Hence, there exists a run of $\mathcal{S}'$ that visits infinitely often $Q \times F$, if and only if there exists runs w of $\mathcal{S}$ and w' of $\mathcal{B}$ such that $actions(w) = actions(w')$ and w' visits infinitely often F, which means that $actions(w) \models \varphi$.

4   Decidability of Repeated Control State Reachability

Let us introduce the order $\leq_0$ as $x \leq_0 y \iff x \leq y \wedge x[0] = y[0]$. We have the following monotony property for $VASS_0$:

Proposition 2. *Let $q \in Q$ and $x, y \in \mathbb{N}^d$ with $x \leq_0 y$. If a sequence of transitions can be fired from (q, x), it can be fired from (q, y).*

Our idea is to make an equivalence between repeated control state reachability and the existence of an increasing loop going through this state.

Definition 5. *Let $\mathcal{V}$ be a $VASS_0$ and $\mathcal{S}$ its associated transition system.*
Given ℓ in $\mathbb{N} \times \mathbb{N}_\omega^{d-1}$, we say that $(x, u, y) \in \mathbb{N}^d \times A^+ \times \mathbb{N}^d$ is a ℓ-increasing loop on q in $\mathcal{V}$ if we have $(q, x) \xrightarrow{u}_\mathcal{S} (q, y)$, $x \leq_0 y$ and $x \leq_0 \ell$.

Our proof is in three steps : First we show that if we have the restriction that $\ell[0] = 0$, we can decide the existence of an ℓ-increasing loop. Then we show that, assuming the run we are looking for goes infinitely through the zero-test, the existence of a run visiting infinitely often a control state reduces to the existence of a ℓ-increasing loop with $\ell[0] = 0$. We conclude by taking also care of runs visiting the zero-test only a finite number of times.

We will fix a normed $VASS_0$ $\mathcal{V} = \langle Q, A, a_Z, T, s_{in} \rangle$ of dimension d and $\mathcal{S}$ its associated transition system. Unless otherwise specified, all lemmas refer to this $VASS_0$.

Lemma 1. *Let $q_f \in Q$ and $\ell \in \{0\} \times \mathbb{N}_\omega^{d-1}$. The existence of an ℓ-increasing loop on q_f is decidable.*

Proof. Let us take $\ell \in \{0\} \times \mathbb{N}_\omega^{d-1}$. Without loss of generality (by reordering counters), we have $\ell = (0, \omega^m, b)$, $b \in \mathbb{N}^n$, with $d = 1 + m + n$.

We will build a VASS$_0$ $\mathcal{V}'$ inducing a transition system $\mathcal{S}'$ that will mimic $\mathcal{S}$ in the following sense (x represents a state $\mathcal{S}$, and x' the associated state in $\mathcal{S}'$):

- The counter that can be tested for zero is preserved.
- A counter $x[i]$ for $1 \leq i \leq m$ (the ones for which $\ell[i] = \omega$) is replaced by two counters $x'[i]$ and $x'[i+m]$, such that $x'[i+m] - x'[i] \leq x[i]$. This simulates a counter that can go arbitrarily below its initial value, and that can leak non-deterministically.
- A counter $x[i]$ for $m+1 \leq i \leq m+n$ (the ones for which $\ell[i] \neq \omega$) is replaced by one counter $x'[i+m]$ such that $x'[i+m] \leq x[i]$. This simulates a counter that can leak non-deterministically.

Note that states of $\mathcal{S}$ will be represented as (x, v, z) with $x \in \mathbb{N}$, $v \in \mathbb{N}^m$ and $z \in \mathbb{N}^n$ while states of $\mathcal{S}'$ will be represented as (x, v, w, z) with $x \in \mathbb{N}$, $(v, w) \in \mathbb{N}^m \times \mathbb{N}^m$ and $z \in \mathbb{N}^n$.

Formally, we define $\mathcal{V}' = \langle Q, A, a_Z, T', s'_{in} \rangle$ of dimension $d' = 1 + 2m + n$ by:

$$s'_{in} = (0, 0, 0, b)$$

$$T' = \begin{aligned} &\{(q, a, (x, 0^m, v, w), q') \mid (q, a, (x, v, w), q') \in T\} \cup &\text{(T1)}\\ &\{(q, \varepsilon, (0, 0^m, 0^m, -e_i^n), q) \mid q \in Q \wedge 1 \leq i \leq n\} \cup &\text{(T2)}\\ &\{(q, \varepsilon, (0, e_i^m, e_i^m, 0^n), q) \mid q \in Q \wedge 1 \leq i \leq m\} \cup &\text{(T3)}\\ &\{(q, \varepsilon, (0, -e_i^m, -e_i^m, 0^n), q) \mid q \in Q \wedge 1 \leq i \leq m\} \cup &\text{(T4)}\\ &\{(q, \varepsilon, (0, 0^m, -e_i^m, 0^n), q) \mid q \in Q \wedge 1 \leq i \leq m\} &\text{(T5)} \end{aligned}$$

(T1) is the traduction of the transition of $\mathcal{S}$. (T2) makes the counters of index from $1 + 2*m + 1$ to $1 + 2*m + n$ (we recall these counters represent the counters of index $1 + m + 1$ to $1 + m + n$ in $\mathcal{S}$) lossy. (T3) + (T4) imply that only the relative value of the counters i and $i + m$ matters (for $1 \leq i \leq m$). This simulates a counter living in $\mathbb{Z}$. Finally, (T5) makes the previous counter lossy.

We will show that the existence of $x, y \in \{0\} \times \mathbb{N}^{d-1}$ and $u \in A^+$ such that $x \leq \ell$, $(q_f, x) \xrightarrow{u}_{\mathcal{S}} (q_f, y)$ and $x \leq y$ is equivalent to the reachability in $\mathcal{S}'$ of $(0, 0^m, 0^m, b)$ from itself using at least one non-epsilon transition. Note that reachability by using at least one non-epsilon transition is reducible to reachability by adding a lossy counter, starting at zero, that is increased when a non-epsilon transition is fired.

$\Rightarrow$ Let us assume the existence of $x, y \in \{0\} \times \mathbb{N}^{d-1}$ and $u \in A^+$ such that $x \leq \ell$, $x \xrightarrow{u} y$ and $x \leq y$.
Let $x = (0, \alpha_1, \beta_1)$, $y = (0, \alpha_2, \beta_2)$ and $\ell = (0, \omega^m, b)$ with $\alpha_1 \leq \alpha_2$, $\beta_1 \leq \beta_2$ and $\beta_1 \leq b$.
Because $(q_f, 0, \alpha_1, \beta_1) \xrightarrow{u}_{\mathcal{S}} (q_f, 0, \alpha_2, \beta_2)$, we have $(q_f, 0, \alpha_1, \alpha_1, \beta_1) \xrightarrow{u}_{\mathcal{S}'} (q_f, 0, \alpha_1, \alpha_2, \beta_2)$. Because $\beta_1 \leq b$, we also have that $(q_f, 0, \alpha_1, \alpha_1, b) \xrightarrow{u}_{\mathcal{S}'} (q_f, 0, \alpha_1, \alpha_2, \beta_2 + b - \beta_1)$.

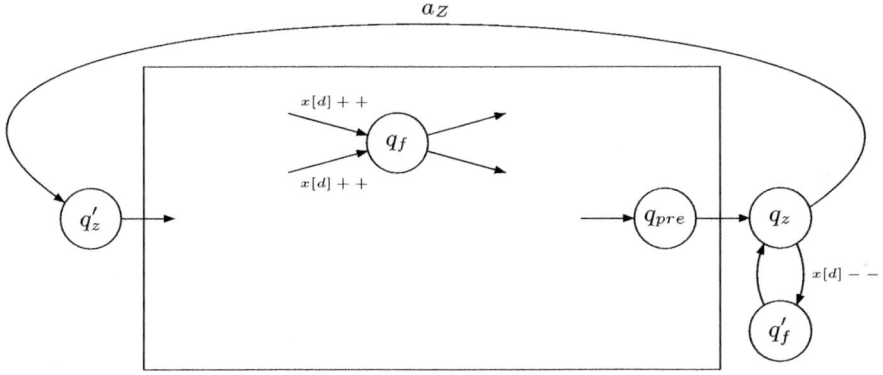

Fig. 1. Schema of the reduction

Then, we have :

$$(q_f, 0, 0^m, 0^m, b) \xrightarrow{\varepsilon}_{\mathcal{S}'} (q_f, 0, \alpha_1, \alpha_1, b)$$
$$\xrightarrow{u}_{\mathcal{S}'} (q_f, 0, \alpha_1, \alpha_2, b + \beta_2 - \beta_1)$$
$$\xrightarrow{\varepsilon}_{\mathcal{S}'} (q_f, 0, \alpha_1, \alpha_1, b)$$
$$\xrightarrow{\varepsilon}_{\mathcal{S}'} (q_f, 0, 0^m, 0^m, b)$$

⇐ Assume that we have $(q_f, 0, 0^m, 0^m, b) \xrightarrow{u} (q_f, 0, 0^m, 0^m, b)$. We will show there exist $x, y \in \{0\} \times \mathbb{N}^{d-1}$ with $x \leq \ell$ such that $(q_f, x) \xrightarrow{u}_{\mathcal{S}} (q_f, y)$.
Let $(t_i, v_i, w_i, z_i)_{0 \leq i \leq k}$ such that $(t_i, v_i, w_i, z_i) \to_{\mathcal{S}'} (t_{i+1}, v_{i+1}, w_{i+1}, z_{i+1})$ and $(t, v_0, w_0, z_0) = (t, v_k, w_k, z_k) = (0, 0^m, 0^m, b)$.
Let α be the vector defined by $\alpha[i] = max_{0 \leq j \leq k} \{v_j[i]\}$. We define μ from $\mathbb{N}^{d'}$ to $\mathbb{N}^d$ by $\mu(t, v, w, z) = (t, w - v + \alpha, z)$. Then, an induction on the length of the transition sequence gives that $(q, s_1) \xrightarrow{u}_{\mathcal{S}'} (q', s_2) \implies \exists s_3 \in \mathbb{N}^d, s_3 \leq_1 \mu(s_2) \wedge (q, \mu(s_1)) \xrightarrow{u}_{\mathcal{S}} (q', s_3)$.
This gives the result.

Note that we can treat a VASS as a $VASS_0$ where the component tested for zero is unused. We get the following corollary of lemma 1 (a similar result can be found in [5] and [4]) that we will also need to use:

Corollary 1. *Let* $\mathcal{V}' = \langle Q, A, T, s_{in} \rangle$ *be a VASS,* $q_f \in Q$ *and* $\ell \in \mathbb{N}_\omega^d$. *It is possible to decide whether there exists a ℓ-increasing loop on q_f in* $\mathcal{V}'$.

Lemma 2. *Let* q_f *be a control state.*
Testing whether there is a run of $\mathcal{S}$ visiting infinitely often q_f and on which the zero-test is fired infinitely often is decidable.

Proof. We reduce this problem to the one of lemma 1. Because $\mathcal{S}$ is normed, there is a single transition labelled by a_Z in T: $(q_z, \delta_z, a_Z, q_z')$. We define $\mathcal{S}' = \langle Q', A, a_Z, T', s_{in}' \rangle$ of dimension $d + 1$ (schematized in figure 1) by:

$$Q' = Q \cup \{q_{pre}, q'_f\}$$
$$s'_{in} = (q_{in}, (x_{in}, 0))$$
$$T' = \begin{array}{l} \{(q, (\delta, 0), a, q') \mid (q, \delta, a, q') \in T \wedge q' \notin \{q_f, q_z\}\} \cup \\ \{(q, (\delta, 1), a, q_f) \mid (q, \delta, a, q_f) \in T\} \cup \\ \{(q, (\delta, 0), a, q_{pre}) \mid (q, \delta, a, q_z)\} \cup \\ \left\{(q_{pre}, 0^{d+1}, \varepsilon, q_z), (q_z, (0^d, -1), \varepsilon, q'_f), (q'_f, 0^{d+1}, \varepsilon, q_z)\right\} \end{array}$$

Note that in S', the last component of the state contains the difference between the number of times the system visited q_f and the number of times the system visited q'_f.

First, let us show that there is a run visiting infinitely often q_f and going through the zero-test infinitely often in S if and only if there is a run visiting infinitely often q'_f in S'.

$\Rightarrow$ Let us assume there is a run in S that visits infinitely often q_f and that goes infinitely often through the zero-test. This run is also a valid run in S' because we only added places and a counter that is only incremented by actions of S. Now, we alter this run by inserting as many loops $q_z \rightleftharpoons q_f$ as possible before each zero-test. This new run fulfills $x[d] = 0$ infinitely often, and because this counter marks the difference between the number of passages in q_f and the number of passages in q'_f, this means q'_f is visited infinitely often.

$\Leftarrow$ Let us assume there is a run in S' that visits q'_f infinitely often. Because of the $x[d]$ counter, this run visits q_f infinitely often. Moreover, because q'_f can only be reached by q_z, that can only go to q'_f or through the zero-test, and that the loop $q_z \rightleftharpoons q_f$ can only be done a finite number of times, if q'_f is visited infinitely often on a run, then the zero-test is also fired an infinite number of times. Hence, we have a run of S' that visits infinitely often q_f and on which the zero-test is fired infinitely often. Now, if we remove in this run the loops $q_z \rightleftharpoons q_f$, we get a run using only transitions of S, and removing the additionnal counter can't make this run non-fireable, so we get a run of S that visits infinitely often q_f and the zero-test.

Now, assume we have a run visiting infinitely often q'_f. We have an infinite sequence $(x_i)_i$, $x_i \in \mathbb{N}^{d+1}$ such that for all $i \in \mathbb{N}$, $(q_f, x_i) \xrightarrow{*} (q_f, x_{i+1})$. By well-order of $\mathbb{N}^{d+1}$, there exists $i < j$ such that $x_i \le x_j$. Also, because the zero-test is fired after the iterations $q'_f \rightleftharpoons q_z$, this means that $x_i[0] = 0$. So, we have a run visiting infinitely often q'_f if and only if there exists (q_f, x) reachable state with $x[0] = 0$, y with $x \le_0 y$ and $u \in A^+$ such that $(q_f, x) \xrightarrow{u} (q_f, y)$ (the "if" part is immediate).

Because the first counter is necessarily 0 on the q'_f control state (assuming an infinite run) and because our system is monotonic with respect to $\le_0$ (proposition 2), we can replace "(q_f, x) reachable state" by "(q_f, x) coverable state" in the previous equivalence. Hence, our problems reduce to decide whether there exists a ℓ-increasing loop on q_f, for ℓ a maximal element of $Cover(S)$.

By [3], we can compute the maximal elements of $Cover(\mathcal{S})$. Then, for each such maximal element, we can use lemma 1 to get our result.

Lemma 3. *Let q_f be a control state.*
Testing whether there is a run of $\mathcal{S}$ visiting infinitely often q_f and on which the zero-test is not fired infinitely often is decidable.

Proof. Let us consider a run visiting q_f infinitely often. Because the zero-test is fired only a finite number of times, after some point, we have a run visiting q_f infinitely often without firing the zero-test. Hence, we reduce our problem to repeated control state reachability in VASS.

We make the intersection of $Cover(\mathcal{S})$ (computed through [3]) with $(\{q_f\} \times \mathbb{N}^d)$. By well-order, if q_f is visited infinitely often, then there exists $x, x' \in \mathbb{N}^d$ and $u \in (A \backslash \{a_z\})^+$ such that $(q_f, x) \xrightarrow{u} (q_f, x')$, $x \leq x'$. Detecting such an increasing loop in a VASS can be seen as a special case of lemma 1 (corollary 1), and by testing the presence of an increasing loop for each maximal element of the cover, we get our result.

Finally, we can combine lemmas 2 and 3 to get:

Theorem 3. *Let q_f be a control state.*
Testing whether there is a run of $\mathcal{S}$ visiting infinitely often q_f is decidable.

And by proposition 1,

Corollary 2. *Model-Checking LTL is decidable on $VASS_0$.*

5   Conclusion

We have shown that despite $VASS_0$ looking more expressive than VASS, another decidability result of VASS is preserved. Between the numerous decidability results that have recently been shown for $VASS_0$ and this new one, a rule of thumb seems to be that $VASS_0$ and VASS enjoy the same decidability properties, and counter-examples have yet to be found. One can wonder if the few problems (regularity of the recognized language for example) that are decidable for VASS and remain unknown for $VASS_0$ follow this rule.

However, it is interesting to note that, despite repeated control state reachability being independent from reachability for Vector Addition Systems [6], our proof requires both reachability and place-boundedness on $VASS_0$. This makes the complexity of our procedure unknown. One might wonder a proof might exist without using reachability and/or place-boundedness, or whether reachability and place-boundedness can actually be reduced to LTL. We leave these questions for future work.

References

1. Abdulla, P., Cerans, K., Jonsson, B., Tsay, Y.-K.: General decidability theorems for infinite-state systems. In: Symposium on Logic in Computer Science, p. 313 (1996)
2. Abdulla, P., Mayr, R.: Minimal cost reachability/coverability in priced timed petri nets. In: de Alfaro, L. (ed.) FOSSACS 2009. LNCS, vol. 5504, pp. 348–363. Springer, Heidelberg (2009)
3. Bonnet, R., Finkel, A., Leroux, J., Zeitoun, M.: Place-boundedness for vector addition systems with one zero-test. In: Lodaya, K., Mahajan, M. (eds.) Proceedings of the 30th Conference on Foundations of Software Technology and Theoretical Computer Science (FSTTCS 2010). Leibniz International Proceedings in Informatics, vol. 8, pp. 192–203. Leibniz-Zentrum für Informatik, Dagstuhl, Germany (2010)
4. Emerson, E.A., Namjoshi, K.S.: On model checking for non-deterministic infinite-state systems. In: Proceedings of the 13th Annual IEEE Symposium on Logic in Computer Science LICS 1998, p. 70. IEEE Computer Society, Washington, DC, USA (1998)
5. Esparza, J.: On the decidability of model checking for several μ-calculi and petri nets. In: Tison, S. (ed.) CAAP 1994. LNCS, vol. 787, pp. 115–129. Springer, Heidelberg (1994), 10.1007/BFb0017477
6. Esparza, J.: Decidability and complexity of petri net problems: An introduction. In: Reisig, W., Rozenberg, G. (eds.) APN 1998. LNCS, vol. 1491, pp. 374–428. Springer, Heidelberg (1998)
7. Finkel, A., Goubault Larrecq, J.: Forward analysis for WSTS, Part I: Completions. In: Albers, S., Marion, J.-Y. (eds.) 26th International Symposium on Theoretical Aspects of Computer Science - STACS 2009, pp. 433–444. IBFI Schloss Dagstuhl, Germany (2009)
8. Finkel, A., Sangnier, A.: Mixing coverability and reachability to analyze vass with one zero-test. In: van Leeuwen, J., Muscholl, A., Peleg, D., Pokorný, J., Rumpe, B. (eds.) SOFSEM 2010. LNCS, vol. 5901, pp. 394–406. Springer, Heidelberg (2010)
9. Finkel, A., Schnoebelen, P.: Well-structured transition systems everywhere! Theoretical Computer Science 256(1-2), 63–92 (2001)
10. Gastin, P., Oddoux, D.: Fast LTL to Büchi automata translation. In: Berry, G., Comon, H., Finkel, A. (eds.) CAV 2001. LNCS, vol. 2102, pp. 53–65. Springer, Heidelberg (2001)
11. Gerth, R., Peled, D., Vardi, M.Y., Wolper, P.: Simple on-the-fly automatic verification of linear temporal logic. In: Proceedings of the Fifteenth IFIP WG6.1 International Symposium on Protocol Specification, Testing and Verification XV, pp. 3–18. Chapman & Hall, Ltd., London (1996)
12. Habermehl, P.: On the complexity of the linear-time μ-calculus for petri nets. In: Azéma, P., Balbo, G. (eds.) ICATPN 1997. LNCS, vol. 1248, pp. 102–116. Springer, Heidelberg (1997)
13. Hack, M.: The equality problem for vector addition systems is undecidable. Theoretical Computer Science 2(1), 77–95 (1976)
14. Kosaraju, S.R.: Decidability of reachability in vector addition systems. In: Proceedings of the Fourteenth Annual ACM Symposium on Theory of Computing, STOC 1982, pp. 267–281. ACM, New York (1982)
15. Leroux, J.: The general vector addition system reachability problem by presburger inductive invariants. In: Symposium on Logic in Computer Science, pp. 4–13 (2009)

16. Leroux, J.: Vector addition system reachability problem: a short self-contained proof. SIGPLAN Not. 46, 307–316 (2011)
17. Mayr, E.W.: An algorithm for the general petri net reachability problem. In: Proceedings of the Thirteenth Annual ACM Symposium on Theory of Computing, STOC 1981, pp. 238–246. ACM, New York (1981)
18. Reinhardt, K.: Reachability in petri nets with inhibitor arcs. Electronic Notes in Theoretical Computer Science 223, 239–264 (2008); Proceedings of the Second Workshop on Reachability Problems in Computational Models (RP 2008)

A   Additionnal Reductions

Definition 6. *Let* $\mathcal{S}_1 = \langle Q_1, A, a_Z, T_1, s_{in1} \rangle$ *and* $\mathcal{S}_2 = \langle Q_2, A, a_Z, T_2, s_{in2} \rangle$ *be two VASS$_0$ of respective dimensions d_1 and d_2. $\mathcal{S}_1$ and $\mathcal{S}_2$ are weakly bisimilar if there exists a relation $\sim \subseteq (Q_1 \times \mathbb{N}^{d_1}) \times (Q_2 \times \mathbb{N}^{d_2})$ such that:*

- $s_{in1} \sim s_{in2}$
- $\begin{cases} s_1 \sim s_2 \\ s_1 \xrightarrow{a}_{\mathcal{S}_1} s_1' \end{cases} \implies \exists s_2' \in Q \times \mathbb{N}^{d_2} \begin{cases} s_2 \xrightarrow{a}_{\mathcal{S}_2} s_2' \\ s_1' \sim s_2' \end{cases}$
- $\begin{cases} s_1 \sim s_2 \\ s_2 \xrightarrow{a}_{\mathcal{S}_2} s_2' \end{cases} \implies \exists s_1' \in Q \times \mathbb{N}^{d_2} \begin{cases} s_1 \xrightarrow{a}_{\mathcal{S}_1} s_1' \\ s_1' \sim s_2' \end{cases}$

Note that we are using weak bisimilarity because of the presence of epsilon-transitions. Satisfiability of a LTL formula is stable by weak bisimilarity[1].

We provide here a quick proof of a well known reduction of VASS$_0$.

Proposition 3. *Let $\mathcal{S}$ be a VASS$_0$. There exists a VASS$_0$ $\mathcal{S}'$ weakly bisimilar to $\mathcal{S}$ such that there exists a unique $(q_z, a_Z, q_z', \delta_z) \in T$.*

Proof. If $\mathcal{S}$ has no such transition, we can simply add new unreachable control states and add the required transition, so we will only consider the case of $\mathcal{S}$ having more than one transition.

Let $\mathcal{S} = \langle Q, A, a_Z, T, (q_{in}, x_{in}) \rangle$ be a VASS$_0$ of dimension d.

Let $T_z = \{(q_{z,i}, a_Z, q_{z,i}', \delta_{z,i}) \mid 0 \le i \le p\}$ be the transitions of $\mathcal{S}$ using the zero-test. Let T_0 be the other transitions. $T = T_0 \uplus T_z$. We define $\mathcal{S}' = \langle Q', A, a_Z, T', s_{in}' \rangle$ of dimension $d + 2$ by:

$$Q' = Q \uplus \{q_z, q_z'\}$$

$$T' = \begin{array}{l} \{(q, a, q', (\delta, 0, 0)) \mid (q, a, q', \delta) \in T_0)\} \cup \\ \{(q_{z,i}, \varepsilon, q_z, (\delta_{z,i}, i, p - i)) \mid 1 \le i \le p\} \cup \\ \{(q_z', \varepsilon, q_{z,i}', (0^d, -i, -(p - i))) \mid 1 \le i \le p\} \cup \\ \{(q_z, a_Z, q_z', 0^{d+2})\} \end{array}$$

$$s_{in}' = (q_{in}, (x_{in}, 0, 0))$$

We note that we have the invariant that the last two components are always zero in all states of Q. Bisimilarity comes easily from that.

---

[1] For a survey of weak bisimilarity and other notions of system equivalence, one might look at "The linear time-branching time spectrum II: The semantics of sequential processes with silent moves", by R.J. van Glabbeek.

Complexity Analysis of the Backward Coverability Algorithm for VASS

Laura Bozzelli[1] and Pierre Ganty[2,*]

[1] UPM Facultad de Informática, Madrid, Spain
[2] IMDEA Software Institute, Madrid, Spain

Abstract. By using the known lower and upper complexity bounds of the coverability problem for VASS, we characterize the complexity of the classical backward algorithm for VASS coverability, and provide optimal bounds on the size of the symbolic representation it computes.

1   Introduction

In [3, 4, 15, 10] checking safety properties for concurrent systems like multithreaded programs, communication protocols, or asynchronous programs is reduced to the coverability problem of VASS (Vector Addition System with States), turning it into a central problem in verification of concurrent systems. Given a VASS G and two configurations s_0 and s_f, the *coverability problem* asks whether s_f is coverable from s_0, i.e. there is a computation in G starting at s_0 and leading to a configuration s which *covers* s_f; that is, s and s_f are in the same control state and the counters of s are pointwise greater or equal than those of s_f (this is noted $s_f \trianglelefteq s$). The complexity of the coverability problem, which is complete for EXPSPACE, was settled in the late 70's (Lipton [13] for the lower bound and Rackoff [14] for the matching upper bound). However, rather surprisingly, the complexity analysis of the algorithms that have been implemented to solve the coverability problem have received little or no attention.[1]

   In this work, we propose to characterize the complexity of the so-called *backward algorithm* which has been implemented in several tools and whose definition can be attributed to [1, 9] and to some extent [2]. Given a VASS G and a target configuration s_f, the backward algorithm iteratively computes the configurations from which s_f is coverable in 0 steps, 1 step, ... until the set of configurations is saturated. More precisely, the algorithm symbolically computes an increasing (w.r.t set inclusion) sequence of sets of configurations starting from the set of configurations which cover s_f. Let us call each element of the computed sequence an *iterate* which is given by a set of configurations closed by above for $\trianglelefteq$. Since such upward closed sets are infinite, each iterate is finitely represented and manipulated by its *basis*, that is the *finite* set of its *minimal elements* (w.r.t $\trianglelefteq$). First, let us recall that the minimal elements yields a decidable, finite, and

---

* This author was sponsored by Comunidad de Madrid's Program PROMETIDOS-CM (S2009TIC-1465), PEOPLE-COFUND's program AMAROUT (PCOFUND-2008-229599), and by the Spanish Ministry of Science and Innovation (TIN2010-20639).

[1] As far as we know, no implementation of Rackoff's algorithm exists.

G. Delzanno and I. Potapov (Eds.): RP 2011, LNCS 6945, pp. 96–109, 2011.

canonical representation of each iterate, and second, because $\trianglelefteq$ is a well-quasi ordering on the set of configurations, it follows that the algorithm is guaranteed to reach a fixpoint $B(G, s_f)$ after finitely many steps. Since $B(G, s_f)$ is the basis of the set of configurations from which s_f is coverable in G, we obtain a decision procedure for the coverability problem: (G, s_0, s_f) is a positive instance of the coverability problem iff $s_{min} \trianglelefteq s_0$ for some $s_{min} \in B(G, s_f)$. Note that $B(G, s_f)$ can be used to solve other coverability related problems such as checking whether from each G-configuration, s_f is coverable.

Our contribution. In this paper, we show that the "backward algorithm" is optimal to solve the coverability problem. Using Rackoff's and Lipton's results [14, 13], respectively, we give upper and lower bounds on the number of iterations of the backward algorithm as well as its execution time. Moreover, our complexity analysis allows us to derive upper bounds on the cardinality of $B(G, s_f)$ and the maximal size of the single elements of $B(G, s_f)$, which are doubly exponential in the dimension of G (the number of counters). Furthermore, we provide matching lower bounds by a readaptation of the Lipton's proof [13].

Besides the backward algorithm, VASS analysis tools often implement a *forward algorithm* whose definition is due to Karp and Miller [12]. The forward algorithm returns a finite representation (the *covering set*) of an overapproximation (the *coverability set*) of the set of configurations reachable from the given initial configuration s_0. Such an overapproximation is sound and also complete for certain problems like the coverability problem. By using the covering set, one can solve, for instance, the coverability problem (by asking whether the target configuration s_f belongs to the coverability set) but also the boundedness problem which asks whether the set of reachable configurations from the given initial configuration is finite. From a complexity standpoint, it is mentioned in [7] that the algorithm of Karp and Miller requires non-primitive recursive space. Let us also cite [8] which gives a more refined complexity analysis of the forward algorithm.

Related work. The closest works to our are [17] which provide an upper bound on the size of $B(G, s_f)$. However, the algorithm to compute $B(G, s_f)$ (originally given in [16]) differs from the backward algorithm and does not yield any conclusion about the complexity of the backward algorithm. Moreover, contrary to us the authors do not provide lower bounds on the size of $B(G, s_f)$.[2]

2 Preliminaries

2.1 Notations and Definitions

Let $\mathbb{Z}$ be the set of integers, $\mathbb{N}$ be the set of nonnegative integers, and $\mathbb{N}^+$ be the set of positive integers. For each $k \in \mathbb{N}^+$ and *vector* $v \in \mathbb{Z}^k$, $v[i]$ denotes the ith component of v, for $i \in \{1, \ldots, k\}$. If $v_1, v_2 \in \mathbb{Z}^k$, then $v_1 + v_2$ denotes that vector $v \in \mathbb{Z}^k$ such that $v[i] = v_1[i] + v_2[i]$ for all $i \in \{1, \ldots, k\}$; $v_1 - v_2$ is defined similarly. Let $v \in \mathbb{Z}^k$, define $\|v\| = \max(\{abs(v[i]) \mid i \in \{1, \ldots, k\}\})$, where $abs(v[i])$ is the absolute value of $v[i]$. Finally, for a finite set Q, $|Q|$ denotes the cardinality of Q.

---

[2] Similarly to Rackoff's algorithm we do not know of any implementation of the algorithm of [16].

2.2   Well-Quasi Orderings

Recall that for a set S, a *partial order* $\preceq$ over S is a reflexive, transitive and antisymmetric binary relation on S. We say that $\preceq$ is a *well-quasi ordering* (wqo, for short) if additionally, for each infinite sequence $s_0, s_1, \ldots$ of elements of S there are indices $i < j$ such that $s_i \preceq s_j$. Given a partial order $\preceq$ over S, a subset U of S is *upward-closed* (w.r.t. $\preceq$) if for all $s, s' \in S$, $s \in U$ and $s \preceq s'$ entail $s' \in U$. A *basis* of U (w.r.t. $\preceq$) is a subset B of U satisfying the following: (1) for each $s \in U$, there is $s' \in B$ such that $s' \preceq s$, and (2) for all $s, s' \in B$, $s \preceq s'$ implies $s = s'$ (i.e., distinct elements of B are incomparable w.r.t. $\preceq$). The following is a well-known result.

Lemma 1. *[11] Let S be a set and $\preceq$ be a partial order over S which is wqo. Then, each upward-closed subset U of S (w.r.t. $\preceq$) admits a unique basis, which is finite and consists of the minimal elements of U (w.r.t. $\preceq$). Moreover, for each monotone infinite sequence of upward-closed sets $U_0 \subseteq U_1 \subseteq \ldots$, there is $i \geq 0$ such that $U_{i+1} = U_i$.*

Let $k \in \mathbb{N}^+$. We consider the partial order over $\mathbb{N}^k$, written $\trianglelefteq$, which is the componentwise extension of $\leq$ over $\mathbb{N}$: let $v, v' \in \mathbb{N}^k$, $v \trianglelefteq v'$ iff $v[i] \leq v'[i]$ for each $1 \leq i \leq k$. Moreover, for a *finite* set Q, we consider the partial order over $Q \times \mathbb{N}^k$, which (with a little abuse of notation) is again denoted by $\trianglelefteq$, defined as: $\langle q, v \rangle \trianglelefteq \langle q', v' \rangle$ iff $q = q'$ and $v \trianglelefteq v'$. It is well-known that $\trianglelefteq$ is a wqo over $\mathbb{N}^k$ (this result is known as Dickson's Lemma [5]). Hence, it easily follows that $\trianglelefteq$ is a wqo over $Q \times \mathbb{N}^k$, for each *finite* set Q. For $s \in Q \times \mathbb{N}^k$, we denote by $s{\uparrow}$ the upward-closed set given by $\{s' \in Q \times \mathbb{N}^k \mid s \trianglelefteq s'\}$. In the rest of this paper, if we say that some set $U \subseteq Q \times \mathbb{N}^k$ is upward-closed, we mean that U is upward-closed set w.r.t. $\trianglelefteq$. For $X \subseteq Q \times \mathbb{N}^k$, $min(X)$ denotes the set of minimal elements in X (w.r.t. $\trianglelefteq$). Note that according to Lemma 1, $min(X)$ is the unique (finite) basis of X if X is upward-closed.

2.3   Vector Addition Systems with States (VASS)

Let $d \in \mathbb{N}^+$. A d-VASS G is a pair $\langle Q, \Delta \rangle$, where Q is a non-empty *finite* set of *control points* and $\Delta \subseteq Q \times \mathbb{Z}^d \times Q$ is a *finite* set of transitions in $Q \times \mathbb{Z}^d \times Q$. The d-VASS G induces an infinite directed graph $[\![G]\!] = \langle Q \times \mathbb{N}^d, \rightarrow \rangle$ whose set of vertices is given by $Q \times \mathbb{N}^d$ and the set of edges is defined as: $\langle q, v \rangle \rightarrow \langle q', v' \rangle$ iff there is $\langle q, u, q' \rangle \in \Delta$ such that $v' = v + u$. Vertices of $[\![G]\!]$ are called *G-states* or simply *states* when G is clear from the context. A *run* $\pi = s_1, \ldots, s_n$ of G is a finite path in the graph $[\![G]\!]$. The length $|\pi|$ of π is n. We define $\|\Delta\| = \max(\{\|v\| \mid \langle q, v, q' \rangle \in \Delta\})$. Moreover, for a state $s = \langle q, v \rangle$ and a *finite* set S of states, define $\|s\| = \|v\|$ and $\|S\| = \max(\{\|s\| \mid s \in S\})$.

For each set S of G-states, $Pre^*(G, S)$ denotes the set of G-states s such that there is a run of G from s to some state in S. Moreover, $Pre(G, S)$ denotes the set of G-states s such that $s \rightarrow s'$ is an edge of $[\![G]\!]$ for some $s' \in S$. It is well-known (see e.g. [1, 9]) that if S is upward-closed, then $Pre^*(G, S)$ and $Pre(G, S)$ are upward-closed as well (this can be easily checked).

2.4   Coverability Problem and Rackoff's Upper Bound

Given a d-VASS $G = \langle Q, \Delta \rangle$ and two G-states s_0 and s_f, a *covering in G of s_f w.r.t. s_0* is a run of G from s_0 which leads to a state s satisfying $s_f \trianglelefteq s$. If such a covering exists, i.e.,

$s_0 \in Pre^*(G, s_f \uparrow)$, we say that s_f *is coverable from* s_0 *in* G. The *coverability problem* asks whether s_f is coverable from s_0 in G for a given d-VASS G and G-states s_0 and s_f. By a straightforward adaptation of the Rackoff's algorithm for the coverability problem [14], we obtain the following result.

Theorem 1. *Let* $G = \langle Q, \Delta \rangle$ *be a* d-VASS *and* s_f *be a state. For each state* s, *if* s_f *is coverable from* s *in* G, *then there is a covering in* G *of* s_f *w.r.t.* s *whose length is independent on* $\|s\|$ *and is at most* $[|Q| \cdot (\|\Delta\| + \|s_f\| + 2)]^{(3d)!+1}$.

Proof of Theorem 1. We need additional definitions. Let $d \in \mathbb{N}^+$ and $I \subseteq \{1, \ldots, d\}$. For $u \in \mathbb{Z}^d$, u^I denotes the vector in $\mathbb{Z}^d$ defined as $u^I[i] = u[i]$ if $i \in I$, and $u^I[i] = 0$ otherwise. For a d-VASS $G = \langle Q, \Delta \rangle$, G^I denotes the d-VASS $G^I = \langle Q, \{\langle q, u^I, q' \rangle \mid \langle q, u, q' \rangle \in \Delta\} \rangle$. Note that $G^{\{1, \ldots, d\}} = G$. Let $s = \langle q, v \rangle$ be a G-state, we denote by s^I the G-state given by $\langle q, v^I \rangle$, and for a run π, we denote by π^I the sequence of G-states obtained from π by replacing each state s along π with s^I. Note that π^I is a run in G^I. For $B \in \mathbb{N}$, a vector $v \in \mathbb{N}^d$ is B-bounded if $v[i] \leq B$ for each $i \in \{1, \ldots, d\}$. A run π of G is B-bounded if for each state $\langle q, v \rangle$ occurring along π, v is B-bounded.

Fix a d-VASS $G = \langle Q, \Delta \rangle$ and a state $s_f = \langle q_f, v_f \rangle$. For each $I \subseteq \{1, \ldots, d\}$ and G-state s, define $dist(I, s)$ to be the length of the shortest covering in G^I of $(s_f)^I$ w.r.t. s^I, if $(s_f)^I$ is coverable from s^I in G^I (note that $dist(I, s) \geq 1$), and $dist(I, s) = 0$ otherwise. Moreover, for each $k \in \{0, 1, \ldots, d\}$, define $f(k) = sup\{dist(I, s) \mid |I| = k$ and s is a G-state$\}$ (note that $f(k) \geq 1$ since s_f is coverable from itself in G). Then:

Lemma 2. *For all* $k \in \{0, 1, \ldots, d\}$, *the following inequalities hold:*
$$f(k) \leq \begin{cases} |Q| & \text{if } k = 0 \\ |Q| \cdot ((\|\Delta\| + \|s_f\|) \cdot f(k-1))^k + f(k-1) & \text{if } k > 0 \end{cases}$$

Proof. The case $k = 0$ is trivial. Now, assume that $k > 0$. By ind. hyp., $f(k-1)$ is finite. Let s be a G-state and $I \subseteq \{1, \ldots, d\}$ s.t. $|I| = k$ and there is a covering π in G^I of $(s_f)^I$ w.r.t. s^I. We need to show that there is a covering in G^I of $(s_f)^I$ w.r.t. s^I of length bounded by $|Q| \cdot ((\|\Delta\| + \|s_f\|) \cdot f(k-1))^k + f(k-1)$. Let $B = \|\Delta\| \cdot f(k-1) + \|s_f\|$. We distinguish two cases:

Case 1: π is B-bounded. Let s' be the last state of π. Then, there is a B-bounded run π' in G^I from s^I to s' such that the states visited by π' are mutually distinct. It is routine to check that the length of π' is at most $|Q| \cdot B^k$. By hypothesis $(s_f)^I \trianglelefteq s'$, hence π' is also a covering in G^I of $(s_f)^I$ w.r.t. s^I. Thus, since $|Q| \cdot B^k \leq |Q| \cdot ((\|\Delta\| + \|s_f\|) \cdot f(k-1))^k$, the result holds in this case.

Case 2: π is *not* B-bounded. Then, there is a G-state s_2 s.t. π can be written in the form $\pi = \pi_1 \cdot \pi_2$ so that π_1 is either empty or B-bounded, π_2 starts at state $(s_2)^I = \langle q_2, v_2 \rangle$, and v_2 is not B-bounded. Hence, there is $i \in I$ such that $v_2[i] > B$. Assume that π_1 is not empty and B-bounded (the other case being simpler). Let s_1 be the last state of π_1. As in case 1, we can replace π_1 with a run π'_1 in G^I from s^I to s_1 of length at most $|Q| \cdot B^k$. Let $J = I \setminus \{i\}$ (hence, $|J| = k - 1$). Since $(\pi_2)^J$ is a covering in G^J of $(s_f)^J$ w.r.t. $(s_2)^J$, by the ind. hyp., there is a covering π'_2 in G^J of $(s_f)^J$ w.r.t. $(s_2)^J$ of length at most $f(k-1)$. Note that at each step of a run of G, any component of a G-state can decrease at most by $\|\Delta\|$. Thus, since π'_2 has length at most $f(k-1)$, $(s_2)^I = \langle q_2, v_2 \rangle$, and

$v_2[i] > B = \|\Delta\| \cdot f(k-1) + \|s_f\|$, it follows that there exists a covering π_2'' in G^I of $(s_f)^I$ w.r.t. $(s_2)^I$ of length at most $f(k-1)$. Hence, $\pi_1' \cdot \pi_2''$ is a covering in G^I of $(s_f)^I$ w.r.t. s^I of length at most $|Q| \cdot B^k + f(k-1)$. Since $|Q| \cdot B^k \leq |Q| \cdot ((\|\Delta\| + \|s_f\|) \cdot f(k-1))^k$, we are done.   □

By solving the recurrence in Lemma 2, we obtain the following result. Hence, Theorem 1 directly follows.

Lemma 3. *For all* $k \in \{0, 1, \ldots, d\}$, $f(k) \leq (|Q| \cdot (\|\Delta\| + \|s_f\| + 2))^{(3k)!+1}$.

Proof. By induction on k. The base case $k = 0$ directly follows from Lemma 2. Now, assume that $k > 0$. Let $C = \|\Delta\| + \|s_f\| + 2$. Then,

$$
\begin{aligned}
f(k) &\leq |Q| \cdot (C \cdot f(k-1))^k + f(k-1) && \text{by Lemma 2}\\
&\leq |Q| \cdot [(C \cdot f(k-1))^k + f(k-1)]\\
&\leq |Q| \cdot (C \cdot f(k-1))^{k+1} && \text{since } C \cdot f(k-1) \geq 2\\
&\leq (|Q| \cdot C \cdot f(k-1))^{k+1}\\
&\leq ((|Q| \cdot C)^{(3(k-1))!+2})^{k+1} && \text{by induction hypothesis}\\
&\leq (|Q| \cdot C)^{(3k)!+1}
\end{aligned}
$$

□

Note that $min(Pre^*(G, s_f\uparrow))$ constitutes a finite canonical representation of the possibly infinite set $Pre^*(G, s_f\uparrow)$, for which the membership problem (and other basic questions) are decidable.[3] It is well-known that $min(Pre^*(G, s_f\uparrow))$ can be computed by a least fixpoint algorithm [1, 9] refered to as the backward algorithm. However, no elementary upper bound is known on the execution time of this algorithm. By using Theorem 1, we provide in the next section such an upper bound. As a consequence, we derive an upper bound on the cardinality of $min(Pre^*(G, s_f\uparrow))$, which is doubly exponential in the dimension d of G. In Section 4, we show that this double exponential blow-up cannot be avoided.

3   Complexity of the Backward Algorithm for Coverability

First, we recall the standard backward algorithm for coverability [1, 9]. Fix a d-VASS $G = \langle Q, \Delta \rangle$ and a state s_f. We define a monotone infinite sequence $U_0 \subseteq U_1 \subseteq \ldots$ of upward-closed sets of states as: $U_0 = s_f\uparrow$, and $U_{i+1} = U_i \cup Pre(G, U_i)$ for each $i \geq 0$. Since $\trianglelefteq$ (over $Q \times \mathbb{N}^d$) is a wqo, $U_i \subseteq U_{i+1}$ for each $i \geq 0$, and $U_i = U_{i+1}$ iff $min(U_i) = min(U_{i+1})$, by Lemma 1 and definition of the sets U_i, we obtain the following.[4]

Remark 1. For each $i \geq 0$, U_i is the set of states s such that there is a covering of s_f w.r.t. s of length less or equal to i. Moreover, there is $i \geq 0$ such that $min(U_{i+1}) = min(U_i)$. Also, whenever $min(U_{i+1}) = min(U_i)$ for some $i \geq 0$, then $Pre^*(G, s_f\uparrow) = U_i$.

---

[3] Given $min(U)$ for an upward-closed set U of G-states, one can decide if a given state is in U (membership problem).

[4] Note that $Pre^*(G, s_f\uparrow)$ is the least fixpoint of $\mu X.(s_f\uparrow) \cup Pre(G, X)$.

Remark 2. [1, 9] Given a G-state s, one can compute $min(Pre(G,s\uparrow))$. Hence, for each $i \geq 0$, given $min(U_i)$, one can compute $min(U_{i+1})$ as follows:
$$min(U_{i+1}) = min(min(U_i) \cup \bigcup_{s \in min(U_i)} min(Pre(G,s\uparrow))) .$$

Then, the backward algorithm at ith step computes $min(U_i)$. If $min(U_i) = min(U_{i+1})$, then the algorithm terminates and outputs $min(U_i)$. By Remark 1, the algorithm terminates and outputs the basis of $Pre(G,s_f\uparrow)$. Now, we analyze its complexity. Let H be the upper bound in Theorem 1 for G and s_f, i.e., $H = [|Q| \cdot (\|\Delta\| + \|s_f\| + 2)]^{(3d)!+1}$.

Lemma 4. *The sequence* $min(U_0), min(U_1), \ldots$ *is stable at* H, *i.e.* $min(U_H) = min(U_{H+1})$.

Proof. By contradiction. Assume that $min(U_H) \neq min(U_{H+1})$. Then, $U_H \neq U_{H+1}$ and since $U_H \subseteq U_{H+1}$, there must be $s \in U_{H+1} \setminus U_H$. By Remark 1, it follows that each covering in G of s_f w.r.t. s has length at least $H + 1$. Since $s \in U_{H+1} \subseteq Pre^*(G,s_f\uparrow)$, s_f is coverable from s. Thus, by definition of H and Theorem 1, there must be a covering of s_f w.r.t. s of length at most H, which is a contradiction. $\square$

Lemma 5. *Let S be a* finite *set of states. Then, one can compute a* finite *set B_S of states such that* $min(S\uparrow \cup Pre(G,S\uparrow)) \subseteq B_S \subseteq S\uparrow \cup Pre(G,S\uparrow)$, $|B_S|$ *is at most* $O(|\Delta| \cdot |S|)$, *and* $\|B_S\|$ *is at most* $O(\|\Delta\| + \|S\|)$. *Moreover, B_S can be computed in time* $O(d \cdot |\Delta| \cdot |S| \cdot \log(\|\Delta\| + \|S\| + 2))$.

Proof. For $v \in \mathbb{Z}^d$, $pos(v)$ denotes the vector in $\mathbb{N}^d$ defined as: $pos(v)[i] = v[i]$ if $v[i] \in \mathbb{N}$, and $pos(v)[i] = 0$ otherwise. Then, $B_S = S \cup A_S$, where A_S is given by

$$A_S = \{\langle q, pos(v' - v)\rangle \mid \langle q,v,q'\rangle \in \Delta \text{ and } \langle q',v'\rangle \in S \text{ for some } q' \in Q\}$$

We show the following, hence, the result easily follows:

1. $A_S \subseteq Pre(G,S\uparrow)$
2. For each $s \in Pre(G,S\uparrow)$, there is $s' \in A_S$ such that $s \trianglerighteq s'$.

Proof of Property 1: let $s \in A_S$. By construction there are $\langle q,v,q'\rangle \in \Delta$ and $\langle q',v'\rangle \in S$ such that $s = \langle q, pos(v' - v)\rangle$. Evidently, it suffices to show that $pos(v' - v) + v \trianglerighteq v'$. Since $pos(v' - v) \trianglerighteq v' - v$, the result follows.
Proof of Property 2: let $s \in Pre(G,S\uparrow)$, where $s = \langle q,v\rangle$ for some $q \in Q$ and $v \in \mathbb{N}^d$. Then, there is $\langle q,v',q'\rangle \in \Delta$ such that $\langle q',v + v'\rangle \in S\uparrow$. Hence, there is $\langle q', v_{min}\rangle \in S$ such that $v + v' \trianglerighteq v_{min}$. Hence, $v \trianglerighteq v_{min} - v'$. Since $v \in \mathbb{N}^d$, we obtain that $v \trianglerighteq pos(v_{min} - v')$. Let $s' = \langle q, pos(v_{min} - v')\rangle$. Note that $s' \in A_S$. Thus, since $s \trianglerighteq s'$, we are done. $\square$

Note that given a *finite* set S of G-states, $min(S)$ can be easily computed in time $O(d \cdot |S|^2 \cdot \log(\|S\| + 2))$. Hence, by Lemma 5, we obtain the following.

Corollary 1. *Let S be a* finite *set of G-states and* $S_{min} = min(S\uparrow \cup Pre(G,S\uparrow))$. *Then,* $\|S_{min}\|$ *is at most* $O(\|\Delta\| + \|S\|)$. *Moreover, S_{min} can be computed in time* $O(d \cdot |\Delta|^2 \cdot |S|^2 \cdot \log(\|\Delta\| + \|S\| + 2))$.

By Lemma 4 and Remark 1, $min(Pre^*(G, s_f\uparrow)) = min(U_H)$. Then, Corollary 1 shows that the backward algorithm terminates in time

$$O(H \cdot d \cdot |\Delta|^2 \cdot max_{0 \leq i \leq H}|min(U_i)|^2 \cdot \log(\|\Delta\| + max_{0 \leq i \leq H}\|min(U_i)\| + 2))$$

Note that, by Corollary 1, for each $i \geq 0$, $\|min(U_i)\| = O(i \cdot \|\Delta\| + \|s_f\|)$. Hence, $|min(U_i)|$ is at most $O(|Q| \cdot (i \cdot \|\Delta\| + \|s_f\|)^d)$. Also, $max_{0 \leq i \leq H}\|min(U_i)\| = O(H \cdot \|\Delta\| + \|s_f\|)$ and $max_{0 \leq i \leq H}|min(U_i)|^2 = O(|Q|^2 \cdot (H \cdot \|\Delta\| + \|s_f\|)^{2d})$. Therefore, since $H = (|Q| \cdot (\|\Delta\| + \|s_f\| + 2))^{2^{O(d \cdot \log d)}}$, we obtain the following.

Theorem 2. *The backward algorithm terminates in time* $(|Q| \cdot (\|\Delta\| + \|s_f\| + 2))^{2^{O(d \cdot \log d)}}$, $\|min(Pre^*(G, s_f\uparrow))\|$ *and* $|min(Pre^*(G, s_f\uparrow))|$ *are at most* $(|Q| \cdot (\|\Delta\| + \|s_f\| + 2))^{2^{O(d \cdot \log d)}}$.

4   Lower Bound

In this section, we prove the following result by an adaptation of Lipton's proof of EXPSPACE-hardness for reachability in VASS [13].

First, we need the following notation. Let $G = \langle Q, \Delta \rangle$ be a d-VASS and $q \in Q$. We denote by $q\uparrow$ the upward-closed set $\{q\} \times \mathbb{N}^d$ of G-states. Also, for a set S of G-states, we denote by $[Pre^*(G, S)]_q$ the subset of $\mathbb{N}^d$ given by $\{v \in \mathbb{N}^d \mid \langle q, v \rangle \in Pre^*(G, S)\}$.

Theorem 3. *For each* $n \in \mathbb{N}$, *one can build a* $O(n)$-*VASS* $G_n = \langle Q_n, \Delta_n \rangle$ *and* $q_n \in Q_n$ *s.t.* $|Q_n| = O(n)$, $|\Delta_n| = O(n)$, $\|\Delta_n\| = 1$, *and the following holds:* (1) $|min(Pre^*(G_n, q_n\uparrow))|$ *is at least* 2^{2^n} *(hence,* $\|min(Pre^*(G_n, q_n\uparrow)))\|$ *is at least* $2^{2^{\Omega(n)}}$), *and* (2) *there are states* $s \in min(Pre^*(G_n, q_n\uparrow))$ *s.t. each run from* s *to a state in* $q_n\uparrow$ *has length at least* 2^{2^n}.

By Property 2 in Theorem 3 and the results in the previous section, we easily deduce the following.

Corollary 2. *Let* $n \in \mathbb{N}$, $G_n = \langle Q_n, \Delta_n \rangle$ *and* $q_n \in Q_n$ *as in Theorem 3. Then, the number of iterations of the backward algorithm with input* G_n *and* $\langle q_n, \langle 0, \ldots, 0 \rangle \rangle$ *is at least* 2^{2^n}.

To make clear the proof of Theorem 3, we consider an high-level variant of VASS, called *net Programs* [6], corresponding to a subclass of nondeterministic counter machines with nonrecursive subroutines. Then, we show that in order to prove Theorem 3, it is sufficient to prove a similar result for net programs. Finally, in Section 4.2, we prove the variant of Theorem 3 for net programs.

For $m, k \in \mathbb{N}^+$ s.t. $k \leq m$ and $U \subseteq \mathbb{N}^m$, $\Pi_k(U)$ denotes the subset of $\mathbb{N}^k$ given by $\{v \in \mathbb{N}^k \mid \langle v[1], \ldots, v[k], 0, \ldots, 0 \rangle \in U\}$. Note that $\Pi_k(U)$ is upward-closed if U is upward-closed. Moreover, the following holds.

Lemma 6. *Let* $m, k \in \mathbb{N}^+$ *such that* $k \leq m$ *and* U *be an upward-closed subset of* $\mathbb{N}^m$. *Then,* $|min(U)| \geq |min(\Pi_k(U))|$.

Proof. For $v \in \mathbb{N}^k$, we denote by $v \cdot \underline{0}$ the vector in $\mathbb{N}^m$ given by $\langle v[1], \ldots, v[k], 0, \ldots, 0 \rangle$. Let $v \in min(\Pi_k(U))$. We show that $v \cdot \underline{0} \in min(U)$, hence, the result follows. Since $v \cdot \underline{0} \in U$, there is $v' \in min(U)$ such that $v' \trianglelefteq v \cdot \underline{0} \in U$. Hence, $v' = v'' \cdot \underline{0}$, $v'' \in \Pi_k(U)$, and $v'' \trianglelefteq v$. Since $v \in min(\Pi_k(U))$, it follows that $v'' = v$, hence $v \cdot \underline{0} = v' \in min(U)$, and we are done. $\square$

4.1 Net Programs

A net program is similar to a nondeterministic Minsky counter machine, but does not have the ability to test a (counter) variable for zero. However, it has the possibility of transferring control to a subroutine (or subprogram). Formally, a net program P on a finite set $\{x_1,\ldots,x_d\}$ of (counter) variables is a tuple $P = \langle ID_1,\ldots,ID_n,\mathsf{Code}\rangle$, where $ID_1,\ldots,ID_n$ are pairwise distinct *subprogram identifiers*, and Code assigns to each $1 \leq p \leq n$, the *code* $\mathsf{Code}(ID_p)$ of subprogram ID_p, which is a sequence of the form
$$\mathsf{Code}(ID_p) = 1_1 : I_1; \ldots 1_{k-1} : I_{k-1}; 1_k : \mathbf{return};$$
where $k \geq 1$, $1_1,\ldots,1_k$ are pairwise distinct (instruction) *labels*, 1_1 (resp., 1_k) is the initial (resp., final) label of subprogram ID_p, and each I_j is an instruction of the form:

- *increment*: $x_i := x_i + 1$ (where $1 \leq i \leq d$),
- *decrement*: $x_i := x_i - 1$ (where $1 \leq i \leq d$),
- *unconditional jump*: **goto** 1 (where $1 \in \{1_1,\ldots,1_k\}$),
- *nondeterministic jump*: **goto** 1 **or goto** $1'$ (where $1,1' \in \{1_1,\ldots,1_k\}$),
- *subprogram call*: **call** ID_i (where $i > p$).[5]

Additionally, we require that labels of distinct subprograms are distinct as well. The subprogram ID_1 is called the *main subprogram* of P, and the initial (resp., final) label of P is the initial (resp., final) label of the main subprogram. For each (instruction) label 1 of P, we denote by $ID(1)$ the identifier of the unique subprogram having 1 as label. Moreover, if 1 is the label of a call instruction, we denote by $called(1)$ the identifier of the called subprogram. Now, we formally define the semantics of net programs. An *extended label* of the net program P above is a pair of the form $\langle C,1\rangle$, where 1 is a label of P and C is a *caller context*, i.e., a (possibly empty) sequence of P-labels $C = 1_1 \ldots 1_k$ such that the following holds: (i) each 1_i is the label of a call instruction, and (ii) if C is nonempty, then $ID(1_{i+1}) = called(1_i)$ for each $1 \leq i \leq k$, where $1_{k+1} = 1$. Note that the set of extended labels of P, written $EL(P)$, is finite. A *P-state* is a pair $\langle\langle C,1\rangle,v\rangle$, where $\langle C,1\rangle \in EL(P)$ and $v \in \mathbb{N}^d$ is a valuation of variables $\{x_1,\ldots,x_d\}$ assigning to each variable x_i, the value $v[i]$. The net program P induces a transition relation $\rightarrow$ over P-states, as follows $\langle\langle C,1\rangle,v\rangle \rightarrow \langle\langle C',1'\rangle,v'\rangle$ iff:

- if 1 is the label of an increment (resp., decrement, resp., jump) instruction, then $\langle\langle C',1'\rangle,v'\rangle$ is as expected (note that $C' = C$ and if 1 is the label of a decrement, then $\langle\langle C,1\rangle,v\rangle$ has a successor iff the value in v of the decremented variable is greater than 0);
- if 1 is the label of a call instruction "**call** ID_j", then $v' = v$, $C' = C \cdot 1$, and $1'$ is the initial label of subprogram ID_j;
- if 1 is the label of a return instruction, then $v' = v$, $C = C' \cdot 1''$ for some $1''$, and $1'$ is the label which follows the call instruction label $1''$ in $\mathsf{Code}(ID(1''))$.

A run or execution of P is a finite sequence $s_1,\ldots,s_h$ of P-states such that $s_i \rightarrow s_{i+1}$ for each $1 \leq i < h$. For a set S of P-states, let $Pre^*(P,S)$ be the set of P-states s such that there is a run of P from s leading to some P-state in S. For each label 1, we denote by $[Pre^*(P,S)]_1$ the set $\{v \in \mathbb{N}^d \mid \langle\langle \varepsilon,1\rangle,v\rangle \in Pre^*(P,S)\}$, and by $1\uparrow$ the set of P-states

---

[5] The requirement $i > p$ ensures that there are no recursive calls.

$\{\langle \varepsilon, 1 \rangle\} \times \mathbb{N}^d$. It is easy to show that if S is an upward-closed set of P-states, then $Pre^*(P, S)$ is upward-closed as well. The following result allows us to reduce the proof of Theorem 3 to its variant for the class of net programs.

Theorem 4. *Let P be a net program on $\{x_1, \ldots, x_d\}$, k be the number of call instructions of P, and* start *and* end *be the initial and final labels of P. Then, one can build in linear-time a $(d+k)$-VASS $G = \langle Q, V \rangle$ such that Q is the set of P-labels, $\|\Delta\| = 1$, $|\Delta| \leq 2 \cdot N$, where N is the number of P-instructions, and $|min([Pre^*(G, end\uparrow)]_{start})| \geq |min([Pre^*(P, end\uparrow)]_{start})|$. Moreover, for each $s \in min(Pre^*(P, end\uparrow))$, there is a G-state $s' \in min(Pre^*(G, end\uparrow))$ such that for each run π in G from s' to a G-state in end$\uparrow$, there is a run of P from s to a P-state in end$\uparrow$ of length $|\pi|$.*

Proof. Let $L = \{l_1, \ldots, l_k\}$ be the set of call instruction labels of P. The $(d+k)$-VASS $G = \langle Q, \Delta \rangle$ is defined as follows (intuitively, we use an additional dimension for each call instruction label of P): Q is the set of P-labels and the set of transitions Δ is obtained in the following way:

- for each increment "$l : x_i := x_i + 1; l' : I; \ldots$", we add the transition $\langle l, v, l' \rangle$, where $v[i] = 1$ and $v[j] = 0$ for $j \neq i$;
- for each decrement "$l : x_i := x_i - 1; l' : I; \ldots$", we add the transition $\langle l, v, l' \rangle$, where $v[i] = -1$ and $v[j] = 0$ for $j \neq i$;
- for each unconditional jump "$l : \textbf{goto } l'; \ldots$", we add transition $\langle l, \underline{0}^{d+k}, l' \rangle$;
- for each nondeterministic jump "$l : \textbf{goto } l' \textbf{ or goto } l''; \ldots$", we add two transitions given by $\langle l, \underline{0}^{d+k}, l' \rangle$ and $\langle l, \underline{0}^{d+k}, l'' \rangle$;
- for each call instruction "$l_i : \textbf{call } ID_p; l : I; \ldots$" (where $1 \leq i \leq k$), we add two transitions $\langle l_i, v_+, l_0 \rangle$ and $\langle l_f, v_-, l \rangle$, where: (i) l_0 (resp., l_f) is the initial (resp., final) label of subprogram ID_p, (ii) $v_+[d+i] = 1$ and $v_+[j] = 0$ for $j \neq d+i$, and (iii) $v_-[d+i] = -1$ and $v_-[j] = 0$ for $j \neq d+i$.

Note that $\|\Delta\| = 1$ and $|\Delta| \leq 2 \cdot N$, where N is the number of P-instructions. Now, we establish the correspondence between the runs of P and the runs of G. Let H be the mapping assigning to each state s of P of the form $\langle\langle l_{i_1} \ldots l_{i_p}, 1 \rangle, v \rangle$, the G-state $H(s)$ defined as follows (note that $i_1, \ldots, i_p \in \{1, \ldots, k\}$ and are pairwise distinct)[6]: $H(s) = \langle l, v_{ext} \rangle$, where for each $1 \leq j \leq d+k$, $v_{ext}[j] = v[j]$ if $j \leq d$, $v_{ext}[j] = 1$ if $j = d + i_h$ for some $1 \leq h \leq p$, and $v_{ext}[j] = 0$ otherwise. By construction, we obtain the following:

Claim: let $s_0, s_1, \ldots, s_n$ be a sequence of states of P. Then, $s_0, s_1, \ldots, s_n$ is a run of P if and only if $H(s_0), H(s_1), \ldots, H(s_n)$ is a run of G. Moreover, for each state s_0' of P, each run of G from $H(s_0')$ has the form $H(s_0'), H(s_1'), \ldots, H(s_m')$ for some sequence $s_1', \ldots, s_m'$ of P-states.

By the claim above, it follows that $\Pi_d([Pre^*(G, end\uparrow)]_{start}) = [Pre^*(P, end\uparrow)]_{start}$. Thus, by Lemma 6 and the claim above, Theorem 4 easily follows.   □

4.2   Proof of Theorem 3

Theorem 3 directly follows from Theorem 4 and the following result.

---

[6] Moreover, note that $called(l_{i_1}), \ldots, called(l_{i_k})$ are pairwise distinct and $called(l_{i_k}) = ID(1)$.

Theorem 5. *For each $n \in \mathbb{N}$, one can build a net program P_n with initial (resp., final) label* start *(resp.,* end*), $O(n)$ instructions, and $O(n)$ variables such that $|min([Pre^*(P_n, \text{end}\uparrow)]_{start})| \geq 2^{2^n}$. Also, there exists $v \in min([Pre^*(P_n, \text{end}\uparrow)]_{start})$ such that each run from $\langle\langle\varepsilon, \text{start}\rangle, v\rangle$ to a state in* end$\uparrow$ *has length at least 2^{2^n}.*

In the rest of this section, we prove Theorem 5.

Construction of P_n. Let $n \in \mathbb{N}$, define $Var_n = \{w_1, w_2, y_n, \bar{y}_n\} \cup \bigcup_{i=0}^{n-1}\{y_i, \bar{y}_i, z_i, \bar{z}_i\}$. The net program P_n has set of variables Var_n and is given by

$$\langle Main_n, Lipton_n, Init_0, \ldots, Init_{n-1}, Dec_n(\bar{y}_n), Dec_{n-1}(\bar{y}_{n-1}), Dec_{n-1}(\bar{z}_{n-1}), \ldots$$
$$Dec_0(\bar{y}_0), Dec_0(\bar{z}_0), Set_n, \text{Code}\rangle$$

where Code is given in Figures 1–3.[7]

The construction of P_n ensures the following: if initially (i.e., at call time of the main subprogram $Main_n$) each variable in $Var_n \setminus \{w_1, w_2\}$ has value 0, then the main subprogram $Main_n$ *can return*[8] if and only if the sum of the initial values of w_1 and w_2 is greater or equal to 2^{2^n}. Now, we proceed with the description of the various subprograms of P_n. The main subprogram $Main_n$ simply calls the subprograms Set_n and $Lipton_n$ (in the given order) and returns. It is easy to check (see Figure 1) that the subprogram Set_n ensures the following.

$\underline{Main_n:}$
start : **call** Set_n;
       call $Lipton_n$;
end : **return**.

$\underline{Lipton_n:}$
start : **call** $Init_0$;
       ......
       call $Init_{n-1}$;
       call $Dec_n(\bar{y}_n)$;
end : **return**.

$\underline{Set_n:}$
start : **goto** 0 **or goto** end;
    0 : **goto** 1 **or goto** 2;
    1 : $w_1 := w_1 - 1; \bar{y}_n = \bar{y}_n + 1$;
       goto start;
    2 : $w_2 := w_2 - 1; y_n = \bar{y}_n + 1$;
       goto start;
end : **return**.

Fig. 1. The subprograms $Main_n$, $Lipton_n$, and Set_n of P_n

Lemma 7. *Assume that Set_n is called with the value of $\bar{y}_n$ being 0. Then: (1) whenever Set_n returns, the value of $\bar{y}_n$ is less or equal to the sum of the initial values (at call time of Set_n) of variables w_1 and w_2, and (2) there is an execution such that Set_n returns with the value of $\bar{y}_n$ being exactly the sum of the initial values of w_1 and w_2.*

---

[7] In Figures 1–3, for clarity, some instruction labels are omitted, and some labels of distinct subprograms are equal (we tacitely assume that they are prefixed by the ID of the associated subprogram).

[8] i.e., there is a run leading to a state whose label is the final label of subprogram $Main_n$.

The subprogram $Lipton_n$ (see Figure 1), whose construction corresponds to a variant of that given by Lipton in [13] (see also [6]), ensures the following: if initially (i.e., at call time of $Lipton_n$) all the variables in $Var_n \setminus \{w_1, w_2, \bar{y}_n\}$ have value 0, then $Lipton_n$ can return if and only if the initial value of $\bar{y}_n$ is greater or equal to 2^{2^n}. The implementation of $Lipton_n$ is based on subprograms $Init_i$, $Dec_i(\bar{z}_i)$, and $Dec_j(\bar{y}_j)$ (where $0 \leq i \leq n - 1$ and $0 \leq j \leq n$). The subprograms $Dec_i(\bar{z}_i)$ and $Dec_j(\bar{y}_j)$ (see Figure 2) ensure the following.

$Dec_0(\bar{x}_0)$:
_____

* x_0 is either y_0 or z_0 *

start : $\bar{x}_0 := \bar{x}_0 - 1$;
$\quad\quad\quad \bar{x}_0 := \bar{x}_0 - 1$;
$\quad\quad\quad x_0 := x_0 + 1$;
$\quad\quad\quad x_0 := x_0 + 1$;
end : **return**.

$Dec_{i+1}(\bar{x}_{i+1})$:
_____

* x_{i+1} is either y_{i+1} or z_{i+1} *
* Initially, $y_i = z_i = 2^{2^i}$ and $\bar{y}_i = \bar{z}_i = 0$ *

$\quad$ out-loop : $y_i := y_i - 1; \bar{y}_i := \bar{y}_i + 1$;
$\quad$ in-loop : $z_i := z_i - 1; \bar{z}_i := \bar{z}_i + 1$;
$\quad\quad\quad\quad\quad \bar{x}_{i+1} := \bar{x}_{i+1} - 1; x_{i+1} := x_{i+1} + 1$;
$\quad\quad\quad\quad\quad$ **goto** in-continue **or goto** in-exit;
in-continue : $z_i := z_i - 1; z_i := z_i + 1$; **goto** in-loop;
$\quad$ in-exit : **call** $Dec_i(\bar{z}_i)$;
$\quad\quad\quad\quad\quad$ **goto** out-continue **or goto** out-exit;
out-continue : $y_i := y_i - 1; y_i := y_i + 1$; **goto** out-loop;
$\quad$ out-exit : **call** $Dec_i(\bar{y}_i)$;
$\quad\quad\quad$ end : **return**.

Fig. 2. The subprograms $Dec_0(\bar{x}_0)$ and $Dec_{i+1}(\bar{x}_{i+1})$ of P_n

Lemma 8. *Let $0 \leq j \leq n$ and $x_j \in \{y_j, z_j\}$ such that $x_j = y_j$ if $j = n$. Assume that $Dec_j(\bar{x}_j)$ is called with the values of $\bar{y}_h$ and $\bar{z}_h$ being 0 and the values of y_h and z_h being 2^{2^h} for each $0 \leq h < j$. Then, the following holds:*

- *$Dec_j(\bar{x}_j)$ can return iff the initial value of $\bar{x}_j$ (at call time of $Dec_j(\bar{x}_j)$) is at least 2^{2^j}. Moreover, if the initial value of $\bar{x}_j$ is exactly 2^{2^j} and the initial value of x_j is 0, then whenever $Dec_j(\bar{x}_j)$ returns, the values of $\bar{x}_j$ and x_j (at return time) are swapped (i.e., x_j has value 2^{2^j} and $\bar{x}_j$ has value 0).*
- *Whenever $Dec_j(\bar{x}_j)$ returns, there are no side-effects on the variables $x \in Var_n \setminus \{\bar{x}_j, x_j\}$ (the values of x at call and return times are the same).*
- *Whenever $Dec_j(\bar{x}_j)$ returns, the number of computational steps from the call time to the return time is at least 2^{2^j}.*

Proof. The proof is by induction on j. The base case ($j = 0$) is trivial (see Figure 2). Now, assume that $j = i+1$ for some $0 \leq i < n$. Let us consider the code of $Dec_{i+1}(\bar{x}_{i+1})$ in Figure 2, which consists of two nested loops: the inner loop is associated with the counter variable z_i, while the outer loop is associated with the counter variable y_i. Note that the body of the inner loop decrements $\bar{x}_{i+1}$. Essentially, since the initial values of y_i and z_i are 2^{2^i}, each of two nested loops can be executed 2^{2^i}-times. Since $2^{2^i} \cdot 2^{2^i} = 2^{2^{i+1}}$, it follows that $\bar{x}_{i+1}$ can be decreased by $2^{2^{i+1}}$. Fix $x_i \in \{y_i, z_i\}$. First, note that at each step the invariant $x_i + \bar{x}_i = 2^{2^i}$ is preserved. Moreover, for the loop associated with counter variable x_i, $Dec_{i+1}(\overline{x}_{i+1})$ can *guess* that the continuation (resp., exit) condition is satisfied, i.e., $x_i > 0$ (resp., $x_i = 0$), by a nondeterministic jump instruction. The continuation condition is implemented by decrementing and then incrementing x_i, while the exit condition is implemented by a call to $Dec_i(\overline{x}_i)$. By the induction hypothesis, $Dec_i(\overline{x}_i)$ can return if and only if $\overline{x}_i$ has value 2^{2^i}, i.e., x_i has value 0. Thus, if the *guess* is not correct, the subprogram $Dec_{i+1}(\overline{x}_{i+1})$ stops without returning. Moreover, by the induction hypothesis, whenever $Dec_i(\overline{x}_i)$ returns, the values of x_i and $\overline{x}_i$ are swapped. This ensures that the inner loop can be re-initialized correctly, and whenever $Dec_{i+1}(\overline{x}_{i+1})$ returns, the values of x_i and $\overline{x}_i$ correspond to the initial ones. Thus, it follows that $Dec_{i+1}(\overline{x}_{i+1})$ can return if and only if $\overline{x}_{i+1}$ can be decreased by $2^{2^{i+1}}$ (i.e., the initial value of $\overline{x}_{i+1}$ is at least $2^{2^{i+1}}$). Finally, if the initial value of $\overline{x}_{i+1}$ is $2^{2^{i+1}}$ and the initial value of x_{i+1} is 0, then the body of the inner loop of $Dec_{i+1}(\overline{x}_{i+1})$ ensure that at return time, the values of x_{i+1} and $\overline{x}_{i+1}$ are swapped. □

Finally, for each $0 \leq i \leq n-1$, the subprogram $Init_i$ (see Figure 3) is used to set the values of y_i and z_i to 2^{2^i}. More precisely, $Init_i$ ensures the following.

$\underline{Init_0:}$

$\qquad$ start : $y_0 := y_0 + 1;$
$\qquad\qquad\quad$ $y_0 := y_0 + 1;$
$\qquad\qquad\quad$ $z_0 := z_0 + 1;$
$\qquad\qquad\quad$ $z_0 := z_0 + 1;$
$\qquad$ end $\;$: **return**.

$\underline{Init_{i+1}:}$

$\qquad$ out-loop : $y_i := y_i - 1; \overline{y}_i := \overline{y}_i + 1;$
$\qquad$ in-loop $\;$: $z_i := z_i - 1; \overline{z}_i := \overline{z}_i + 1;$
$\qquad\qquad\qquad$ $y_{i+1} := y_{i+1} + 1; z_{i+1} := z_{i+1} + 1;$
$\qquad\qquad\qquad$ **goto** in-continue **or goto** in-exit;
$\qquad$ in-continue : $z_i := z_i - 1; z_i := z_i + 1;$ **goto** in-loop;
$\qquad$ in-exit $\;$: **call** $Dec_i(\overline{z}_i);$
$\qquad\qquad\qquad$ **goto** out-continue **or goto** out-exit;
$\qquad$ out-continue : $y_i := y_i - 1; y_i := y_i + 1;$ **goto** out-loop;
$\qquad$ out-exit : **call** $Dec_i(\overline{y}_i);$
$\qquad\qquad$ end $\;$: **return**.

Fig. 3. The subprograms $Init_0$ and $Init_{i+1}$ of P_n

Lemma 9. *Let $0 \leq j \leq n-1$. Assume that $Init_j$ is called with the following condition being satisfied at call time: (i) the values of $y_j, z_j, \overline{y}_j, \overline{z}_j$ are 0, and (ii) the values of $\overline{y}_h, \overline{z}_h$ are 0 and the values of y_h and z_h are 2^{2^h} for each $0 \leq h < j$. Then, $Init_i$ can return. Moreover, whenever $Init_i$ returns, y_i and z_i have value 2^{2^i} and there are no-side effects for the other variables $x \in Var_n \setminus \{y_i, z_i\}$ (i.e., the values of x at call and return times are the same).*

Proof. The proof is by induction on j. The base case ($j = 0$) is trivial (see Figure 3). Now, assume that $j = i+1$ for some $0 \leq i < n-1$. Let us consider the code of $Init_{i+1}$ in Figure 3, which is the same as $Dec_{i+1}(\overline{x}_{i+1})$, with the unique difference that the body of the inner loop increments the two variables y_{i+1} and z_{i+1}. Hence, reasoning as in the proof of Lemma 8, the result easily follows (in particular, under the considered assumptions, whenever $Init_{i+1}$ returns, the values of y_{i+1} and z_{i+1} are increased exactly by $2^{2^{i+1}}$). □

Assume that at call time of subprogram $Lipton_n$, each variable in $Var_n \setminus \{w_1, w_2, \bar{y}_n\}$ has value 0. Then, By Lemmata 8 and 9, $Lipton_n$ can return iff at call time $\bar{y}_n$ has value at least 2^{2^n}. Moreover, whenever $Lipton_n$ returns, then the number of computational steps from the call time to the return time is at least 2^{2^n}. Thus, by Lemma 7, we obtain the following.

Lemma 10. *Assume that at call time of $Main_n$ each variable in $Var_n \setminus \{w_1, w_2\}$ has value 0. Then, $Main_n$ can return iff the sum of the values of w_1 and w_2 at call time is at least 2^{2^n}. Moreover, whenever $Main_n$ returns, then the number of computational steps from the call time to the return time is at least 2^{2^n}.*

Proof of Theorem 5. First, we need an additional result. For all $n \in \mathbb{N}$, we denote by Λ_n and $\Upsilon_n \subseteq \Lambda_n$ the subsets of $\mathbb{N}^2$ given by

$$\Lambda_n = \{v \in \mathbb{N}^2 \mid v[1] + v[2] \geq 2^{2^n}\} \text{ and } \Upsilon_n = \{v \in \mathbb{N}^2 \mid v[1] + v[2] = 2^{2^n}\}$$

Lemma 11. *Let $n, m \in \mathbb{N}$ and U be an upward-closed subset of $\mathbb{N}^{m+2}$ such that $\Pi_2(U) = \Lambda_n$. Then, $|min(U)| \geq 2^{2^n}$ and $min(U) \supseteq \{v \cdot \underline{0}^m \mid v \in \Upsilon_n\}$.*

Proof. For $v \in \mathbb{N}^2$, we denote by $v \cdot \underline{0}$ the vector in $\mathbb{N}^{m+2}$ given by $\langle v[1], v[2], 0, \ldots, 0 \rangle$. First, we show the following.

Claim 1: $\Upsilon_n \subseteq min(\Lambda_n)$

Proof of Claim 1: Let $v \in \Upsilon_n$. Since $v \in \Lambda_n$, there must be $v_{min} \in min(\Lambda_n)$ such that $v_{min} \trianglelefteq v$ (note that Λ_n is upward-closed). By definition of Υ_n, it follows that $v_{min} \in \Upsilon_n$. Thus, since all elements in Υ_n are pairwise incomparable, we obtain that $v_{min} = v$. Hence, $v \in min(\Lambda_n)$, and the result follows. □

Moreover, by the proof of Lemma 6, the following holds

Claim 2: $min(U) \supseteq \{v \cdot \underline{0} \mid v \in min(\Pi_2(U))\}$.

Evidently, Υ_n has cardinality 2^{2^n}. By hypothesis, $\Pi_2(U) = \Lambda_n$. Thus, by Claims 1 and 2, the result follows. □

Fix $n \in \mathbb{N}$ and an ordering of Var_n such that w_1 and w_2 precede all the other variables. Let start (resp., end) be the initial (resp., final) label of P_n. By construction, P_n has $O(n)$ instructions and $O(n)$ variables. Thus, by Lemmata 10 and 11, Theorem 5 easily follows.

Acknowledgement. We would like to thank Rupak Majumdar for asking about the complexity of the backward algorithm and Javier Esparza for encouraging us.

References

[1] Abdulla, P., Čerāns, K., Jonsson, B., Tsay, Y.: General Decidability Theorems for Infinite-State Systems. In: LICS 1996, pp. 313–321. IEEE Computer Society Press, Los Alamitos (1996)

[2] Arnold, A., Latteux, M.: Recursivite et cones rationnels fermes par intersection. Calcolo 15, 381–394 (1978)

[3] Delzanno, G., Raskin, J.-F.: Symbolic representation of upward-closed sets. In: Graf, S. (ed.) TACAS 2000. LNCS, vol. 1785, pp. 426–440. Springer, Heidelberg (2000)

[4] Delzanno, G., Raskin, J.-F., Van Begin, L.: Attacking symbolic state explosion. In: Berry, G., Comon, H., Finkel, A. (eds.) CAV 2001. LNCS, vol. 2102, pp. 298–310. Springer, Heidelberg (2001)

[5] Dickson, L.E.: Finiteness of the odd perfect and primitive abundant numbers with n distinct prime factors. American Journal of Mathematics 35, 413–422 (1913)

[6] Esparza, J.: Decidability and Complexity of Petri Net Problems - An Introduction. In: Reisig, W., Rozenberg, G. (eds.) APN 1998. LNCS, vol. 1491, pp. 374–428. Springer, Heidelberg (1998)

[7] Esparza, J., Nielsen, M.: Decidability issues for Petri nets - a survey. Journal of Informatik Processing and Cybernetics 30(3), 143–160 (1994)

[8] Figueira, D., Figueira, S., Schmitz, S., Schnoebelen, P.: Ackermannian and primitive-recursive bounds with Dickson's lemma. In: LICS 2011: Proc. 26th Annual IEEE Symp. on Logic in Computer Science, pp. 269–278. IEEE Computer Society Press, Los Alamitos (2011)

[9] Finkel, A., Schnoebelen, P.: Well-structured transition systems everywhere! Theoretical Computer Science 256(1-2), 63–92 (2001)

[10] Ganty, P., Majumdar, R.: Algorithmic verification of asynchronous programs. CoRR, abs/1011.0551 (2010)

[11] Higman, G.: Ordering by divisibility in abstract algebras. Proceedings of the London Mathematical Society (3) 2(7), 326–336 (1952)

[12] Karp, R.M., Miller, R.E.: Parallel program schemata. Journal of Comput. Syst. Sci. 3(2), 147–195 (1969)

[13] Lipton, R.: The Reachability Problem Requires Exponential Space. Technical Report 62, Yale University (1976)

[14] Rackoff, C.: The Covering and Boundedness Problems for Vector Addition Systems. Theoretical Computer Science 6, 223–231 (1978)

[15] Sen, K., Viswanathan, M.: Model checking multithreaded programs with asynchronous atomic methods. In: Ball, T., Jones, R.B. (eds.) CAV 2006. LNCS, vol. 4144, pp. 300–314. Springer, Heidelberg (2006)

[16] Valk, R., Jantzen, M.: The residue of vector sets with applications to decidability problems in Petri nets. Acta Informatica 21, 643–674 (1985)

[17] Yen, H.-C., Chen, C.-L.: On minimal elements of upward-closed sets. Theoretical Computer Science 410(24-25), 2442–2452 (2009)

Automated Termination in Model Checking Modulo Theories

Alessandro Carioni[1], Silvio Ghilardi[1], and Silvio Ranise[2]

[1] Università degli Studi di Milano, Milano, Italia
[2] FBK-Irst, Trento, Italia

Abstract. We use a declarative SMT-based approach to model-checking of infinite state systems to design a procedure for automatically establishing the termination of backward reachability by using well-quasi-orderings. Besides showing that our procedure succeeds in many instances of problems covered by general termination results, we argue that it could predict termination also on single problems outside the scope of applicability of such general results.

1 Introduction

Infinite state model checking is nowadays a mature field and many successful attempts have been made to verify disparate problems, using various kinds of methodologies. Still, termination of search (both forward and backward) is a major problem and it is sometimes difficult to predict *in advance* whether a given problem will be solved in a finite amount of time or some source of divergence is hidden somewhere and appropriate techniques (such as acceleration or abstraction) should be employed in order to make the search terminating. In the literature, several results for the termination of backward reachability are known covering entire classes of infinite state systems [13,11,5,2,6]. In most cases, following the seminal work in [1], these results are based on the use of well-quasi-orders on configurations, which are (finite) symbolic representations of infinite sets of backward reachable states. Thus, their applicability crucially rely on the human ability to reformulate a given specification so that it fits in one of the classes of systems for which termination is guaranteed.

In this paper, we propose an *automated* technique capable (when successful) of predicting termination from the static analysis of a given verification problem that is amenable to backward reachability. We develop our ideas in the model checking modulo theories framework [14,18] where array-based (guarded assignment) transition systems are used to symbolically specify a wide range of systems by using certain classes of first-order formulae and background theories. There are two main ingredients. The former is that of a *wqo-theory* W, which is the declarative counterpart of a well-quasi-order on configurations used in the main arguments for termination in [1]. The second ingredient is the standard notion in first-order logic textbooks (see, e.g., [12]) of *syntactic interpretation*. Then, we cook these two ingredients together to design a method for establishing the

G. Delzanno and I. Potapov (Eds.): RP 2011, LNCS 6945, pp. 110–124, 2011.

termination of backward reachability on a given verification problem by checking for the *existence* of a syntactic translation from W, satisfying certain conditions. Such conditions refer to a search space *restricted to formulae describing small models* and the possibility of such restriction is indeed the essential content of our main result (Theorem 4.5 below). Interestingly, the conditions of Theorem 4.5 can all be checked trough proof obligations that can be efficiently discharged by using SMT solving techniques. We shall turn to an informal discussion on our main issue at the beginning of Section 4.

For space constraints, we shall supply here only high level explanations and formal statements of our results: the interested reader is referred to the online available extended version [8] for worked out examples and full proofs.

2   Preliminaries

We assume the usual syntactic (e.g., signature, variable, term, ground term, atom, literal, formula, and sentence) and semantic (e.g., structure, sub-structure, embedding, assignment, truth, and validity) notions of many-sorted first-order logic (see, e.g., [12,10]). The equality symbol $=$ is included in all signatures considered below. We use $L, M, \ldots$ for literals and $\phi, \psi, \ldots$ for formulae. A signature is *relational* iff it does not contain function symbols and it is *quasi-relational* iff the only function symbols it contains are constants. A formula is *open* (or *quantifier-free*) iff it does not contain quantifiers; it is *universal* (resp. *existential*) iff it is obtained from an open formula by prefixing it a finite sequence of universal (resp. existential) quantifiers. If $\phi(\underline{x})$ is a formula with free variables included in the tuple $\underline{x} = x_1, \ldots, x_n$ and $\underline{a} = a_1, \ldots, a_n$ is a (sort-conforming) tuple of elements of the support $|\mathcal{M}|$ of a structure $\mathcal{M}$, we write $\mathcal{M} \models \phi(\underline{a})$ to denote that $\phi(\underline{x})$ is valid in $\mathcal{M}$ under the assignment $\{x_1 \mapsto a_1, \ldots, x_n \mapsto a_n\}$.

SMT. Following [22], a *theory* T is a pair $(\Sigma, \mathcal{C})$, where Σ is a signature and $\mathcal{C}$ is a class of Σ-structures; the structures in $\mathcal{C}$ are the *models* of T. Below, let $T = (\Sigma, \mathcal{C})$. A Σ-formula ϕ is *T-satisfiable* if there exists a Σ-structure $\mathcal{M}$ in $\mathcal{C}$ such that ϕ is true in $\mathcal{M}$ under a suitable assignment to the free variables of ϕ (in symbols, $\mathcal{M} \models \phi$); it is *T-valid* (in symbols, $T \models \varphi$) if its negation is *T*-unsatisfiable. Two formulae φ_1 and φ_2 are *T-equivalent* if $\varphi_1 \leftrightarrow \varphi_2$ is *T*-valid. The *quantifier-free satisfiability modulo the theory T ($SMT(T)$) problem* amounts to establishing the *T*-satisfiability of quantifier-free Σ-formulae. A theory is said to be *syntactically specified* if we are given a set of Σ-sentences (called the *axioms* of T): in this case, the class $\mathcal{C}$ of the models of T is formed by the Σ-structures in which all the axioms of T are true. A theory T is *universal* iff it is syntactically specified and its axioms are universal sentences.

Diagrams. Given a (finite, in our applications) Σ-structure $\mathcal{M}$, take a free variable x_a for every a in the support of $\mathcal{M}$ and call $\underline{x}$ the set of all x_a (varying a). The *Σ-diagram* $\delta_{\mathcal{M}}$ *of* $\mathcal{M}$ [10] is the set of all Σ-literals $L(\underline{x})$ such that $\mathcal{M}, a \models L$, where a is the assignment mapping x_a to a. By abuse of notation, we shall confuse the variable x_a with the element a. Intuitively, we can view $\delta_{\mathcal{M}}$

as a sort of 'multiplication table' of the structure $\mathcal{M}$. Notice that the diagram of a finite structure is also finite and can be seen as the formula obtained by the conjunction of its literals.

Interpretations. Informally, an interpretation $(-)^*$ of a Σ-theory T into a Σ'-theory T' is a mapping from the expressions of T to the expressions of T' which preserves the validity of sentences. We will consider a special class of interpretations, generalizations of our definition exist—see, e.g., [12]—but we do not need them here. Formally, $(-)^*$ is a mapping associating (i) a sort S^* of Σ' with each sort S of Σ, (ii) a Σ'-formula $R^*(x_1, \ldots, x_n)$ with each predicate symbol R of Σ in such a way that the variables $x_1, \ldots, x_n$ occurring free in R^* match the translations of the arity sorts of R (implicitly we assume that identity of sort S is translated into identity of sort S^*), and (iii) a Σ'-term $f^*(x_1, \ldots, x_n)$ with each function symbol f of Σ (with the same condition on $x_1, \ldots, x_n$ as for predicate symbols). Then, $(-)^*$ can be extended inductively to formulae in the obvious way: $(R(t_1, \ldots, t_n))^* = R^*(t_1^*/x_1, \ldots, t_n^*/x_n)$ for atomic formulae and $(A \wedge B)^* = A^* \wedge B^*, (\neg A)^* = \neg A^*, (\forall x A)^* = \forall x A^*$ for non-atomic formulae. The last requirement that $(-)^*$ is supposed to satisfy is the following: (iv) for every sentence ϕ in the signature of T, if $T \models \phi$ then $T' \models \phi^*$. Notice that, if T is specified syntactically, it is sufficient to check (iv) only for the axioms of T. In this paper, we shall limit ourselves to *quantifier-free* translations, i.e. to translations in which the formulae R^* mentioned in (ii) above are quantifier-free.

3   Array-Based Systems and Backward Reachability

Array-based transition systems [14,19,16,18] have been proved useful for the verification of several classes of infinite state systems, such as broadcast protocols, lossy channel systems, timed networks, parametric and distributed systems. The Model Checker Modulo Theories (MCMT) tool [20] implements symbolic backward reachability for array-based systems (its executable, several benchmark problems, the documentation, and related papers can be downloaded at `http://homes.dsi.unimi.it/~ghilardi/mcmt`). The state variables of array-based systems are arrays "connecting" the theories T_I and T_E: the former describes the topology and the latter the data structures of the system. We fix for the whole paper the following conventions:

(T1) $T_I = (\Sigma_I, \mathcal{C}_I)$ is a relational mono-sorted theory whose unique sort is named INDEX), its $SMT(T_I)$-problem is decidable, and $\mathcal{C}_I$ is closed under substructures (meaning that $\mathcal{C}_I$ contains every substructure of any $\mathcal{M} \in \mathcal{C}_I$);

(T2) T_E is a multi-sorted theory whose sorts are $\mathtt{ELEM}_1, \ldots, \mathtt{ELEM}_S$ for some $S \geq 1$ and its $SMT(T_E)$-problem is decidable;

(T3) A_I^E is a compound theory, intended to describe 'arrays with indexes in T_I and elements in T_E.' Formally, the signature of A_I^E contains the sort symbols of $\Sigma_I \cup \Sigma_E$, together with a new sort symbol $\mathtt{ARRAY}_\ell$ for each sort $\mathtt{ELEM}_\ell$ of Σ_E; it contains also all the function and predicate symbols in $\Sigma_I \cup \Sigma_E$ together with a new function symbol $\_[\_]_\ell$: $\mathtt{ARRAY}_\ell, \mathtt{INDEX} \longrightarrow \mathtt{ELEM}_\ell$ for each sort $\mathtt{ELEM}_\ell$ of Σ_E. The models $\mathcal{M}$ of A_I^E are the structures whose Σ_I- and Σ_E-reducts are models of T_I and T_E, respectively; the sort $\mathtt{ARRAY}_\ell$ is interpreted as the set of total functions $\mathtt{INDEX}^{\mathcal{M}} \longrightarrow \mathtt{ELEM}_\ell^{\mathcal{M}}$ and the symbol $\_[\_]_\ell$ as function application.

So, $a[i]_\ell$ denotes the element of sort $\mathtt{ELEM}_\ell$ stored in the array a of sort $\mathtt{ARRAY}_\ell$ at index i; the subscript ℓ is dropped whenever it is clear from the context.

Definition 3.1. *An* array-based (transition) system (*for* (T_I, T_E)) *is a triple* $\mathcal{S} = (a, I, \tau)$ *where* (i) $a = a_1, \ldots, a_s$ *is a tuple of free constants of array sorts (these are to be thought as* state variables *storing data of sorts* $\mathtt{ELEM}_1, \ldots, \mathtt{ELEM}_s$, *respectively);* (ii) $I(a)$ *is the* initial *formula;* (iii) $\tau(a, a')$ *is the* transition *formula, where* a' *contains the renamed copies of the variables in* a. *The formula* I *is assumed to be a* $\forall^I$-*formula and the formula* τ *is a disjunction* $\bigvee_{h=1}^r \tau_h$ *of guarded assignments in functional form.*

To give the definition of a $\forall^I$-formula and of a guarded assignment in functional form, we preliminarily introduce the following notational convention and definitions. Below, d, e range over variables of a sort $\mathtt{ELEM}_\ell$ of Σ_E, $i, j, k, z, \ldots$ over variables of sort $\mathtt{INDEX}$. An underlined variable name abbreviates a tuple of variables of unspecified (but finite) length and, if $\underline{i} := i_1, \ldots, i_n$, the notation $a[\underline{i}]$ abbreviates the $s * n$-tuple of terms $a_1[i_1], \ldots, a_s[i_1], \ldots, a_1[i_n], \ldots, a_s[i_n]$. Possibly sub/super-scripted expressions of the form $\phi(\underline{i}, \underline{e}), \psi(\underline{i}, \underline{e})$ denote *quantifier-free* $(\Sigma_I \cup \Sigma_E)$-*formulae in which at most the variables* $\underline{i} \cup \underline{e}$ *occur.* Also, $\phi(\underline{i}, \underline{t}/\underline{e})$ (or simply $\phi(\underline{i}, \underline{t})$) abbreviates the substitution of the Σ-terms $\underline{t}$ for the variables $\underline{e}$. Thus, for instance, $\phi(\underline{i}, a[\underline{i}])$ denotes the formula obtained by replacing $\underline{e}$ with $a[\underline{i}]$ in the quantifier-free formula $\phi(\underline{i}, \underline{e})$.

Given a theory T (in our case, T will be A_I^E), a T-*partition* is a finite set $C_1(\underline{x}), \ldots, C_n(\underline{x})$ of quantifier-free formulae (with free variables contained in the tuple $\underline{x}$) such that $T \models \forall \underline{x} \bigvee_{i=1}^n C_i(\underline{x})$ and $T \models \bigwedge_{i \neq j} \forall \underline{x} \neg (C_i(\underline{x}) \wedge C_j(\underline{x}))$. The formulae $C_1, \ldots, C_k$ are called the *components* of the T-partition. A *case-definable extension* $T' = (\Sigma', \mathcal{C}')$ of a theory $T - (\Sigma, \mathcal{C})$ is obtained from T by applying (finitely many times) the following procedure: (i) take a T-partition $C_1(\underline{x}), \ldots, C_n(\underline{x})$ together with Σ-terms $t_1(\underline{x}), \ldots, t_n(\underline{x})$; (ii) let Σ' be $\Sigma \cup \{F\}$, where F is a "fresh" function symbol (i.e. $F \notin \Sigma$) whose arity is equal to the length of $\underline{x}$; (iii) take as $\mathcal{C}'$ the class of Σ'-structures $\mathcal{M}$ whose Σ-reduct is a model of T and such that $\mathcal{M} \models \bigwedge_{i=1}^n \forall \underline{x} (C_i(\underline{x}) \rightarrow F(\underline{x}) = t_i(\underline{x}))$. Thus a case-definable extension T' of a theory T contains finitely many additional function symbols, called case-defined functions. By abuse of notation, below, we shall identify T with its case-definable extensions T'.

A formula $\forall \underline{i}. \phi(\underline{i}, a[\underline{i}])$ is a $\forall^I$-*formula*, one of the form $\exists \underline{i}. \phi(\underline{i}, a[\underline{i}])$ is an $\exists^I$-*formula*, and a sentence $\exists a \, \exists \underline{i} \, \forall \underline{j} \, \psi(\underline{i}, \underline{j}, a[\underline{i}], a[\underline{j}])$ is an $\exists^{A,I} \forall^I$-*sentence*. A *guarded assignment in functional form* is a formula of the form

$$\exists e \exists \underline{k} \, (\phi_L(e, \underline{k}, a[\underline{k}]) \wedge a' = \lambda j. F(e, \underline{k}, a[\underline{k}], j, a[j])) \tag{1}$$

where: (i) $F = F_1, \ldots, F_s$ is a tuple of case-defined functions; (ii) the existentially quantified data variable e ranges over a sort $\mathtt{ELEM}_\ell$ such that T_E admits quantifier elimination with respect to quantified variables of sort $\mathtt{ELEM}_\ell$.

Given an array-based system $\mathcal{S} = (a, I, \tau)$ and an $\exists^I$-formula $U(a)$ describing a set of *unsafe* states (also called *error* states), the symbolic backward reachability procedure iteratively computes the set of backward reachable states $BR(a)$ as follows. (Below, we give a very high-level description of the symbolic backward reachability procedure implemented in MCMT. For a description of the techniques and heuristics used to make the procedure effective in practice, the reader is pointed to [19,17,16,18]. In particular, [18] reports an extensive experimental evaluation of the tool.) Preliminarily, define (for $n \geq 0$) the *n-pre-image* of a formula $K(a)$ as $Pre^0(\tau, K) := K$ and $Pre^{n+1}(\tau, K) := Pre(\tau, Pre^n(\tau, K))$, where $Pre(\tau, K) := \exists a'.(\tau(a, a') \wedge K(a'))$. Intuitively, $Pre^n(\tau, U)$ describes the set of backward reachable states in $n \geq 0$ steps. At the n-th iteration, the *backward reachability procedure* computes the formula $BR^n(\tau, U) := \bigvee_{i=0}^n Pre^i(\tau, U)$ representing the set of states which are backward reachable from the states in U with at most n steps. While computing $BR^n(\tau, U)$, the procedure also checks whether the system is unsafe by establishing if the formula $I \wedge Pre^n(\tau, U)$ is A_I^E-satisfiable (*safety test*) or whether a fix-point has been reached by checking if $(BR^n(\tau, U) \rightarrow BR^{n-1}(\tau, U))$ is A_I^E-valid or, equivalently, if the formula $BR^n(\tau, U) \wedge \neg BR^{n-1}(\tau, U)$ is A_I^E-unsatisfiable (*fix-point test*).

To mechanize this procedure, it is mandatory to identify a class of formulae for representing sets of backward reachable states which is closed under pre-image computation and such that the safety and fix-point checks are decidable. Using $\exists^I$-formulae for representing unsafe states, this is indeed the case as stated in the following theorem, a corollary of more general results in [14,18].

Theorem 3.2. *Let $\mathcal{S} = (a, I, \tau)$ be an array-based system; we have that*
(RF1) *if K is an $\exists^I$-formula, then $Pre(\tau, K)$ is equivalent to an (effectively computable) $\exists^I$-formula;*
(RF2) *if the set U of unsafe states is represented by an $\exists^I$-formula, then the A_I^E-satisfiability checks for safety and fix-point of the backward reachability procedure are effective.*

As shown in [14,19,18,9,7], this theorem allows the automated verification of reachability properties for several classes of systems (e.g., parameterised systems, timed networks, or fault-tolerant algorithms). For several of these problems, a (declarative reformulation of) the *approximate model technique* (see, e.g., [4]) is required as explained in [15]. **(RF2)** is a special case of the decidability of the A_I^E-satisfiability problem for $\exists^{A,I}\forall^I$-sentences in [18]. Assumptions **(T1)** on T_I are essential for **(RF2)** because, if they are dropped, undecidability of $\exists^{A,I}\forall^I$-sentences arises [18]. The proof of **(RF2)** in [18] consists of a decision procedure integrating quantifier-free SMT solving and quantifier instantiation. Powerful heuristics [17] are also crucial for implementation.

An important refinement of the backward reachability procedure above is to exploit invariants whenever they are available. An invariant $J(a)$ for the array-based system $\mathcal{S} = (a, I, \tau)$ is a $\forall^I$-formula such that (a) $A_I^E \models I(a) \rightarrow J(a)$ and (b) $A_I^E \models Pre(\tau, J) \rightarrow J(a)$. The requirement that $J(a)$ is a $\forall^I$-formula allows us to reduce conditions (a) and (b) to the A_I^E-satisfiability of $\exists^{A,I}\forall^I$-sentences, which is decidable because of **(T1)**. Techniques for invariant synthesis are discussed in [18] and are also implemented in MCMT. Whenever an invariant J is known, we can replace A_I^E with $A_I^E \cup \{J\}$ in our satisfiability tests (e.g. for fix-point in the backward reachability procedure). The presence of invariants in such tests is often crucial either to greatly speed up the performances of MCMT or to obtain termination (see again [18] for details).

3.1 Closure under Pre-image Computation

The proof of **(RF1)** (Theorem 3.2) in, e.g., [14,18] consists of simple logical manipulations (this is a distinguishing feature of our approach). Here, we briefly discuss a variant of such proofs, whose details are needed to state the main result of this paper (see Theorem 4.5 below).

Let $\mathcal{S} = (a, I, \tau)$ be an array-based system, where τ is a finite disjunction of formulae $\tau_1, ..., \tau_r$ of the form (1). We consider only $Pre(\tau_h, K)$ since $Pre(\tau, K)$ is easily seen to be equivalent to the disjunction of $Pre(\tau_1, K), ..., Pre(\tau_r, K)$. Let us now focus on the definition of $F = F_1, ..., F_s$ in (1). Without loss of generality (since partitions admit common refinements), we assume that the A_I^E-partition $\{C_1(e, \underline{k}, a[\underline{k}], j, a[j]), ..., C_m(e, \underline{k}, a[\underline{k}], j, a[j])\}$ is the same for each F_l ($l \in \{1, ..., s\}$). Thus, each case-defined function F_l can be written as

$$F_l(j, a[j]) := \mathtt{case\ of}\{C_1(j, a[j]) : t_{l1}(j, a[j]);\ \cdots\ C_m(j, a[j]) : t_{lm}(j, a[j])\}, \quad(2)$$

for $l = 1, \ldots, s$. (According to the definition of case-definable extension, the logical reading of the `case of` construct is the conjunction of the formulae $\bigwedge_{z=1}^{m} \forall \underline{j}.(C_z(j, a[j]) \to F_l(j, a[j]) = t_{lz})$ for each $l = 1, ..., s$.) Notice that F_l, C_1, $..., C_m, t_{l1}, ..., t_{lm}$ depend not only on $j, a[j]$ but also on $e, \underline{k}, a[\underline{k}]$; to simplify notation, we omit these dependences in the rest of this paper. If $K(a)$ is the $\exists^I$-formula $\exists \underline{i}\, \psi(\underline{i}, a[\underline{i}])$, then $Pre(\tau_h, K)$ is logically equivalent to the formula obtained from

$$\phi_L \wedge \psi(\underline{i}, F[\underline{i}]) \quad(3)$$

by prefixing it with the existential quantifiers $\exists \underline{i}\, \exists \underline{k}\, \exists e$; here the notation $F[\underline{i}]$ abbreviates the $(n * s)$-tuple of terms $F_l(i_z, a[i_z])$, varying $l = 1, \ldots, s$ and $z = 1, \ldots, n$ when $\underline{i} = i_1, \ldots, i_n$. We can further manipulate the formula (3) in order to eliminate the defined symbols $F_1, ..., F_s$. To do this, we consider the functions $f : \underline{i} \to \{1, \ldots, m\}$ and rewrite the formula (3) as the disjunction (varying f) of the formulae

$$\begin{aligned}\tau_h[\psi, f] :=\ \ &\phi_L \wedge C_{f(i_1)}(i_1, a[i_1]) \wedge \cdots \wedge C_{f(i_n)}(i_n, a[i_n]) \wedge \\ &\wedge \psi(\underline{i}, t_{1f(i_1)}(i_1, a[i_1]), \ldots, t_{sf(i_n)}(i_n, a[i_n]))\ .\end{aligned}$$

At this point, it is clear that $Pre(\tau_h, K)$ is equivalent to $\bigvee_f \exists \underline{i}\, \exists \underline{k}\, \exists e\, \tau_h[\psi, f]$, where the functions f indexing the disjunction will be called *case-marking functions* and their purpose is to mark each index in $\underline{i}$ with the case that formally applies to it. This concludes the proof of **(RF1)**.

4   Wqo-Theories, QE-Degrees, and Termination

The assumptions of Theorem 3.2 guarantee that the backward reachability procedure described in the previous section can be mechanized but they are not sufficient to guarantee termination. This is because the symbolic representation of pre-images (in our case, $\exists^I$-formulae) may not be expressive enough to represent a fix-point of the set of backward reachable states. Termination can be achieved only under additional assumptions. The classical (non declarative) method for obtaining termination (see, e.g., [1,5,6]) consists in endowing the states of the system with a preorder relation $\preceq$ and in assuming that (1) pre-images are monotonic w.r.t. $\preceq$ and (2) $\preceq$ is a well-quasi-ordering (wqo). Now, (1) implies that the pre-image of an upward-closed (w.r.t. $\preceq$) set is still upward-closed and (2) implies that every upward-closed set can be characterized by a finite set of minimal (w.r.t. $\preceq$) elements. Thus, starting from an upward-closed set U of states, the iterative computation of the backward reachable configurations from U necessarily terminates because the fixpoint is upward closed and hence its minimal elements are rechable in finitely many steps. Obviously, this requires that relevant upward-closed sets can be effectively represented and manipulated. Our goal here is to recast this argument in our declarative framework underlying the backward reachability procedure of Section 3. Roughly, our plan is as follows: without loss of generality (see [18], Section 4), states of the system can be identified with the values assigned to the array constants a in a model $\mathcal{M}$ having finitely many generators of sort INDEX. Since we represent backward reachable states by $\exists^I$-formulae, we replace $\preceq$ by the embeddability relation between such finitely generated models. The fact that existential formulae are preserved by super-structures guarantees that they describe upward closed sets of states. However, it would be too strong to require embeddability among finitely generated models to be a wqo: instead, we make an abstraction of the system, through a wqo-theory W and a syntactic translation into A_I^E. This will replace embeddability between finitely generated models of A_I^E by *embeddability between the abstract states*, i.e. between the finitely generated models of W. The definition of a wqo theory just says that embeddability between such abstract states is a wqo. The key step of our plan consists of checking whether every pre-image P_i computed by the backward reachability procedure is a translation of an existential formula β_i in W for $i \geq 0$ where $P_0 := U$. The sequence $\beta_0, \beta_1, \dots$ is finite because it describes increasingly larger upsets of a wqo, hence the sequence $P_0, P_1, \dots$ must be finite too. In fact, if there exists $b \geq 1$ such that β_b is a fix-point, i.e. $W \models \beta_b \rightarrow \beta_{b-1}$, by using the syntactic translation we must also have that $A_I^E \models P_b \rightarrow P_{b-1}$ where P_b and P_{b-1} are the translations of β_b and β_{b-1}, respectively. Hence P_b is a fixed point of our backward reachability procedure. Thus we must find a condition that guarantees that the backward reachability procedure generates only $\exists^I$-formulae which are translations of existential formulae of W: to this aim, we reduce the general case to finitely many cases involving formulae of the kind $\exists e \, \exists \underline{i} \, \exists \underline{k} \, \tau_h[\psi, f]$, *where ψ is the translation of the diagram of a "small" model of W.* Finally, since the elimination of the existentially quantified data variable e is required, we need also assumptions on the quantifier elimination algorithm.

Wqo-Theories. A wqo $(P, \leq)$ is a set P endowed with a binary reflexive and transitive relation $\leq$ such that for every infinite sequence $p_1, p_2, \ldots$ of elements from P there are $i < j$ such that $p_i \leq p_j$.

Definition 4.1. *A wqo-theory is a universal theory whose finitely generated models[1] are a well-quasi-order with respect to the embeddability relation.*

Simple examples of wqo-theories can be obtained by taking vector spaces on a fixed field, torsion-free abelian groups, etc. The reason why we get a wqo in these cases is that an embedding always exists whenever the dimension is lower. Examples of wqo-theories which are more relevant to this paper can be obtained by re-interpreting declaratively some special cases of Kruskal theorem or Higman lemma, as sketched in the following example.

Example 4.2. Consider a signature contaning one sort, finitely many 0-ary and 1-ary predicates and a single binary predicate $\leq$ (besides equality). We get a wqo theory W by Higmann lemma if we syntactically specify W through the following set of axioms

$$\forall x\, (x \leq x), \quad \forall x, y, z\, (x \leq y \wedge y \leq z \rightarrow x \leq z), \text{ and } \forall x, y\, (x \leq y \vee y \leq x),$$

(the axioms say that $\leq$ is to be interpreted as a total pre-order relation).

QE-Degree. A theory T admits *quantifier elimination* (relative to a sort S) iff for every formula $\varphi(\underline{x})$ containing only quantified variables of sort S, there exists a quantifier-free formula $\varphi'(\underline{x})$ such that $T \models \forall \underline{x}(\varphi(\underline{x}) \leftrightarrow \varphi'(\underline{x}))$.

A set $\mathcal{P}$ of Σ-predicates is said to be *T-representative* for a Σ-theory T iff for every Σ-literal $L(\underline{x})$ one can compute a formula $\psi(\underline{x})$ which is a positive combination (i.e. a disjunction of conjunctions) of atoms whose root predicate symbol is in $\mathcal{P}$ such that $T \models \forall \underline{x}\,(L \leftrightarrow \psi)$. The atoms whose root symbol is a predicate in $\mathcal{P}$ are called *T-representative literals*. The set of T-representative literals is denoted by $\mathcal{L}_{\mathcal{P}}$. Notice that T-representative literals are closed under taking substitutions of terms for variables, by definition.

Definition 4.3. *Let T be a theory eliminating quantifiers for a sort S. We say that T has QE-degree N with respect to a set of representative predicates $\mathcal{P}$ iff for every finite family $\{L_i\}_{i \in I}$ of literals from $\mathcal{L}_{\mathcal{P}}$, the formula*

$$\exists x \left(\bigwedge_{i \in I} L_i \right) \leftrightarrow \bigwedge_{I_0 \subseteq_N I} \exists x \left(\bigwedge_{i \in I_0} L_i \right) \tag{4}$$

is T-valid (here the variable x is of sort S and $I_0 \subseteq_N I$ means that I_0 is a subset of I having cardinality at most N).

Notice that the left-to-right implication in (4) is universally valid.

---

[1] Recall that a model $\mathcal{M}$ is finitely generated iff there is a finite subset X of the support $|\mathcal{M}|$ of $\mathcal{M}$ such that for every $c \in |\mathcal{M}|$, there are $\underline{b} \subseteq X$ and a term $t(\underline{x})$ such that $\mathcal{M} \models t(\underline{b}) = c$.

Example 4.4. Real linear arithmetic has the signature $\{0, 1, -, +, =, <, \leq\}$ and the single structure $\mathbb{R}$ (endowed with the natural interpretation) as class of models. If we take the predicates in $\{<, \leq, =\}$ as representatives, an inspection of the formulae produced by the Fourier-Motzkin quantifier elimination procedure shows that this theory has QE-degree equal to 2. A similar result holds for the so-called real and integer 'difference logic.'

We remark that Definition 4.3 does not allow negative literals to be T-representatives. However, there is an obvious way to circumvent this limitation (when needed) by expanding the signature in an inessential way. For instance, if we want negated equations to be T-representative, it is sufficient to expand the signature with a new binary predicate Neq, to let the formula $\forall x \forall y (Neq(x, y) \leftrightarrow x \neq y)$ be T-valid (e.g., by adding it to the axioms of T), and then to include Neq into the set $\mathcal{P}$ of representative predicates. Because of this, we prefer to speak of 'T-representative literals' rather than of 'T-representative atoms'. Finally, notice that, in a multi-sorted context, we might be interested in quantifier elimination over just one sort (e.g., the sort representing time, in timed networks); in this case, it is convenient to take all predicates not involving such sort *and their negations* as representative, by using the above trick.

4.1   Automated Termination

Let us fix from now on a wqo theory W and an array-based system $\mathcal{S} = (a, I, \tau)$ (built up over theories T_I, T_E). We make the following extra assumptions:

(E1) T_E has QE-degree N for all sorts occurring in τ as sorts of an existentially quantified data variable (this is the variable e in (1));

(E2) for each disjunct of the form (1) in τ, ϕ_L and the partition components in F are conjunctions of representative literals;

(E3) W has a finite relational signature Σ_W with unique sort S and there is a syntactic interpretation $(-)^*$ from W into A_I^E such that $S^* = \text{INDEX}$.

The fact that a translation map $(-)^*$ is a syntactic interpretation from W into A_I^E is decidable since the axioms for W are universal and their translations modulo A_I^E generate $\exists^{A,I} \forall^I$-sentences, whose satisfiability is decidable. As a consequence, if the signature of A_I^E is finite (which is always the case in practical examples), the syntactic interpretations $(-)^*$ can be *enumerated*. In theory, this guarantees the possibility to find the right translation (if it exists) for the termination argument of Theorem 4.5 below to work, relatively to the wqo W, the array-based system $\mathcal{S}$, and the set of unsafe states U under consideration. In practice, however, the search for the right translation could be driven by user provided hints.

We are ready to state the main results of the paper. We use $\alpha(\underline{i}), \beta(\underline{i}), \dots$ to denote quantifier-free Σ_W-formulae in which at most the variables $\underline{i}$ occur. We say that an $\exists^I$-formula $\exists \underline{i} \, \psi(\underline{i}, a[\underline{i}])$ is a *translation* iff ψ is equivalent (modulo A_I^E) to a formula α^* for some $\alpha(\underline{i})$. Being a translation is clearly a decidable notion: this is because the signature of W is finite and relational, so the search space for the suitable $\alpha(\underline{i})$ is finite (the required validity tests fall within the decidability result for $\exists^{A,I}\forall^{I}$-sentences).

Theorem 4.5. *Assume (E1)-(E3). Suppose that the $\exists^{I}$-formula $U(a)$ describing unsafe states is a translation and let M be the maximum arity of the predicate symbols in Σ_W. The backward reachability procedure terminates if the following conditions are satisfied by every finite model $\mathcal{M}$ of W ($\sharp\mathcal{M}$ indicates the cardinality of the support of $\mathcal{M}$):*

(i) *if $\sharp\mathcal{M} \leq M$, then the the translation $\delta^*_{\mathcal{M}}$ of the diagram of $\mathcal{M}$ is A^E_I-equivalent to a conjunction of representative literals $L^{\mathcal{M}}_1 \wedge \cdots \wedge L^{\mathcal{M}}_{k_{\mathcal{M}}}$;*

(ii) *if $\sharp\mathcal{M} \leq M$, then for every substructure $\mathcal{M}[\underline{i_0}]$ of $\mathcal{M}$, for every $r = 1,\ldots,k_{\mathcal{M}}$, we have that if $L^{\mathcal{M}}_r$ is of the kind $L(\underline{i_0}, a[\underline{i_0}])$ (i.e. if it mentions at most the elements $\underline{i_0}$ of the support of $\mathcal{M}[\underline{i_0}]$), then $A^E_I \models \delta^*_{\mathcal{M}[\underline{i_0}]} \to L^{\mathcal{M}}_r$;*

(iii) *if $\sharp\mathcal{M} \leq M * N$, then the formulae obtained after the elimination of the quantifier $\exists e$ from the formulae $\exists e\, \tau_h[\delta^*_{\mathcal{M}}, f]$ (varying τ_h among the disjuncts of τ and f under the suitable case-marking functions) are all translations.*

In practice, N and M have very small values, typically $N = M = 2$. Thus, only small models must be inspected in order to apply the theorem. When there is no existentially quantified data variable in all transitions (i.e. variable e does not occur neither in the ϕ_L's nor in the F's of (1)), the proof of Theorem 4.5 shows that we can assume $N = 1$ and condition (i) is not needed. In general, condition (i) might not be made fully automated; however, if it holds, an enumerative search can always effectively checks it. In practice, the situation is simple because the following straightforward procedure succeeds and is sufficient to guarantee (i). Consider $\delta^*_{\mathcal{M}}(\underline{i})$: if we replace in it all literals by their Boolean combinations of representative literals and put the result in disjunctive normal form, we get a formula of the form $\theta_1(\underline{i}, a[\underline{i}]) \vee \cdots \vee \theta_k(\underline{i}, a[\underline{i}])$ where the θ_j's are conjunctions of representative literals. Condition (i) of Theorem 4.5 is certainly guaranteed if $k = 1$ or if all θ_j but one are A^E_I-inconsistent. Condition (ii) is technical but usually holds trivially (we can figure that only pathological examples may violate it). The significant condition to verify is just (iii); it can be effectively checked because "being a translation" is decidable.

Theorem 4.5 covers termination of the backward reachability procedure described in Section 3 for several classes of systems including broadcast protocols [13,11], lossy channels systems [5], timed networks with integer clocks [3], and timed networks with a single real clock [6].

The complexity of the procedure of Theorem 4.5 is hard to evaluate, because too many parameters contribute to it (not only M, N, but also the size of τ, of Σ_W, and of the translated atoms, the complexity of quantifier elimination, the number and the complexity of the involved SMT tests, etc.); however, it is arguable that we are well below the lower bounds known for deciding problems which are nevertheless covered by Theorem 4.5 (see e.g. [21]). Thus it might be convenient to apply our termination test before directly running backward search.

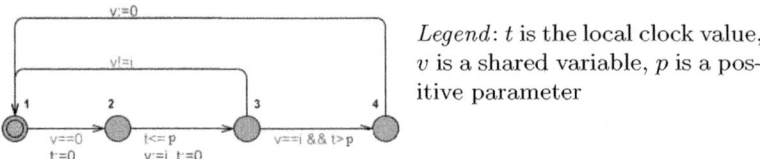

Fig. 1. Automaton for one process of the Fischer's protocol

4.2   An Application of Theorem 4.5: The Fischer Protocol

The goal of the Fischer protocol is to ensure mutual exclusion in a network of processes, using a clock and a shared variable v. Each process has a local clock and a control state variable ranging over $\{1, 2, 3, 4\}$. Each process is identified by a natural number > 0 and can read/update a shared variable whose values is either 0 or the index of one of the processes. A process wishing to enter the critical section 4 starts in 1. If $v = 0$, the process goes to 2 and resets its local clock. From 2, the process can go to state 3 if the clock is $< p$ time unit (where p is a positive parameter), sets v to its own index, and again resets its clock. From 3, the process can go to 4 if the clock is $> p$ time unit and v is still equal to the index of the process performing the transition. When exiting 4, the process sets v to 0. The set of unsafe states, i.e. those states violating the mutual exclusion property, can be characterized by the presence of at least two processes entering 4 at the same time. This specification is depicted in Figure 1. Before applying Theorem 4.5, we make few observations to simplify the technical development. As implemented in MCMT, the shared variable v is modelled as a constant array which is updated uniformly so that the invariant $\forall i \, \forall j \, (v[i] = v[j])$ holds. The invariant will be taken into consideration during backward search as explained in Section 3. For simplicity, below, we do not indicate the redundant dependency on the index i and write just v instead of $v[i]$. A second remark concerns *processes identifiers (id's)*, which are integers and the sort of integers is not the same as the sort INDEX. Formally, we view the id's as an array $id : \text{INDEX} \to \mathbb{Z}$ that is never updated; we implicitly add the conjunct $\forall i \, \forall j \, (id[i] = id[j] \to i = j)$ to the initial formula I and use this too as an invariant (this invariant just says that different processes have different id's); again, for simplicity however, below we write just i for $id[i]$. Since the id's are positive, we also implicitly add the conjunct $\forall i \, (id[i] > 0)$ to the initial formula and use it as an invariant. A third remark is about the use of *additional trivial invariants* that are a substantial part of the specification of the problem and do not capture any deep insight into the system. In our case, for local clocks, stored in the real-valued array t, we use the invariant $\forall i \, (t[i] \geq 0)$ in order to specify that clock values are non-negative.

The theory A_I^E is composed of the theory of pure equality for T_I and the theory T_E is the union of three theories $T_{E_1}, T_{E_2}, T_{E_3}$, where T_{E_1} is linear real arithmetic, T_{E_2} is the theory of the single finite structure $Q = \{1, 2, 3, 4\}$ (the signature of T_{E_2} has just four constants and the identity predicate), and T_{E_3} is linear integer arithmetic. Since the specification contains the positive parameter

p, the theory T_{E_1} is extended with a further constant p constrained by the axiom $p > 0$. Quantifier elimination is assumed only for T_{E_1} and we take as representative predicates all the predicates of T_{E_2}, T_{E_3} together with their negations; we pick just $<, \leq, =$ as representative predicates from T_{E_1} (we get QE-degree 2 with this choice). The following array-based system (a, I, τ) formalizes the Fischer protocol:

- a is a 4-tuple of array variables $\langle l, t, v, id \rangle$, whose target sorts are those of T_{E_1}, T_{E_2} and T_{E_3} (twice), respectively (l is the array of locations, t is the array of clocks, v is the shared register, and id is the array of the id's);
- the formula I is $\forall i \, (l[i] = 1 \wedge t[i] = 0 \wedge v = 0)$, which constrains the initial locations to be equal to 1 and that clocks and the shared variable to 0;
- τ is the disjunction of the time elapsing transition τ_1 and the discrete transitions $\tau_2, ..., \tau_6$ listed below above (identical updates of the array id are not displayed for the sake of conciseness):

$$\tau_1 := \exists c. \left(c > 0 \wedge v' = v \wedge l' = l \wedge t' = \lambda j.(t[j] + c) \right)$$

$$\tau_2 := \exists i. \begin{pmatrix} l[i] = 1 \wedge v = 0 \ \wedge v' = v \wedge \\ l' = \lambda j. \text{ (if } j = i \text{ then 2 else } l[j]) \ \wedge \\ t' = \lambda j. \text{ (if } j = i \text{ then 0 else } t[j]) \end{pmatrix}$$

$$\tau_3 := \exists i. \begin{pmatrix} l[i] = 2 \wedge t[i] \leq p \ \wedge v' = i \ \wedge \\ l' = \lambda j. \text{ (if } j = i \text{ then 3 else } l[j]) \ \wedge \\ t' = \lambda j. \text{ (if } j = i \text{ then 0 else } t[j]) \end{pmatrix}$$

$$\tau_4 := \exists i. \begin{pmatrix} l[i] = 3 \wedge \ v \neq i \ \wedge v' = v \wedge t' = t \ \wedge \\ l' = \lambda j. \text{ (if } j = i \text{ then 1 else } l[j]) \end{pmatrix}$$

$$\tau_5 := \exists i. \begin{pmatrix} l[i] = 3 \wedge \ v = i \ \wedge t[i] > p \ \wedge v' = v \wedge t' = t \ \wedge \\ l' = \lambda j. \text{ (if } j = i \text{ then 4 else } l[j]) \end{pmatrix}$$

$$\tau_6 := \exists i. \begin{pmatrix} l[i] = 4 \ \wedge v' = 0 \ \wedge t' = t \ \wedge \\ l' = \lambda j. \text{ (if } j = i \text{ then 1 else } l[j]) \end{pmatrix}$$

This completes the array-based specification for the Fischer protocol. Finally, the $\exists^I$-formula U describing the set of unsafe states in which the mutual exclusion property for location 4 is violated is $\exists i_1 \exists i_2 \, (i_1 \neq i_2 \wedge l[i_1] = 4 \wedge l[i_2] = 4)$.

To apply Theorem 4.5, we use the theory W in Example 4.2, relatively to the set of predicates indicated below, together with their syntactic translations.

- The unary predicates $Q_1, ..., Q_4$ are translated as the formulae $l[i] = 1, ..., l[i] = 4$, respectively;
- the unary predicates $P_{=0}, P_{>0}, P_{<p}, P_{=p}, P_{>p}$ are translated as $t[i] = 0, t[i] > 0, t[i] < p, t[i] = p, t[i] > p$, respectively;
- the unary predicate F is translated as $v = i$;
- the 0-ary predicate f is translated as $v = 0$;
- the binary predicate $\leq$ is translated as $t[i_1] \leq t[i_2]$.

Clearly, the translations of the axioms of W (namely reflexivity, transitivity and linearity axioms for $\leq$) are valid modulo A_I^E. The unsafety formula is a translation because $i_1 \neq i_2 \wedge l[i_1] = 4 \wedge l[i_2] = 4$ is the translation of $i_1 \neq i_2 \wedge Q_4(i_1) \wedge Q_4(i_2)$. It remains to check conditions (i)-(ii)-(iii) of Theorem 4.5. For lack of space, we outline how to do this for (iii) by analysing the models of W having at most four elements. Instead of examining them one by one, we use a powerful heuristics. All representative literals coming from the translations of the diagrams of the four-elements models are included in the following list:

$$(L) \quad k \neq k', l[k] = 1, l[k] = 2, l[k] = 3, l[k] = 4, v = k, v \neq k, v = 0, v \neq 0,$$
$$t[k] = 0, t[k] > 0, t[k] < p, t[k] = p, t[k] > p, t[k] \leq t[k'], t[k] < t[k'],$$

where $k, k' \in \{i_1, i_2, i_3, i_4\}$ - the literal $t[k] < t[k']$ is the translation of $k' \not\leq k$. (Notice that to get the list above, it is sufficient to consider models on a support with at most two elements, because there are no function symbols and at most binary predicates in Σ_W, so any model has a diagram which is the conjunction of the diagrams of all submodels of cardinality at most two.) If we succeed in proving that the formulae $\exists e \, \tau_1[L_1 \wedge L_2, f]$ and $\tau_h[L_1 \wedge L_2, f]$ ($2 \leq h \leq 6$) are translations for every pair of literals L_1, L_2 coming from the above list (with possibly $L_1 \equiv L_2$), we actually proved more than what is required by condition (iii) (the limitation to at most two literals is due to the fact the QE-degree is 2).

The case of the discrete transitions $\tau_2, ..., \tau_6$ is trivial. It remains to analyze the time elapsing transition τ_1 where $\exists e \, \tau_1[L_1 \wedge L_2, f]$ does not have case-marking function and is $\exists e \, (e > 0 \wedge L_1^{+e} \wedge L_2^{+e})$ where L_i^{+e} is L_i after the substitution of the terms $t[k]$ with $t[k] + e$. The relevant cases to be analyzed are 28 and in all of them we get that $\exists e \, \tau_1[L_1 \wedge L_2, f]$ is a translation. For example, $\exists e \, (e > 0 \wedge t[k_1] + e > 0 \wedge t[k_2] + e = p)$ gives $p > t[k_2] \wedge t[k_1] + p - t[k_2] > 0$ which is equivalent to $p > t[k_2]$ (i.e. to $P_{<p}(k_2)^*$), taking into account the invariant saying that clocks are non-negative.

Thus, all conditions from Theorem 4.5 have been checked and we can predict termination of backward search for Fischer protocol. We emphasize that *the above arguments can be fully mechanized: they consist just in satisfiability checks that can be automatically generated and quickly discharged by suitable tools.*

5   Conclusions

We identified a sufficient condition for the termination of a symbolic backward reachability procedure encompassing many results from the literature in a uniform and declarative framework. We believe that the statement of Theorem 4.5 could be seen as a paradigm for a declaratively-oriented approach to termination; the statement itself needs to be further investigated and exploited in connection to more examples of wqo-theories and syntactic translations arising from encoding termination arguments based on Kruskal theorem.

An interesting direction for future work consists in applying the methods of this paper in connection to abstraction techniques: our results could be profitably employed to predict whether a proposed abstraction of a system yields a terminating search.

References

1. Abdulla, P.A., Cerans, K., Jonsson, B., Tsay, Y.-K.: General decidability theorems for infinite-state systems. In: Proc. of LICS, pp. 313–321 (1996)
2. Abdulla, P.A., Delzanno, G., Henda, N.B., Rezine, A.: Regular model checking without transducers. In: Grumberg, O., Huth, M. (eds.) TACAS 2007. LNCS, vol. 4424, pp. 721–736. Springer, Heidelberg (2007)
3. Abdulla, P.A., Deneux, J., Mahata, P.: Multi-clock timed networks. In: Proc. of LICS 2004, the 18th IEEE Int. Symp. on Logic in Computer Science (2004)
4. Abdulla, P.A.: Forcing monotonicity in parameterized verification: From multisets to words. In: van Leeuwen, J., Muscholl, A., Peleg, D., Pokorný, J., Rumpe, B. (eds.) SOFSEM 2010. LNCS, vol. 5901, pp. 1–15. Springer, Heidelberg (2010)
5. Abdulla, P.A., Jonsson, B.: Verifying programs with unreliable channels. Information and Computation 127(2), 91–101 (1996)
6. Abdulla, P.A., Jonsson, B.: Model checking of systems with many identical timed processes. Theoretical Computer Science, 241–264 (2003)
7. Alberti, F., Ghilardi, S., Pagani, E., Ranise, S., Rossi, G.P.: Brief Announcement: Automated Support for the Design and Validation of Fault Tolerant Parameterized Systems—a case study. In: Lynch, N.A., Shvartsman, A.A. (eds.) DISC 2010. LNCS, vol. 6343, pp. 392–394. Springer, Heidelberg (2010)
8. Carioni, A., Ghilardi, S., Ranise, S.: Automated Termination in Model Checking Modulo Theories - extended version,
   http://homes.dsi.unimi.it/~ghilardi/allegati/CGR_RP11_extended.pdf
9. Carioni, A., Ghilardi, S., Ranise, S.: MCMT in the Land of Parametrized Timed Automata. In: Proc. of VERIFY 2010 (2010)
10. Chang, C.-C., Keisler, J.H.: Model Theory, 3rd edn. North-Holland, Amsterdam (1990)
11. Delzanno, G., Esparza, J., Podelski, A.: Constraint-based analysis of broadcast protocols. In: Flum, J., Rodríguez-Artalejo, M. (eds.) CSL 1999. LNCS, vol. 1683, pp. 50–66. Springer, Heidelberg (1999)
12. Enderton, H.B.: A Mathematical Introduction to Logic. Academic Press, New York (1972)
13. Esparza, J., Finkel, A., Mayr, R.: On the verification of broadcast protocols. In: Proc. of LICS, pp. 352–359. IEEE Computer Society, Los Alamitos (1999)
14. Ghilardi, S., Nicolini, E., Ranise, S., Zucchelli, D.: Towards SMT Model-Checking of Array-based Systems. In: Armando, A., Baumgartner, P., Dowek, G. (eds.) IJCAR 2008. LNCS (LNAI), vol. 5195, pp. 67–82. Springer, Heidelberg (2008)
15. Ghilardi, S., Ranise, S.: A Note on the Stopping Failures Models, Unpublished Draft, mcmt web site (2009)
16. Ghilardi, S., Ranise, S.: Goal Directed Invariant Synthesis for Model Checking Modulo Theories. In: Giese, M., Waaler, A. (eds.) TABLEAUX 2009. LNCS, vol. 5607, pp. 173–188. Springer, Heidelberg (2009)
17. Ghilardi, S., Ranise, S.: Model Checking Modulo Theory at work: the integration of Yices in MCMT. In: AFM 2009 (co-located with CAV 2009) (2009)
18. Ghilardi, S., Ranise, S.: Backward reachability of array-based systems by SMT-solving: termination and invariant synthesis. LMCS 6(4) (2010)
19. Ghilardi, S., Ranise, S., Valsecchi, T.: Light-Weight SMT-based Model-Checking. In: Proc. of AVOCS 2007-2008, ENTCS (2008)

20. Ghilardi, S., Ranise, S.: MCMT: A Model Checker Modulo Theories. In: Giesl, J., Hähnle, R. (eds.) IJCAR 2010. LNCS, vol. 6173, pp. 22–29. Springer, Heidelberg (2010)
21. Philippe, S.: Verifying lossy channel systems has nonprimitive recursive complexity. Information Processing Letters 83(5), 251–261 (2002)
22. Ranise, S., Tinelli, C.: The SMT-LIB Standard: Version 1.2. Technical report, Dep. of Comp. Science, Iowa (2006), http://www.SMT-LIB.org/papers

Monotonic Abstraction for Programs with Multiply-Linked Structures[*]

Parosh Aziz Abdulla[1], Jonathan Cederberg[1], and Tomáš Vojnar[2]

[1] Uppsala University, Sweden
[2] FIT, Brno University of Technology, Czech Republic

Abstract. We investigate the use of monotonic abstraction and backward reachability analysis as means of performing shape analysis on programs with multiply pointed structures. By encoding the heap as a vertex- and edge-labeled graph, we can model the low level behaviour exhibited by programs written in the C programming language. Using the notion of *signatures*, which are predicates that define sets of heaps, we can check properties such as absence of null pointer dereference and shape invariants. We report on the results from running a prototype based on the method on several programs such as insertion into and merging of doubly-linked lists.

1 Introduction

Dealing with programs manipulating dynamic pointer-linked data structures is one of the most challenging tasks of automated verification since these data structures are of unbounded size and may have the form of complex graphs. As discussed below, various approaches to automated verification of dynamic pointer-linked data structures are currently studied in the literature. One of these approaches is based on using *monotonic abstraction* and *backward reachability* [4,2]. This approach has been shown to be very successful in handling systems with complex graph-structured configurations when verifying parameterized systems [3]. However, in the area of verification of programs with dynamic linked data structures, it has so far been applied only to relatively simple singly-linked data structures.

In this paper, we investigate the use of monotonic abstraction and backward reachability for verification of programs dealing with dynamic linked data structures with multiple selectors. In particular, we consider verification of sequential programs written in a subset of the C language including its common control statements as well as its pointer manipulating statements (apart from pointer arithmetics and type casting). For simplicity, we restrict ourselves to data structures with two selectors. This restriction can, however, be easily lifted. We consider verification of safety properties in the form of absence of null and dangling pointer dereferences as well as preservation of shape invariants of the structures being handled.

We represent heaps in the form of simple vertex- and edge-labeled graphs. As is common in backward verification, our verification technique starts from sets of bad

---

[*] The first two authors were supported by the Swedish UPMARC project, the third author was supported by the COST OC10009 project of the Czech Ministry of Education.

G. Delzanno and I. Potapov (Eds.): RP 2011, LNCS 6945, pp. 125–138, 2011.

configurations and checks whether some initial configurations are backward reachable from them. For representing sets of bad configurations as well as the sets of configurations backward reachable from them, we use the so-called *signatures* which arise from heap graphs by deleting some of their nodes, edges, or labels. Each signature represents an *upward-closed set* of heaps wrt. a special *pre-order* on heaps and signatures. We show that the considered C pointer manipulating statements can be *approximated* such that one can compute predecessors of sets of heaps represented via signatures wrt. these statements.

We have implemented the proposed approach in a light-weight Java-based prototype and tested it on several programs manipulating doubly-linked lists and trees. The results show that monotonic abstraction and backward reachability can indeed be successfully used for verification of programs with multiply-linked dynamic data structures.

Related work. Several different approaches have been proposed for automated verification of programs with dynamic linked data structures. The most-known approaches include works based on monadic second-order logic on graph types [10], 3-valued predicate logic with transitive closure [14], separation logic [12,11,15,6], other kinds of logics [16,9], finite tree automata [5,7], forest automata [8], graph grammars [13], upward-closed sets [4,2], as well as other formalisms.

As we have already indicated above, our work extends the approach of [4,2] from singly-linked to multiply-linked heaps. This extension has required a new notion of signatures, a new pre-order on them, as well as new operations manipulating them. Not counting [4,2], the other existing works are based on other formalisms than the one used here, and they use a forward reachability computation whereas the present paper uses a backward reachability computation. Apart from that, when comparing the approach followed in this work with the other existing approaches, one of the most attractive features of our method is its simplicity. This includes, for instance, a simple specification of undesirable heap shapes in terms of signatures. Each such signature records some bad pattern that should not appear in the heaps, and it is typically quite small (usually with three or fewer nodes). Furthermore, our approach uses local and quite simple reasoning on the graphs in order to compute predecessors of symbolically represented infinite sets of heaps[1]. Moreover, the abstraction used in our approach is rather generic, not specialised for some fixed class of dynamic data structures.

Outline. In Section 2, we give some preliminaries and introduce our model for describing heaps. We present the class of programs we consider in Section 3. In Section 4, we introduce signatures as symbolic representations for infinite sets of configurations. We show how to use signatures for specifying bad heap patterns (that violate safety properties of the considered programs) in Section 5. In Section 6, we describe a symbolic backward reachability analysis algorithm for checking safety properties. Next, we report on experiments with the proposed method in Section 7. Finally, we give some conclusions and directions for future work in Section 8.

---

[1] Approaches based on separation logic and forest automata also use local updates, but the updates used here are still simpler.

2   Heaps

Preliminaries. For a partial function $f : A \to B$ and $a \in A$, we write $f(a) = \perp$ to signify that f is undefined at a. We take $f[a \mapsto b]$ to be the function f' such that $f'(a) = b$ and $f'(x) = f(x)$ otherwise. We define the *restriction* of f to A', written $f|_{A'}$, as the function f' such that $f'(a) = f(a)$ if $a \in A'$, and $f'(a) = \perp$ if $a \notin A'$. Given $b \in B$, we write $f^{-1}(b)$ to denote the set $\{a \in A : f(a) = b\}$.

Heaps. We model the dynamically allocated memory, also known as the *heap*, as a labeled graph. The nodes of the graph represent memory cells, and the edges represent how these nodes are linked by their successor pointers. Each edge is labeled by a color, reflecting which of the possibly many successor pointers of its source cell the edge is representing. In this work, we—for simplicity—consider structures with two selectors, denoted as 1 and 2 (instead of, e.g., next and prev commonly used in doubly-linked lists or left and right used in trees) only. The results can, however, be generalized to any number of selectors.

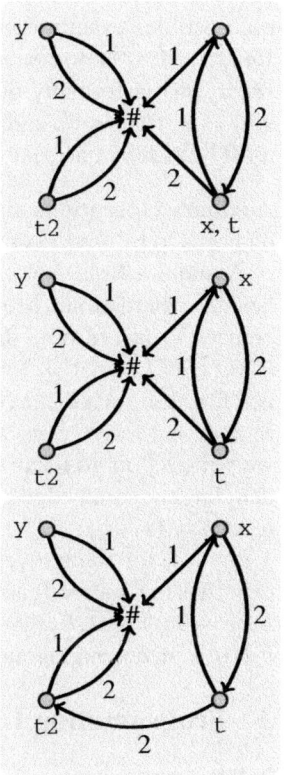

To model *null pointers*, we introduce a special node called the *null node*, written #. Null successors are then modeled by making the corresponding edge point to this node. When allocated memory is relinquished by a program, any pointers previously pointing to that memory become *dangling*. Dangling pointers also arise when memory is freshly allocated and not yet initialized. This situation is reflected in our model by the introduction of another special node called the *dangling node*, denoted as *. In the same manner as for the null node, a pointer being dangling is modeled by having the corresponding edge point to the dangling node.

Furthermore, we model a program variable by labeling the node that a specific variable is pointing to with the variable in question.

Fig. 1. Heaps

Three examples of heaps can be seen in Figure 1 (we will get back to what they represent in Section 3). To avoid unnecessarily cluttering the pictures, the special node * has been left out. We will adopt the convention of omitting any of the special nodes * and # from pictures unless they are labeled or have edges pointing to them.

Assume a finite set of program variables X and a set $C = \{1,2\}$ of edge colors. Formally, a *heap* is a tuple $(\overline{M}, E, s, t, \tau, \lambda)$ where

- $\overline{M} = M \cup \{\#, *\}$ represents the finite set of allocated memory cells, together with the two special nodes representing the null value and the dangling pointer, respectively.
- E is a finite set of edges.

- The *source function* $s : E \to M$ is a total function that gives the source of the edges.
- The *target function* $t : E \to \overline{M}$ is a total function that gives the target of the edges.
- The *type function* $\tau : E \to C$ is a total function that gives the color of the edges.
- $\lambda : X \to \overline{M}$ is a total function that defines the positions of the program variables.

We also require that the heaps obey the following invariant:

$$\forall c \in C \ \forall m \in M : |s^{-1}(m) \cap \tau^{-1}(c)| = 1.$$

The invariant states that among the edges going out from each cell there is exactly one with color 1 and one with color 2. Note that as a consequence of these invariants, each cell has exactly two outgoing edges. Therefore, each heap h induces a function $succ_{h,c} : M \to \overline{M}$ for each $c \in C$, which maps each cell to its c-successor. For $m \in M$, $succ_{h,c}(m)$ is formally defined as the $m' \in \overline{M}$ such that there is an edge $e \in E$ with $s(e) = m$, $t(e) = m'$, and $\tau(e) = c$. This is indeed a function due to the fact that there must be exactly one such edge, according to the specified invariants.

Auxiliary Operations on Heaps. We will now introduce some notation for operations on heaps to be used in the following.

Assume a heap $h = (\overline{M}, E, s, t, \tau, \lambda)$. For $m \in M$, we write $h \ominus m$ to describe the heap h' where m has been deleted together with its two outgoing edges, and any references to m are now dangling references. Formally, $h \ominus m$ is defined as the heap $h' = (\overline{M}', E', s', t', \tau', \lambda')$ where $\overline{M}' = \overline{M} \setminus \{m\}$, $E' = E \setminus s^{-1}(m)$, $s' = s|_{E'}$, $t' : E' \to \overline{M}'$ is a function such that $t'(e) = *$ if $e \in t^{-1}(m)$ and $t'(e) = t(e)$ otherwise, $\tau' = \tau|_{E'}$, and $\lambda'(x) = *$ if $x \in \lambda^{-1}(m)$ and $\lambda'(x) = \lambda(x)$ otherwise. In a similar manner, for $m' \notin M$, we write $h \oplus m'$ to mean the heap where we have added a new cell as well as two new dangling outgoing edges. Formally, $h \oplus m' = (\overline{M}', E', s', t', \tau', \lambda)$ where $\overline{M}' = \overline{M} \cup \{m'\}$, $E' = E \cup \{e_1, e_2\}$, $s' = s[e_1 \mapsto m', e_2 \mapsto m']$, $t' = t[e_1 \mapsto *, e_2 \mapsto *]$ and $\tau' = \tau[e_1 \mapsto 1, e_2 \mapsto 2]$ for some $e_1, e_2 \notin E$. By $h.s[e \mapsto m]$, we mean the heap identical to h, except that the source function now maps $e \in E$ to $m \in M$. This is formally defined as $h.s[e \mapsto m] = (\overline{M}, E, s[e \mapsto m], t, \tau, \lambda)$. The definitions of $h.t[e \mapsto m]$, $h.\tau[e \mapsto m]$, and $h.\lambda[x \mapsto m]$ are analogous.

3 Programming Language

In this section, we briefly present the class of programs which our analysis is designed for. We also formalize the transition systems which are induced by such programs.

In particular, our analysis and the prototype tool implementing it are designed for sequential programs written in a subset of the C language. The considered subset contains the common control flow statements (like `if`, `while`, `for`, etc.) and the C pointer manipulating statements, excluding pointer arithmetics and type casting. As for the structures describing nodes of dynamic data structures, we—for simplicity of the presentation as well as of the prototype implementation—allow one or two selectors to be used only. However, one can easily generalize the approach to more selectors. Statements manipulating data other than pointers (e.g., integers, arrays, etc.) are ignored—or, in case of tests, replaced by a non-deterministic choice. We allow non-recursive functions that can be inlined[2].

---

[2] Alternatively, one could use function summaries, which we, however, not consider here.

Figure 2 contains an example code snippet written in the considered C subset (up to the tests on integer data that will be replaced by a non-deterministic choice for the analysis). In this example, the data structure DLL represents nodes of a doubly-linked list with two successor pointers as well as a data value. The function merge takes as input two doubly-linked lists and combines them into one doubly-linked list[3]. In Figure 1, the result of executing two of the statements in the merge can be seen. From the top graph, the middle is generated by executing the statement at line 20. By then executing the statement at line 25, the bottom graph is generated. (Note that instead of the next and prev selectors, the figure uses selectors 1 and 2, respectively.)

```
1   typedef struct DLL {
2       struct DLL *next;
3       struct DLL *prev;
4       int data;
5   } DLL;
6
7   DLL *merge(DLL *l1, DLL *l2) {
    ...
17      while(!(x==NULL)&&!(y==NULL)) {
18          if(x->data < y->data) {
19              t = x;
20              x = t->next;
21          } else {
22              t = y;
23              y = t->next;
24          }
25          t->prev = t2;
26          t2->next = t;
27          t2 = t;
28      }
    ...
```

Fig. 2. A program for merging doubly-linked lists

Operational Semantics and the Induced Transition System. From a C program, we can extract a *control flow graph* (PC, T) by standard techniques. Here PC is a finite set of *program counters*, and T is a finite set of transitions. A transition t is a tuple of the form (pc, op, pc') where $pc, pc' \in PC$, and op is an operation manipulating the heap. The operation op is of one of the following forms:

- x == y or x != y, which means that the program checks the stated condition.
- x = y, x = y.next(i), or x.next(i) = y, which are assignments functioning in the same way as assignments in the C language[4].
- x = malloc() or free(x), which are allocation and deallocation of dynamic memory, working in the same manner as in the C language.

When firing t, the program counter is updated from pc to pc', and the heap is modified according to op with the usual C semantics formalized below.

---

[3] In fact, if the input lists are sorted, the output list will be sorted too, but this is not of interest for our current analysis—let us, however, note that one can think of extending the analysis to track ordering relations between data in a similar way as in [2], which we consider as one of interesting possible directions for future work.

[4] Here, next(i) refers to the i-th selector of the appropriate memory cell.

The induced transition system. We will now define the transition system $(S, \longrightarrow)$ induced by a control flow graph (PC, T). The states of the transition system are pairs (pc, h) where $pc \in PC$ is the current location in the program, and h is a heap. The transition relation $\longrightarrow$ reflects the way that the program manipulates the heap during program execution.

Given states $s = (pc, h)$ and $s' = (pc', h')$ there is a transition from s to s', written $s \longrightarrow s'$, if there is a transition $(pc, op, pc') \in T$ such that $h \xrightarrow{op} h'$. The condition $h \xrightarrow{op} h'$ holds if the operation op can be performed to change the heap h into the heap h'. The definition of $\xrightarrow{op}$ is found below.

Assume two heaps $h = (\overline{M}, E, s, t, \tau, \lambda)$ and $h' = (\overline{M}', E', s', t', \tau', \lambda')$. We say that $h \xrightarrow{op} h'$ if one of the following is fulfilled:

- op is of the form x == y, $\lambda(x) = \lambda(y) \neq *$, and $h = h'$.[5]
- op is of the form x != y, $\lambda(x) \neq \lambda(y)$, $\lambda(x) \neq *$, $\lambda(y) \neq *$, and $h = h'$.
- op is of the form x = y, $\lambda(y) \neq *$, and $h' = h.\lambda[x \mapsto \lambda(y)]$.
- op is of the form x = y.next(i), $\lambda(y) \notin \{*, \#\}$, $succ_{h,i}(\lambda(y)) \neq *$, and $h' = h.\lambda[x \mapsto succ_{h,i}(\lambda(y))]$.
- op is of the form x.next(i) = y, $\lambda(x) \neq *$, $\lambda(y) \neq *$, and $h' = h.t[e \mapsto \lambda(y)]$ where e is the unique edge in E such that $s(e) = \lambda(x)$ and $\tau(e) = i$.
- op is of the form x = malloc() and there is a heap h_1 such that $h_1 = h \oplus m$ and $h' = h_1.\lambda[x \mapsto m]$ for some $m \notin \overline{M}$.[6]
- op is of the form free(x), $\lambda(x) \neq *$, and $h' = h \ominus \lambda(x)$.

4   Signatures

In this section, we introduce the notion of signatures which is a symbolic representation of infinite sets of heaps.

Intuitively, a signature is a predicate describing a set of minimal conditions that a heap has to fulfill to satisfy the predicate. It can be viewed as a heap with some parts "missing".

Formally, a signature is defined as a tuple $(\overline{M}, E, s, t, \tau, \lambda)$ in the same way as a heap, with the difference that we allow the τ and λ functions to be partial. For signatures, we also require some invariants to be obeyed, but they are not as strict as the invariants for heaps. More precisely, a signature has to obey the following invariants:

1. $\forall c \in C \; \forall m \in M : |s^{-1}(m) \cap \tau^{-1}(c)| \leq 1$,
2. $\forall m \in M : |s^{-1}(m)| \leq 2$.

These invariants say that a signature can have *at most* one outgoing edge of each color in the set $\{1, 2\}$, and at most two outgoing edges in total. Note that heaps are a special case of signatures, which means that each heap is also a signature.

---

[5] Note that the requirement that $\lambda(x)$ and $\lambda(y)$ are not dangling pointers are not part of the standard C semantics. Comparing dangling pointers are, however, bad practice and our tool therefore warns the user.

[6] Although the malloc operation may fail, we assume for simplicity of presentation that it always succeeds.

Operations on Signatures. We formalize the notion of a signature as a predicate by introducing an ordering on signatures. First, we introduce some additional notation for manipulating signatures. Recall that, for a heap $h = (\overline{M}, E, s, t, \tau, \lambda)$ and $m \in M$, $h \ominus m$ is a heap identical to h except that m has been deleted. As the formal definition of $\ominus$ carries over directly to signatures, we will use it also for signatures.

Given a signature $sig = (\overline{M}, E, s, t, \tau, \lambda)$, we define the removal of an edge $e \in E$, written $sig \boxminus e$, as the signature $(\overline{M}, E', s', t', \tau', \lambda)$ where $E' = E \setminus \{e\}$, $s' = s|_{E'}$, $t' = t|_{E'}$, and $\tau' = \tau|_{E'}$. Similarly, given $m_1 \in M$, $m_2 \in \overline{M}$, and $c \in C$, the addition of a c-edge from m_1 to m_2 is written $sig \boxplus (m_1 \xrightarrow{c} m_2)$. This is formalized as $sig \boxplus (m_1 \xrightarrow{c} m_2) = (\overline{M}, E', s', t', \tau', \lambda)$ where $E' = E \cup \{e'\}$ for some $e' \notin E$, $s' = s[e' \mapsto m_1]$, $t' = t[e' \mapsto m_2]$, and $\tau' = \tau[e' \mapsto c]$. Note that the addition of edges might make the result violate the invariants for signatures. However, we will always use it in such a way that the invariants are preserved. Finally, for $m \notin \overline{M}$, we define $sig.(\overline{M} := \overline{M} \cup \{m\})$ as the signature $(\overline{M} \cup \{m\}, E, s, t, \tau, \lambda)$.

Ordering on Signatures. For a signature $sig = (\overline{M}, E, s, t, \tau, \lambda)$ and $m \in \overline{M}$, we say that m is *unlabeled* if $\lambda^{-1}(m) = \emptyset$. We say that m is *isolated* if m is unlabeled and also $s^{-1}(m) = \emptyset$ and $t^{-1}(m) = \emptyset$ both hold. We call m *simple* when m is unlabeled and $s^{-1}(m) = \{e_1\}$, $t^{-1}(m) = \{e_2\}$, $e_1 \neq e_2$, and $\tau(e_1) = \tau(e_2)$ all hold. Intuitively, an isolated cell has no touching edges, whereas a simple cell has exactly one incoming and one outgoing edge of the same color. For $sig_1 = (\overline{M}_1, E_1, s_1, t_1, \tau_1, \lambda_1)$ and $sig_2 = (\overline{M}_2, E_2, s_2, t_2, \tau_2, \lambda_2)$, we write that $sig_1 \lhd sig_2$ if one of the following is true:

- *Isolated cell deletion.* There is an isolated $m \in M_2$ s.t. $sig_1 = sig_2 \ominus m$.
- *Edge deletion.* There is an edge $e \in E_2$ such that $sig_1 = sig_2 \boxminus e$.
- *Contraction.* There is a simple cell $m \in M_2$, edges $e_1, e_2 \in E_2$ with $t_2(e_1) = m$, $s_2(e_2) = m$, $\tau(e_1) = \tau(e_2)$, and a signature sig' such that $sig' = sig_2.t[e_1 \mapsto t(e_2)]$ and $sig_1 = sig' \ominus m$.
- *Edge decoloring.* There is an edge $e \in E_2$ such that $sig_1 = sig_2.\tau[e \mapsto \bot]$.
- *Label deletion.* There is a label $x \in X$ such that $sig_1 = sig_2.\lambda[x \mapsto \bot])$.

We call the above operations *ordering steps*, and we say that a signature sig_1 is smaller than a signature sig_2 if there is a sequence of ordering steps from sig_2 to sig_1, written $sig_1 \sqsubseteq sig_2$. Formally, $\sqsubseteq$ is the reflexive transitive closure of $\lhd$.

The Semantics of Signatures. Using the ordering relation $\sqsubseteq$ defined above, we can interpret each signature as a predicate. As previously noted, the intuition is that a heap h satisfies a predicate sig if h contains *at least* the structural information present in sig. We make this precise by saying that h satisfies sig, written $h \in [\![sig]\!]$, if $sig \sqsubseteq h$. In other words, $[\![sig]\!]$ is the set of all heaps in the *upward closure* of sig with respect to the ordering $\sqsubseteq$. For a set S of signatures, we define $[\![S]\!] = \bigcup_{s \in S} [\![s]\!]$.

5   Bad Configurations

We will now show how to use the concept of signatures to specify *bad states*. The main idea is to define a finite set of signatures characterizing the set of all heaps that are *not* considered correct. Such a set of signatures is called the set of *bad patterns*.

We present the notion on a concrete example, namely, the case of a program that should produce a single acyclic doubly-linked list pointed to by a variable x. In such a case, the following properties are required to hold at the end of the program:

1. Starting from any allocated memory cell, if we follow the next(1) pointer and then immediately the next(2) pointer, we should end up at the original memory cell.
2. Likewise, starting from any allocated cell, if we follow the next(2) pointer and then immediately the next(1) pointer, we should end up at the original cell.
3. If we repeatedly follow a pointer of the same type starting from any allocated cell, we should never end up where we started. In other words, no node is reachable from itself in one or more steps using only one type of pointer.
4. The variable x is not dangling, and there are no dangling next pointers.
5. The variable x points to the beginning of the list.
6. There are no unreachable memory cells.

We call properties 1 and 2 *Doubly-Linkedness*, property 3 is called *Non-Cyclicity*, property 4 is called *Absence of Dangling Pointers*, property 5 is called *Pointing to the Beginning of the List*, and, finally, property 6 is called *Absence of Garbage*.

Doubly-Linkedness. As noted above, the set of bad states with respect to a property p is characterized by a set of signatures such that the union of their upward closure with respect to $\sqsubseteq$ contains all heaps not fulfilling p. The property we want to express is that following a pointer of one color and then immediately following a pointer of the other color gets you back to the same node. The bad patterns are then simply the set $\{b_1, b_2, b_3, b_4\}$, shown to the right, as they describe exactly the property of taking one step of each color and *not* ending up where we started.

$b_1:$ ○—1→○—2→○

$b_2:$ ○—2→○—1→○

$b_3:$ ○—1→○—2→#

$b_4:$ ○—2→○—1→#

Non-Cyclicity. To describe all states that violate the property of not being cyclic, is to describe exactly those states that do have a cycle. Note that all the edges of the cycle has to be of the same color. Therefore, the bad patterns we get for non-cyclicity is the set $\{b_5, b_6\}$, depicted to the right.

$b_5:$

$b_6:$

Absence of Dangling Pointers. To describe dangling pointers, two bad patterns suffice—namely, the pattern b_7 depicted to the right stipulates that the variable x that should point to the resulting list is not dangling, and the pattern b_8 requires that there is no dangling next pointer.

$b_7: \overset{x}{*}$ $b_8:$ ○—→*

Pointing to the Beginning of the List. To describe that the pointer variable x should point to the beginning of a list, one bad pattern suffices—namely, the pattern b_9 depicted to the right (saying that the node pointed by x has a predecessor). Note that the pattern does not prevent the resulting list from being empty.

$b_9:$ ○←2—○ $\overset{x}{}$

Absence of Garbage. To express that there should be no garbage, the patterns b_{10} and b_{11} are needed. The b_{10} pattern says that if the resulting list is empty, there should be no allocated cell. The b_{11} pattern designed for non-empty lists then builds on that we check the *Doubly-Linkedness* property too. When we assume it to hold, the isolated node can never be part of a well-formed list segment: Indeed, since the two edges in b_{11} are both pointing to the null cell, any possible inclusion of the isolated node into the list results in a pattern that is larger either than b_1 or than b_2.

Clearly, the above properties are common for many programs handling doubly-linked lists (the name of the variable pointing to the resulting list can easily be adjusted, and it is easy to cope with multiple resulting lists too). We now describe some more properties that can easily be expressed and checked in our framework.

Absence of Null Pointer Dereferences. The bad pattern used to prove absence of null pointer dereferences is b_{12}. A particular feature of this pattern is that it is duplicated many times. More precisely, for each program statement of the form y = x.next(i) or x.next(i) = y, the pattern is added to the starting set of bad states S_{bad} coupled with the program counter just before the operation. In other words, we construct a state that we know would result in a null pointer dereference if reached and try to prove that the configuration is unreachable. The construction for dangling pointer dereferences is analogous.

Cyclicity. To encode that a doubly-linked list is cyclic, we use b_{13} as a bad pattern. Given that we already have *Doubly-Linkedness*, we only need to enforce that the list is not terminated. This is achieved by the existence of a null pointer in the list since such a pointer will break the doubly-linkedness property. Note that this relies on the fact that the result actually is a doubly-linked list.

Treeness. To violate the property of being a tree, the data structure must have a cycle somewhere, two paths to the same node, or two incoming edges to some node. The bad patterns for trees are thus the set $\{b_{14}, b_{15}, b_{16}\}$ depicted to the right.

A Remark on Garbage. Note that the treatment of garbage presented above is not universal in the sense that it is valid for all data structures. In particular, if the data structure under consideration is a tree, garbage cannot expressed in our present framework. Intuitively, there is only one path in each direction that ends with null in a doubly-linked list, whereas a tree can have more paths to null. Thus a pattern like b_{11} is not sufficient since the isolated node can still be incorporated into the tree in a valid way. One way to solve this problem, which is a possible direction for future work, is to add some concept of *anti-edges* which would forbid certain paths in a structure from arising.

6   Reachability Analysis

In this section, we present the algorithm used for analysing the transition system defined in Section 3. We do this by first introducing an abstract transition system that has the property of being *monotonic*. Given this abstract system, we show how to perform *backward reachability analysis*. Such analysis requires the ability to compute the predecessors of a given set of states, all of which is described below.

Monotonic Abstraction. Given a transition system $T = (S, \longrightarrow)$ and an ordering $\sqsubseteq$ on S, we say that T is monotonic if the following holds. For any states s_1, s_2 and s_3 such that $s_1 \sqsubseteq s_2$ and $s_1 \longrightarrow s_3$, we can always find a state s_4 such that $s_2 \sqsubseteq s_4$ and $s_3 \longrightarrow s_4$.

The transition system defined in Section 3 does not exhibit this property. We can, however, construct an over-approximation of our transition relation in such a way that it becomes monotonic. This new transition system $\longrightarrow_A$ is constructed from $\longrightarrow$ by using the state s_3 above as our required s_4. Formally, $s \longrightarrow_A s'$ iff there is an s'' such that $s''/J \sqsubseteq s$ and $s'' \longrightarrow s'$.

Since our abstraction generates an over-approximation of the original transition system, if it is shown that no bad pattern is reachable under this abstraction, the result holds for the original program too. The inverse does not hold, and so the analysis may generate false alarms, which, however, does not happen in our experiments. Further, the analysis is not guaranteed to terminate in general. However, it has terminated in all the experiments we have done with it (cf. Section 7).

Auxiliary Operations on Signatures. To perform backward reachability analysis, we need to compute the predecessor relation. We show how to compute the set of predecessors for a given signature with respect to the abstract transition relation $\longrightarrow_A$.

In order to compute pre, we define a number of auxiliary operations. These operations consist of *concretizations*; they add "missing" components to a given signature. The first operation adds a variable x. Intuitively, given a signature sig, in which x is missing, we add x to all places in which x may occur in heaps satisfying sig.

Let $M^{\#} = M \cup \{\#\}$ and $sig = (\overline{M}, E, s, t, \tau, \lambda)$. We define the set $sig{\uparrow}(\lambda(x) \notin \{\bot, *\})$ to be the set of all signatures $sig' = (\overline{M}', E', s', t', \tau', \lambda')$ s.t. one of the following is true:

- $\lambda(x) \in M^{\#}$ and $sig = sig'$. The variable is already present in sig, so no changes need to be made.
- $\lambda(x) = \bot$ and there is a cell $m \in M^{\#}$ such that $sig' = sig.\lambda[x \mapsto m]$. We add x to a cell that is explicitly represented in sig.
- $\lambda(x) = \bot$, and there is a cell $m \notin \overline{M}$ and a signature sig_1 such that $sig_1 = sig.(\overline{M} := \overline{M} \cup \{m\})$ and $sig' = sig_1.\lambda[x \mapsto m]$. We add x to a cell that is missing in sig. Note that according to the definition of $[\![sig]\!]$, there may exist cells in $h \in [\![sig]\!]$ that are not explicitly represented in sig.
- $\lambda(x) = \bot$, and there is a cell $m \notin \overline{M}$, edges $e_1 \in E$, $e_2 \notin E$ and signatures sig_1, sig_2 and sig_3 such that $sig_1 = sig.(\overline{M} := \overline{M} \cup \{m\})$, $sig_2 = sig_1 \boxplus (m \overset{\tau(e_1)}{\longrightarrow} t(e_1))$, $sig_3 = sig_2.t[e_1 \mapsto m]$ and $sig' = sig_3.\lambda[x \mapsto m]$. We add x to a cell that is not explicit in sig. The difference to the previous case is that the missing cell lies *between* two explicit cells m_1, m_2 in sig, along an edge between them.

We now define an operation that adds a missing edge between to specific cells in a signature. Given cells $m_1 \in M, m_2 \in \overline{M}$, we say that a signature sig' is in the set $sig{\uparrow}$ $(m_1 \overset{c}{\longrightarrow} m_2)$ if one of the following is true:

- There is an $e \in E$ such that $s(e) = m_1$, $t(e) = m_2$, $\tau(e) = c$ and $sig' = sig$. The edge is already present, so no addition of edge is needed.
- There is an $e \in E$ such that $s(e) = m_1$, $t(e) = m_2$, $\tau(e) = \perp$, there is no $e' \in E$ such that $s(e') = m_1$ and $\tau(e') = c$, and we have and $sig' = sig.\tau[e \mapsto c]$. There is a decolored edge whose color we can update to c. To do this we need to ensure that there is no such edge already.
- There is no $e \in E$ such that $s(e) = m_1$ and $\tau(e) = c$, $|s^{-1}(m_1)| \leq 1$ and $sig' = sig \boxplus (m_1 \xrightarrow{c} m_2)$. The edge is not present, and m_1 does not already have an outgoing edge of color c. We add the edge to the graph.

The third operation adds labels x and y to the signature in such a way that they both label the same cell.

Formally, we say that a signature sig' is in the set $sig \uparrow (\lambda(x) = \lambda(y))$ if one of the following is true:

- $\lambda(x) \in M^{\#}$, $\lambda(x) = \lambda(y)$ and $sig' = sig$. Both labels are already present and labeling the same cell, so no changes are needed.
- $\lambda(y) = \perp$ and there is a $sig_1 \in sig \uparrow (\lambda(x) \notin \{\perp, *\})$ such that $sig' = sig_1.\lambda[y \mapsto \lambda_1(x)]$. The label y is not present, so we add it to a signature where x is guaranteed to be present.
- $\lambda(x) = \perp$, $\lambda_{sig}(y) \in M^{\#}$ and $sig' = sig.\lambda[x \mapsto \lambda(y)]$. The label x is not present, so we add it to the cell that is labeled by y.

Computing Predecessors. We will now describe how to compute the predecessors of a signature sig and an operation op, written $\text{pre}(op)(sig)$.

Assume a signature $sig = (\overline{M}, E, s, t, \tau, \lambda)$. We define $\text{pre}(x = \texttt{malloc()})(sig)$ as the set sig' of signatures such that there are signatures $sig_1 = (\overline{M}_1, E_1, s_1, t_1, \tau_1, \lambda_1)$, sig_2, and sig_3 satisfying

- $sig_1 \in sig \uparrow (\lambda(x) \notin \{\perp, *\})$, there is no $y \in X$ such that $\lambda_1(y) = \lambda_1(x)$ and no $e \in E_1$ such that $t_1(e) = \lambda_1(x)$,
- $sig_2 \in sig_1 \uparrow (\lambda_1(x) \xrightarrow{1} *)$,
- $sig_3 \in sig_2 \uparrow (\lambda_1(x) \xrightarrow{2} *)$, and
- $sig' = sig_3 \ominus \lambda_1(x)$.

We let $\text{pre}(x = y)(sig)$ be the set sig' of signatures such that there is a signature sig_1 satisfying $sig_1 \in sig \uparrow (\lambda(x) = \lambda(y))$ and $sig' = sig_1.\lambda[x \mapsto \perp]$.

Next, we define $\text{pre}(\texttt{x==y})(sig)$ to be the set of all sig' s.t. $sig' \in sig \uparrow (\lambda(x) = \lambda(y))$. On the other hand, we define $\text{pre}(\texttt{x!=y})(sig)$ to be the set of all $sig' = (\overline{M}', E', s', t', \tau', \lambda')$ with $\lambda'(x) \neq \lambda'(y)$ and such that there is a signature $sig_1 \in sig \uparrow (\lambda(x) \notin \{\perp, *\})$ such that $sig' \in sig_1 \uparrow (\lambda(y) \notin \{\perp, *\})$.

Further, $\text{pre}(\texttt{x = y.next(i)})(sig)$ is defined as the set of all signatures $sig' = (\overline{M}', E', s', t', \tau', \lambda')$ such that there are sig_1, $sig_2 = (\overline{M}_2, E_2, s_2, t_2, \tau_2, \lambda_2)$, sig_3 with

- $sig_1 = sig \uparrow (\lambda(x) \notin \{\perp, *\})$,
- $sig_2 = sig_1 \uparrow (\lambda(x) \notin \{\perp, *\})$,

- $sig_3 \in sig\!\uparrow\!(\lambda_2(y) \xrightarrow{i} \lambda_2(x))$, and
- $sig' = sig_3.\lambda[x \mapsto \bot]$.

We let $\mathrm{pre}(\text{x.next(i)} = \text{y})(sig)$ be the set of all $sig' = (\overline{M}', E', s', t', \tau', \lambda')$ such that there are sig_1, sig_2, $sig_3 = (\overline{M}_3, E_3, s_3, t_3, \tau_3, \lambda_3)$, and $e \in E_3$ with

- $sig_1 = sig\!\uparrow\!(\lambda(x) \notin \{\bot, *\})$,
- $sig_2 = sig_1\!\uparrow\!(\lambda(x) \notin \{\bot, *\})$,
- $sig_3 \in sig\!\uparrow\!(\lambda_2(x) \xrightarrow{i} \lambda_2(y))$,
- $s_3(e) = \lambda_3(x)$, $t_3(e) = \lambda_3(y)$, $\tau(e) = i$, and
- $sig' = sig_3 \boxminus e$.

Finally, we define $\mathrm{pre}(\text{free(x)})(sig)$ to be the set of all $sig' = (\overline{M}', E', s', t', \tau', \lambda')$ such that there are $sig_1 = (\overline{M}_1, E_1, s_1, t_1, \tau_1, \lambda_1)$, sig_2, and $m \notin \overline{M}_1$ with

- $\overline{M}_1 = \overline{M}$, $E_1 = E \setminus t^{-1}(*)$, $s_1 = s|_{E_1}$, $t_1 = t|_{E_1}$, $\tau_1 = \tau|_{E_1}$ and $\lambda_1(x) = \bot$ if $\lambda(x) = *$, $\lambda_1(x) = \lambda(x)$ otherwise,
- $sig_2 = sig_1 \oplus m$, and
- $sig' = sig_2.\lambda[x \mapsto m]$.

The Reachability Algorithm. We are now ready to describe the backward reachability algorithm used for checking safety properties. Given a set S_{bad} of bad patterns for the property under consideration, we compute the successive sets S_0, S_1, S_2, ..., where $S_0 = S_{bad}$ and $S_{i+1} = \bigcup_{s \in S_i} \mathrm{pre}(s)$. Whenever a signature s is generated such that there is a previously generated s' with $s' \sqsubseteq s$, we can safely discard s from the analysis. When all the newly generated signatures are discarded, the analysis is finished. The generated signatures at this point denote all the heaps that can reach a bad heap using the approximate transition relation $\longrightarrow_A$. If all the generated signatures characterize sets that are disjoint from the set of initial states, the safety property holds.

Remark. As the configurations of the transition system are pairs consisting of a heap and a control state, the set S_{bad} is a set of pairs where the control state is a given state, typically the exit state in the control flow graph. This extension is straightforward. For a more in depth discussion of monotonic abstraction and backwards reachability, see [1].

7 Implementation and Experimental Results

We have implemented the above proposed method in a Java prototype. To improve the analysis, we combined the backward reachability algorithm with a light-weight flow-based alias analysis to prune the state space. This analysis works by computing a set of necessary conditions on the program variables for each program counter. Whenever we compute a new signature, we check whether it intersects with the conditions for the corresponding program counter, and if not, we discard the signature. Our experience with this was very positive, as the two analyses seem to be complementary. In particular, programs with limited branching seemed to benefit from the alias analysis.

We also used the result of the alias analysis to add additional information to the signatures. More precisely, suppose that, the alias analysis has given us that at a specific program counter pc, x and y must alias. Furthermore, suppose that we compute a signature sig that is missing at least one of x and y at pc. We can then safely replace sig with $sig\uparrow(\lambda(x) = \lambda(y))$.

In Table 1, we show results obtained from experiments with our prototype. We considered programs traversing doubly-linked lists, inserting into them (at the beginning or according to the value of the element being inserted—since the value is abstracted away, this amounts to insertion to a random place), merging ordered doubly-linked lists (the ordering is ignored), and reversing them. We also considered algorithms

Table 1. Experimental results

Program	Struct	Time	#Sig.
Traverse	DLL	11.4 s	294
Insert	DLL	3.5 s	121
Ordered Insert	DLL	19.4 s	793
Merge	DLL	6 min 40 s	8171
Reverse	DLL	10.8 s	395
Search	Tree	1.2 s	51
Insert	Tree	6.8 s	241

for searching an element in a tree and for inserting new leaves into trees. We ran the experiments using a PC with Intel Core 2 Duo 2.2 GHz and 2GB RAM (using only one core as the implementation is completely serial). The table shows the time it took to run the analysis, and the number of signatures computed throughout the analysis. For each program manipulating doubly-linked lists, we used the set $\{b_1, b_2, \ldots, b_{11}\}$ as described in Section 5 as the set of bad states to start the analysis from. For the programs manipulating trees, we used the set $\{b_{14}, b_{15}, b_{16}\}$.

The obtained results show that the proposed method can indeed successfully handle non-trivial properties of non-trivial programs. Despite the high running times for some of the examples, our experience gained from the prototype implementation indicates that there is a lot of space for further optimizations as discussed in the following section.

8   Conclusions and Future Work

We have proposed a method for using monotonic abstraction and backward analysis for verification of programs manipulating multiply-linked dynamic data structures. The most attractive feature of the method is its simplicity, concerning the way the shape properties to be checked are specified as well as the abstraction and predecessor computation used. Moreover, the abstraction used in the approach is rather generic, not specialised for some fixed class of dynamic data structures. The proposed approach has been implemented and successfully tested on several programs manipulating doubly-linked lists and trees.

An important direction for future work is to optimize the operations done within the reachability algorithm. This especially concerns checking of entailment on the heap signatures (e.g., using advanced hashing methods to decrease the number of signatures being compared) and/or minimization of the number of generated signatures (perhaps using a notion of a coarser ordering on signatures that could be gradually refined to reach the current precision only if a need be). It also seems interesting to parallelize the approach since there is a lot of space for parallelization in it. We believe that such improvements are worth the effort since the presented approach should—in principle—be applicable even for checking complex properties of complex data structures such as skip lists which are very hard to handle by other approaches without their significant modifications and/or help from the users. Finally, it is also interesting to think of extending the proposed approach with ways of handling non-pointer data, recursion, and/or concurrency.

References

1. Abdulla, P.A.: Well (and Better) Quasi-Ordered Transition Systems. Bulletin of Symbolic Logic 16, 457–515 (2010)
2. Abdulla, P.A., Atto, M., Cederberg, J., Ji, R.: Automated Analysis of Data-Dependent Programs with Dynamic Memory. In: Liu, Z., Ravn, A.P. (eds.) ATVA 2009. LNCS, vol. 5799, pp. 197–212. Springer, Heidelberg (2009)
3. Abdulla, P.A., Ben Henda, N., Delzanno, G., Rezine, A.: Handling Parameterized Systems with Non-atomic Global Conditions. In: Logozzo, F., Peled, D.A., Zuck, L.D. (eds.) VMCAI 2008. LNCS, vol. 4905, pp. 22–36. Springer, Heidelberg (2008)
4. Abdulla, P.A., Bouajjani, A., Cederberg, J., Haziza, F., Rezine, A.: Monotonic Abstraction for Programs with Dynamic Memory Heaps. In: Gupta, A., Malik, S. (eds.) CAV 2008. LNCS, vol. 5123, pp. 341–354. Springer, Heidelberg (2008)
5. Bouajjani, A., Habermehl, P., Rogalewicz, A., Vojnar, T.: Abstract Regular Tree Model Checking of Complex Dynamic Data Structures. In: Yi, K. (ed.) SAS 2006. LNCS, vol. 4134, pp. 52–70. Springer, Heidelberg (2006)
6. Calcagno, C., Distefano, D., O'Hearn, P.W., Yang, H.: Compositional Shape Analysis by Means of Bi-abduction. In: Proc. of POPL 2009. ACM Press, New York (2009)
7. Deshmukh, J.V., Emerson, E.A., Gupta, P.: Automatic Verification of Parameterized Data Structures. In: Hermanns, H. (ed.) TACAS 2006. LNCS, vol. 3920, pp. 27–41. Springer, Heidelberg (2006)
8. Habermehl, P., Holík, L., Rogalewicz, A., Šimáček, J., Vojnar, T.: Forest Automata for Verification of Heap Manipulation. Technical Report FIT-TR-2011-01, FIT BUT, Czech Republic (2011), http://www.fit.vutbr.cz/~isimacek/pub/FIT-TR-2011-01.pdf
9. Madhusudan, P., Parlato, G., Qiu, X.: Decidable Logics Combining Heap Structures and Data. In: Proc. of POPL 2011. ACM Press, New York (2011)
10. Møller, A., Schwartzbach, M.: The Pointer Assertion Logic Engine. In: Proc. of PLDI 2001. ACM Press, New York (2001)
11. Nguyen, H.H., David, C., Qin, S., Chin, W.N.: Automated Verification of Shape and Size Properties via Separation Logic. In: Cook, B., Podelski, A. (eds.) VMCAI 2007. LNCS, vol. 4349, pp. 251–266. Springer, Heidelberg (2007)
12. Reynolds, J.C.: Separation Logic: A Logic for Shared Mutable Data Structures. In: Proc. of LICS 2002. IEEE CS, Los Alamitos (2002)
13. Rieger, S., Noll, T.: Abstracting Complex Data Structures by Hyperedge Replacement. In: Ehrig, H., Heckel, R., Rozenberg, G., Taentzer, G. (eds.) ICGT 2008. LNCS, vol. 5214, Springer, Heidelberg (2008)
14. Sagiv, S., Reps, T.W., Wilhelm, R.: Parametric Shape Analysis via 3-valued Logic. TOPLAS 24(3) (2002)
15. Yang, H., Lee, O., Berdine, J., Calcagno, C., Cook, B., Distefano, D., O'Hearn, P.W.: Scalable Shape Analysis for Systems Code. In: Gupta, A., Malik, S. (eds.) CAV 2008. LNCS, vol. 5123, pp. 385–398. Springer, Heidelberg (2008)
16. Zee, K., Kuncak, V., Rinard, M.: Full Functional Verification of Linked Data Structures. In: Proc. of PLDI 2008. ACM Press, New York (2008)

Efficient Bounded Reachability Computation for Rectangular Automata

Xin Chen[1], Erika Ábrahám[1], and Goran Frehse[2]

[1] RWTH Aachen University, Germany
[2] Université Grenoble 1 Joseph Fourier - Verimag, France

Abstract. We present a new approach to compute the reachable set with a bounded number of jumps for a rectangular automaton. The reachable set under a flow transition is computed as a polyhedron which is represented by a conjunction of finitely many linear constraints. If the bound is viewed as a constant, the computation time is polynomial in the number of variables.

1 Introduction

Hybrid systems are systems equipped with both continuous dynamics and discrete behavior. A popular modeling formalism for hybrid systems are *hybrid automata*. In this paper, we consider a special class of hybrid automata, called *rectangular automata* [1]. The main restriction is that the derivatives, invariants and guards are defined by lower and upper bounds in each dimension, forming *rectangles* or *boxes* in the value domain. Rectangular automata can be used to model not only simple timed systems but also asymptotically approximate hybrid systems with nonlinear behaviors [2,3,4,5].

Since hybrid automata often model safety-critical systems, their *reachability analysis* builds an active research area. The reachability problem is decidable only for *initialized* rectangular automata [1], which can be reduced to timed automata [6]. The main merit of rectangular automata is that the reachable set under a flow is always a (convex) *polyhedron*. It means that the reachable set in a bounded number of jumps can be exactly computed as a set of polyhedra, unlike for general hybrid automata which need approximative methods such as [7,8,9]. In the past, some geometric methods are proposed for exactly or approximately computing the reachable sets in a bounded number of jumps (see, e.g., [4,3]). There are also tools like HyTech [10] and PHAVer[11] which can compute bounded reachability for rectangular automata in a geometric way.

However, nearly all of the proposed methods compute the exact reachable set under a flow based on the vertices of the initial set and the derivative rectangle. Since a d-dimensional rectangle has 2^d many vertices, those methods are not able to handle high-dimensional cases. In [3], an approximative method is proposed to over-approximate the reachable set by polyhedra which are represented by conjunctions of linear constraints. Since only $2d$ linear constraints are needed to define a d-dimensional rectangle, the computation time of the method is polynomial in d. However, the accuracy degenerates dramatically when d increases.

G. Delzanno and I. Potapov (Eds.): RP 2011, LNCS 6945, pp. 139–152, 2011.

In this paper, we compute the reachable set as polyhedra which are represented by finite linear constraint sets [12], where we need only $2d$ linear constraints to define a d-dimensional rectangle. We show that when the number of jumps is bounded by a constant, the computational complexity of our approach is polynomial in d. We also include the cases that some of the rectangles in the definition of a rectangular automaton are not full-dimensional.

The paper is organized as follows. After introducing some basic definitions in Section 2, we describe our efficient approach for computing the bounded reachable set in Section 3. In Section 4, we compare our approach and PHAVer based on a scalable example. Missing proofs can be found in [13].

2   Preliminaries

2.1   Polyhedra and Their Computation

For a point (or column vector) $v \in \mathbb{R}^d$ in the d-dimensional Euclidean space $\mathbb{R}^d$ we use $v[i]$ to denote its ith component, $1 \leq i \leq d$, and v^T for the row vector being its transpose.

In the following we call linear inequalities $c^T x \leq z$ for some $c \in \mathbb{R}^d$, $z \in \mathbb{R}$ and x a variable, also *constraints*. Given a finite set $\mathcal{L}$ of linear equalities and linear inequalities, we write $S : \mathcal{L}$ for $S = \{x \in \mathbb{R}^d \mid x \text{ satisfies } \bigwedge_{L \in \mathcal{L}} L\}$, and also write $S : L$ instead of $S : \{L\}$. We say that $L \in \mathcal{L}$ is *redundant* in $\mathcal{L}$ if $S : \mathcal{L} = S' : \mathcal{L} \backslash \{L\}$. Redundant (in)equalities can be detected using linear programming [14].

A finite set $\{v_1, \ldots, v_{d'}\} \subseteq \mathbb{R}^d$ of linearly independent vectors span an $(d'-1)$-dimensional *affine subspace* Π of $\mathbb{R}^d$ by the affine combinations of $v_1, \ldots, v_{d'}$:

$$\Pi = \{ \sum_{1 \leq i \leq d'} \lambda_i v_i \mid \sum_{1 \leq i \leq n} \lambda_i = 1, \lambda_i \in \mathbb{R} \}$$

The *affine hull* $aff(S)$ of a set $S \subseteq \mathbb{R}^d$ is the smallest affine subspace $\Pi \subseteq \mathbb{R}^d$ containing S, and we have that $dim(S) = dim(aff(S))$. We call a subset of a vector space *full-dimensional* if its affine hull is the whole space.

A $((d-1)$-dimensional) *hyperplane* in $\mathbb{R}^d$ is a $(d-1)$-dimensional affine subspace of $\mathbb{R}^d$. Each hyperplane H can be defined as $H : c^T x = z$ for some $c \in \mathbb{R}^d$ and $z \in \mathbb{R}$. For $d' < d-1$, a d'-dimensional affine subspace H' of $\mathbb{R}^d$ is called a *lower-* or d'-*dimensional hyperplane* and can be defined as an intersection of $d-d'$ many hyperplanes (see [12]), i.e., as $H' : \bigwedge_{1 \leq i \leq d-d'} c_i^T x = z_i$. Since every linear equation $c^T x = z$ can be expressed by $c^T x \leq z \wedge -c^T x \leq -z$, for $d'' \leq d$, a d''-dimensional hyperplane can be defined by a set of $2(d-d'')$ constraints.

A $(d$-dimensional) *halfspace* S in $\mathbb{R}^d$ is a d-dimensional set $S : c^T x \leq z$ for some $c \in \mathbb{R}^d$ and $z \in \mathbb{R}$. For $d' < d$, a d'-dimensional set $S' \subseteq \mathbb{R}^d$ is a *lower-* or d'-*dimensional halfspace* if it is the intersection of a $(d$-dimensional) halfspace S and a d'-dimensional hyperplane $H' \not\subseteq S$. Note that for $d'' \leq d$, a d''-dimensional halfspace can be defined by a set of $2(d-d'')+1$ constraints.

Given a constraint $c^T x \leq z$, its *corresponding equation* is $c^T x = z$. The *corresponding hyperplane* of a d'-dimensional halfspace S with $d' \leq d$ is the $(d'-1)$-dimensional hyperplane defined by the set of the corresponding equations of the constraints that define S.

For a finite set $\mathcal{L}$ of constraints we call $P : \mathcal{L}$ a *polyhedron*. Polyhedra can also be understood as the intersection of finitely many halfspaces. *Polytopes* are bounded polyhedra.

A constraint $c^T x \leq z$ is *valid* for a polyhedron P if all $x \in P$ satisfy it. For $c^T x \leq z$ valid for P and for $H_F : c^T x = z$, the set $F : P \cap H_F$ is a *face* of P. If $F \neq \emptyset$ then we call H_F a *support hyperplane* of P, and the vectors λc for $\lambda > 0$ are the *normal vectors* of H_F. The hyperplane $H : c^T x = z$ is a support hyperplane of a polyhedron P if and only if for the *support function* $\rho_P : \mathbb{R}^d \rightarrow \mathbb{R} \cup \{\infty\}$, $\rho_P(v) = \sup v^T x$ s.t. $x \in P$, we have that $\rho_P(c) = z$. We call a face F of a polyhedron P *facet* if $dim(F) = dim(P)-1$, and *vertex* if $dim(F) = 0$. For d'-dimensional faces we simply write d'-*faces*. We use $NF(P)$ to denote the number of P's facets. Given a face F of P, the *outer normals* of F are the vectors $v_F \in \mathbb{R}^d$ such that $\rho_P(v_F) = \sup v_F^T x$ for any $x \in F$. We also define $\mathcal{N}(F, P)$ as the set of the outer normals of F in P.

For a d'-dimensional polyhedron $P : \mathcal{L}_P$, every facet F_P of P can be *determined* by some $\mathcal{L}_{F_P} \subseteq \mathcal{L}_P$, that is $\mathcal{L}_{F_P}$ defines a d'-dimensional halfspace which contains P and the corresponding hyperplane is the affine hull of F_P (see [12]).

Lemma 1. *If a constraint set $\mathcal{L}$ defines a d'-dimensional polyhedron $P \subset \mathbb{R}^d$ and there is no redundant constraint in $\mathcal{L}$, then the set $\mathcal{L}$ contains $NF(P)+2(d-d')$ constraints.*

Proof. We need a set $\mathcal{L}'$ of $2(d-d')$ constraints to define $aff(P)$. For every facet F_P of P, we need a constraint L_{F_P} such that $\mathcal{L}' \cup \{L_{F_P}\}$ determines F_P.

For a polyhedron $P : \bigcup_{1 \leq i \leq n} \{c_i^T x \leq z_i\}$ and a scalar $\lambda \geq 0$, the *scaled polyhedron* λP can be computed by $\lambda P : \bigcup_{1 \leq i \leq n} \{c_i^T x \leq \lambda z_i\}$. The *conical hull* of P is the polyhedral cone $cone(P) = \bigcup_{\lambda \geq 0} \lambda P$. If the conical hull of P is d'-dimensional, then P is at least $(d'-1)$-dimensional (see [12]).

Example 1. Figure 1(a) shows a polyhedron $P : x_2 \leq 3 \wedge -x_1 \leq -1 \wedge x_1 - 2x_2 \leq -3$ with three irredundant constraints. The support hyperplanes H_1, H_2, H_3 intersect P at its facets. The fourth hyperplane H_4 is also a support hyperplane of P, but it only intersects P at a vertex, and the related constraint $-x_2 \leq -2$ is redundant. The conical hull of P is shown in Figure 1(b).

Given two polyhedra $P : \mathcal{L}_P$ and $Q : \mathcal{L}_Q$, their *intersection* $P \cap Q$ can be defined by the union of their constraints $P \cap Q : \mathcal{L}_P \cup \mathcal{L}_Q$. The *Minkowski sum* $P \oplus Q$ of P and Q is defined by $P \oplus Q = \{p+q \mid p \in P, q \in Q\}$. It is still a polyhedron, as illustrated in Figure 2. We have the following important theorem for the faces of $P \oplus Q$.

Theorem 1 ([15,16]). *For any polytopes P and Q, each face F of $P \oplus Q$ can be decomposed by $F = F_P \oplus F_Q$ for some faces F_P and F_Q of P and Q respectively. Moreover, this decomposition is unique.*

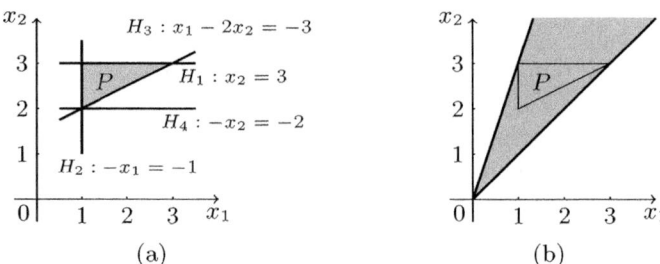

Fig. 1. A 2-dimensional polytope and its conical hull

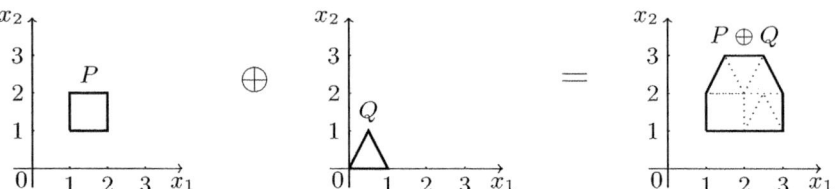

Fig. 2. An example of $P \oplus Q$

2.2   Rectangular Automata

A *box* $B \subseteq \mathbb{R}^d$ is an axis-aligned rectangle which can be defined by a set of constraints of the form $x \leq \bar{a}$ or $-x \leq -\underline{a}$ where x is a variable, and $\bar{a}, \underline{a}$ are rationals. A box $B : \mathcal{L}_B$ is *bounded* if for every variable x there are constraints $x \leq \bar{a}$ and $-x \leq -\underline{a}$ in $\mathcal{L}_B$ for some rationals $\bar{a}, \underline{a}$, otherwise B is *unbounded*. Let $\mathcal{B}^d$ be the set of all boxes in $\mathbb{R}^d$.

Rectangular automata [1] are a special class of hybrid automata [17].

Definition 1. *A d-dimensional rectangular automaton is a tuple* $\mathcal{A} = (Loc, Var, Flow, Jump, Inv, Init, Guard, ResetVar, Reset)$ *where*

- *Loc is a finite set of locations, also called discrete states.*
- *Var = $\{x_1, \ldots, x_d\}$ is a set of d ordered real-valued variables. We denote the variables as a column vector $x = (x_1, \ldots, x_d)^T$.*
- *Flow : Loc → $\mathcal{B}^d$ assigns each location a flow condition which is a box in $\mathbb{R}^d$.*
- *Jump : Loc × Loc is a set of jumps (or discrete transitions).*
- *Inv : Loc → $\mathcal{B}^d$ maps to each location an invariant which is a bounded box.*
- *Init: Loc → $\mathcal{B}^d$ maps to each location a bounded box as initial variable values.*
- *Guard : Jump → $\mathcal{B}^d$ assigns to each jump a guard which is a box.*
- *ResetVar : Jump → 2^{Var} assigns to each jump a set of reset variables.*
- *Reset : Jump → $\mathcal{B}^d$ maps to each jump a reset box such that for all $e \in Jump$ and $x_i \in ResetVar(e)$ the box Reset(e) is bounded in dimension i.*

Example 2. Figure 3 shows an example rectangular automaton. For brevity, we specify boxes by their defining intervals. The location set is $Loc = \{l_0, l_1\}$. The

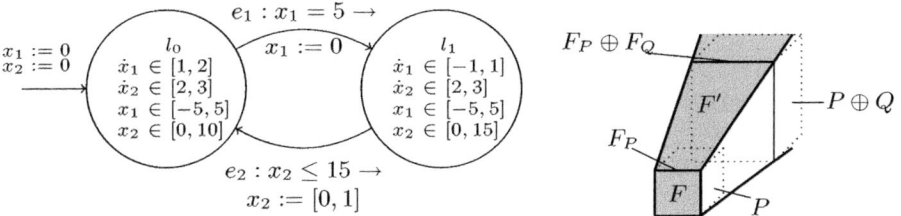

Fig. 3. A rectangular hybrid automaton $\mathcal{A}$ **Fig. 4.** A 3D example of R

initial states are $Init(l_0) = [0,0] \times [0,0]$ and $Init(l_1) = \emptyset$. The flows are defined by $Flow(l_0) = [1,2] \times [2,3]$ and $Flow(l_1) = [-1,1] \times [2,3]$, and the invariants by $Inv(l_0) = [-5,5] \times [0,10]$ and $Inv(l_1) = [-5,5] \times [0,15]$. There are two jumps $Jump = \{e_1, e_2\}$ with $e_1 = (l_0, l_1)$ and $e_2 = (l_1, l_0)$. The guards are $Guard(e_1) = [5,5] \times (-\infty, +\infty)$ and $Guard(e_2) = (-\infty, +\infty) \times (-\infty, 15]$, the reset variable sets $ResetVar(e_1) = \{x_1\}$ and $ResetVar(e_2) = \{x_2\}$, and the reset boxes $Reset(e_1) = [0,0] \times (-\infty, +\infty)$ and $Reset(e_2) = (-\infty, +\infty) \times [0,1]$.

A *configuration* of $\mathcal{A}$ is a pair (l, u) such that $l \in Loc$ is a location and $u \in Inv(l)$ a vector assigning the value $u[i]$ to x_i for $i = 1, \ldots, d$. There are two kinds of transitions between configurations:

- *Flow:* A transition $(l, u) \xrightarrow{t} (l, u')$ where $t \geq 0$, such that there exists $b \in Flow(l)$ such that $u' = u + tb$ and for all $0 \leq t' \leq t$ we have $u + t'b \in Inv(l)$.
- *Jump:* A transition $(l, u) \xrightarrow{e} (l', u')$ such that $e = (l, l') \in Jump$, $u \in Guard(e)$, $u' \in Inv(l') \cap Reset(e)$, and $u[i] = u'[i]$ for all $x_i \in Var \backslash ResetVar(e)$.

An *execution* of $\mathcal{A}$ is a sequence $(l_0, u_0) \xrightarrow{\alpha_0} (l_1, u_1) \xrightarrow{\alpha_1} \cdots$ where $\xrightarrow{\alpha_i}$ is either a flow or a jump for all i. A configuration is *reachable* if it can be visited by some execution. The *reachability computation* is the task to compute the set of the reachable configurations. In this paper we consider bounded reachability with the number of jumps in the considered executions bounded by a positive integer.

3   A New Approach for Reachability Computation

In this section, we present a new approach to compute the reachable set for a rectangular automaton where the number of jumps is bounded.

3.1   Facets of the Reachable Set Under Flow Transitions

For a location l of a rectangular automaton with $Flow(l) = Q$ and $Init(l) = P$, the states reachable from P via the flow can be computed in a geometric way:

$$R_l(P) = (P \oplus cone(Q)) \cap Inv(l). \tag{1}$$

As already mentioned, previously proposed methods compute $R_l(P)$ by considering the evolutions of all vertices of P under the flow condition Q. That means, also all vertices of Q must be considered. Since Q is a bounded box, it has 2^d vertices which make the computation intractable for large d. We present an approach to compute $R_l(P)$ exactly based on three constraint sets which define P, Q and $Inv(l)$ respectively. We show that if we are able to compute a constraint set that defines $P \oplus Q$ in PTIME, then a constraint set which defines $R_l(P)$ can also be computed in PTIME.

We firstly investigate the faces of the set $R = P \oplus cone(Q)$ in the general case that P and Q are polytopes in $\mathbb{R}^d$. From the following two lemmata we derive that the number of R's facets is bounded by $(n_P + n_{P \oplus Q})$ where n_P is the number of the facets of P and $n_{P \oplus Q}$ is the number of the faces from dimension $(dim(R)-2)$ to $(dim(R)-1)$ in $P \oplus Q$.

Lemma 2. *Given a polytope $Q \subseteq \mathbb{R}^d$ and a positive integer d', a d'-face $F_{cone(Q)}$ of the polyhedron $cone(Q)$ can be expressed by $cone(F_Q)$ where F_Q is a nonempty face of Q and it is at least $(d'-1)$-dimensional.*

Proof. The polyhedron $cone(Q)$ can be expressed by $cone(V_Q)$ where V_Q is the vertex set of Q. Then a nonempty face $F_{cone(Q)}$ of $cone(Q)$ can be expressed by $cone(V_Q')$ where $V_Q' \subseteq V_Q$ is nonempty. Assume S is the halfspace whose corresponding hyperplane is $H = aff(F_{cone(Q)})$. Since $cone(Q) \subseteq S$, we also have that $Q \subseteq S$, moreover, we can infer that H is a support hyperplane of Q and $F_Q = H \cap Q$ is a nonempty face of Q whose vertex set is V_Q'. Therefore, the face $F_{cone(Q)}$ can be expressed by $cone(F_Q)$. From the definition of conical hull, if $F_{cone(Q)}$ is d'-dimensional then F_Q is at least $(d'-1)$-dimensional. □

Lemma 3. *Given two polytopes $P, Q \subseteq \mathbb{R}^d$, any d'-face F_R of the polytope $R = P \oplus cone(Q)$ is either a d'-face of P, or the decomposition $F_R = \bigcup_{\lambda \geq 0}(F_P \oplus \lambda F_Q)$ where F_P, F_Q are some nonempty faces of P, Q respectively and $\bar{F}_P \oplus F_Q$ is a face of $P \oplus Q$ which is at least $(d'-1)$-dimensional.*

Proof. We have two cases for a face F_R of R, (1) F_R is a face of P; (2) F_R can be expressed by $F_P \oplus cone(F_Q)$ where F_Q is a *nonempty* face of Q (from Theorem 1 and Lemma 2). In the case (2), we rewrite R and F_R by

$$R = P \oplus cone(Q) = \bigcup_{\lambda \geq 0}(P \oplus \lambda Q) \quad \text{and} \quad F_R = F_P \oplus cone(F_Q) = \bigcup_{\lambda \geq 0}(F_P \oplus \lambda F_Q)$$

Since F_R is a face of R, i.e., it is on the boundary of R, we infer that for all $\lambda \geq 0$ the set $F_P \oplus \lambda F_Q$ is a face of $P \oplus \lambda Q$. Thus $F_P \oplus F_Q$ is a face of $P \oplus Q$. Since F_R is d'-dimensional, the set $F_P \oplus F_Q$ is at least $(d'-1)$-dimensional. □

Example 3. In Figure 4 on page 143, the set F is a facet of both P and R. In contrast, the facet F' can be expressed by $\bigcup_{\lambda \geq 0}(F_P \oplus \lambda F_Q)$.

The facets of R can be found by enumerating all the facets of P, and all the faces from dimension $(dim(R)-2)$ to $(dim(R)-1)$ in $P \oplus Q$.

Lemma 4. *Let* $P : \mathcal{L}_P$ *and* $P \oplus Q : \mathcal{L}_{P \oplus Q}$ *be some polytopes with* $\mathcal{L}_{P \oplus Q} = \{g_j^T x \leq h_j \mid 1 \leq j \leq m\}$ *irredundant. We define the constraint set* $\mathcal{L} = \bigcup_{1 \leq i < j \leq m} \mathcal{L}_{i,j}$ *such that for each* $1 \leq i < j \leq m$, $\mathcal{L}_{i,j} = \{L_{i,j}\}$ *if the intersection of* $H_i : g_i^T x = h_i$, $H_j : g_j^T x = h_j$ *and* $P \oplus Q$ *is nonempty, and* $L_{i,j}$ *is a constraint whose corresponding hyperplane* $H_{i,j}$ *satisfies (1)* $H_{i,j}$ *is a support hyperplane of* P, *(2)* $H_{i,j}$ *is a support hyperplane of* $P \oplus Q$ *and (3)* $H_i \cap H_j \subseteq H_{i,j}$. *Otherwise* $\mathcal{L}_{i,j} = \emptyset$.

Suppose that $\mathcal{L}'$ *is the set of all constraints in* $\mathcal{L}_P$ *and* $\mathcal{L}_{P \oplus Q}$ *that are valid for* R. *Then the polytope* R *can be defined by* $\mathcal{L} \cup \mathcal{L}'$.

Note that $L_{i,j}$ is not unique for each $1 \leq i < j \leq m$, but we only need one of them. Intuitively, for any facet F_R of R, if F_R is also a facet of P then it can be determined by a subset $\mathcal{L}'_P$ of $\mathcal{L}_P$. Since the constraints in $\mathcal{L}'_P$ are also valid for R, we also have that $\mathcal{L}'_P \subseteq \mathcal{L}'$. Otherwise $F_R = \bigcup_{\lambda \geq 0}(F_P \oplus \lambda F_Q)$ for some nonempty faces F_P, F_Q of P, Q respectively. There are two cases, (a) if $F_P \oplus F_Q$ is $(dim(R)-1)$-dimensional, then F_R can be determined by a subset $\mathcal{L}'_{P \oplus Q}$ of $\mathcal{L}_{P \oplus Q}$, it is also included by $\mathcal{L}'$; (b) if $F_P \oplus F_Q$ is $(dim(R)-2)$-dimensional, the facet F_R is determined by a subset of $\mathcal{L}$. Hence, $\mathcal{L} \cup \mathcal{L}'$ defines R.

3.2 Compute the Reachable Set under Flow Transitions

In order to compute the constraint set that defines R, we need to find the hyperplanes $H_{i,j}$ stated in Lemma 4. We determine the $H_{i,j} : c^T x = z$ by solving a feasibility problem for the normal vector $c \in \mathbb{R}^d$ and the value $z \in \mathbb{R}$ as follows. Assume $dim(R) = d_R$, $P : \mathcal{L}_P$ and $P \oplus Q$ is defined by the irredundant set $\mathcal{L}_{P \oplus Q} = \{g_j^T x \leq h_j \mid 1 \leq j \leq m\}$. Firstly, we check if the set $H_i \cap H_j \cap (P \oplus Q)$ with $H_i : g_i^T x = h_i$ and $H_j : g_j^T x = h_j$ is nonempty by solving the following linear program:

$$\text{Find } x_I \in \mathbb{R}^d \quad s.t. \quad g_i^T x_I = h_i \wedge g_j^T x_I = h_j \wedge x_I \in P \oplus Q.$$

If such an x_I is found, then the intersection is nonempty, and there must be a (d_R-2)-face $F_{P \oplus Q}$ of $P \oplus Q$ contained in it since $\mathcal{L}_{P \oplus Q}$ is irredundant. We require that $H_{i,j}$ is a support hyperplane of $P \oplus Q$ and contains $F_{P \oplus Q}$. This can be ensured by finding c in the set $C_{i,j} = \{\alpha g_i + \beta g_j \mid \alpha, \beta \geq 0, \alpha + \beta > 0\}$ and demanding $c^T x_I = z$. An example is shown in Figure 5.

We also require that $H_{i,j}$ is a support hyperplane of P. This can be guaranteed by demanding $\rho_P(c) = z$ and $c = \alpha g_i + \beta g_j$. In order to replace $\rho_P(\alpha g_i + \beta g_j)$ by $\alpha \rho_P(g_i) + \beta \rho_P(g_j)$, we need to ensure their equivalence. This can be done by finding at least one point $p \in P$ such that $g_i^T p = \rho_P(g_i)$ and $g_j^T p = \rho_P(g_j)$. Since the (d_R-2)-face $F_{P \oplus Q}$ is contained in $H_i \cap H_j$, we have that $g_i^T x = \rho_{P \oplus Q}(g_i)$ and $g_j^T x = \rho_{P \oplus Q}(g_j)$ for all $x \in F_{P \oplus Q}$. From Theorem 1, $F_{P \oplus Q}$ can be decomposed by $F_P \oplus F_Q$ for some faces F_P, F_Q of P, Q respectively, and we can infer that for all $x \in F_P$ it holds that $g_i^T x = \rho_P(g_i)$ and $g_j^T x = \rho_P(g_j)$. Hence we can replace $\rho_P(\alpha g_i + \beta g_j)$ by $\alpha \rho_P(g_i) + \beta \rho_P(g_j)$.

Then the vector c can be computed by solving the following problem:

$$\text{Find } c \in \mathbb{R}^d \ s.t. \ \begin{cases} c = \alpha g_i + \beta g_j \ \wedge \ \alpha + \beta > 0 \wedge \alpha \geq 0 \wedge \beta \geq 0 \\ c^T x_I = \alpha \rho_P(g_i) + \beta \rho_P(g_j) \end{cases} \tag{2}$$

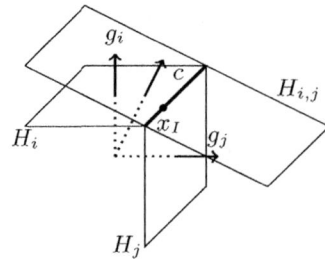

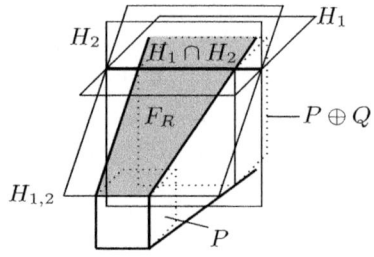

Fig. 5. A 3-dimensional example of the vector c **Fig. 6.** An example of $H_{i,j}$

Algorithm 1. Algorithm to compute R

Input: $P : \mathcal{L}_P$, $Q : \mathcal{L}_Q$
Output: An irredundant constraint set $\mathcal{L}_R$ of R
 1: Compute an irredundant constraint set $\mathcal{L}_{P\oplus Q}$ of $P \oplus Q$; $\mathcal{L}_R := \emptyset$;
 2: **for all** constraints $c^T x \le z$ in $\mathcal{L}_P \cup \mathcal{L}_{P\oplus Q}$ **do**
 3:     **if** $c^T x \le z$ is valid to R **then**
 4:         Add the constraint $c^T x \le z$ into $\mathcal{L}_R$;
 5:     **end if**
 6: **end for**
 7: **for all** constraints $g_i^T x = h_i$ and $g_j^T x = h_j$ in $\mathcal{L}_{P\oplus Q}$ **do**
 8:     Find a hyperplane $H_{i,j} : c^T x = z$ by solving Problem (2);
 9:     **if** $H_{i,j}$ exists **then**
10:         Add the constraint $c^T x \le z$ to $\mathcal{L}_R$;
11:     **end if**
12: **end for**
13: Remove the redundant constraints from $\mathcal{L}_R$;
14: **return** $\mathcal{L}_R$

We set $z = \rho_P(c)$. An example of $H_{i,j}$ is given in Figure 6.

We also need to find the valid constraints for R in $\mathcal{L}_P$ and $\mathcal{L}_{P\oplus Q}$. Given a constraint $L : c^T x \le z$, L is valid for R if and only if $\rho_R(c) \le z$. Since

$$\rho_R(c) = \rho_P(c) + \lambda \rho_Q(c) = \sup c^T x + \lambda \sup c^T y \quad s.t. \quad x \in P, y \in Q, \lambda \ge 0,$$

we compute $\rho_P(c)$ and $\rho_Q(c)$ by linear programming. If $\rho_Q(c) \le 0$ then $\rho_R(c) = \rho_P(c)$, otherwise $\rho_R(c) = \infty$.

If we have the constraints for $P \oplus Q$ then Problem (2) is linear. Algorithm 1 shows the computation of the irredundant constraints of R. Finally, the polytope $\mathcal{L}_{R_l(X)}$ can be defined by the set $\mathcal{L}_R \cup \mathcal{L}_{Inv(l)}$ where $\mathcal{L}_{Inv(l)}$ defines $Inv(l)$.

3.3   Compute the Reachable Set After a Jump

A jump $e = (l, l')$ of a rectangular automaton can update a variable by a value in an interval $[\underline{a}, \overline{a}]$. If the set of the reachable states in l is computed as $(l, R_l(X))$,

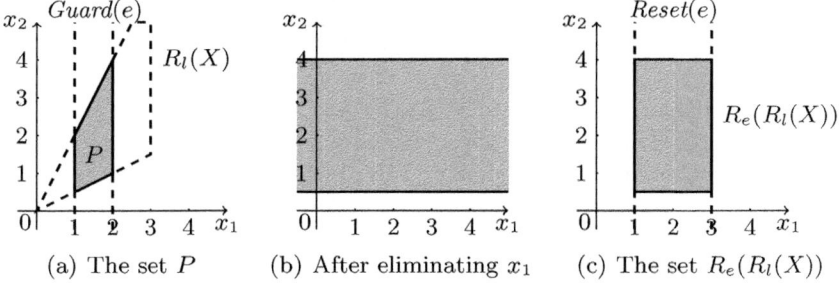

Fig. 7. A 2-dimensional example of resetting x_1 to $[1,3]$

then the set of states at which e is enabled can be computed by $(l, R_l(X) \cap Guard(e))$. Thus the reachable set after the jump e is $(l', R_e(R_l(X)))$ where

$$R_e(R_l(X)) = \{u' \in Inv(l') \cap Reset(e) \mid \exists u \in R_l(X) \cap Guard(e). \\ \forall x_i \in Var \backslash ResetVar(e).u'[i] = u[i]\} \tag{3}$$

The set $R_e(R_l(X))$ can also be computed in a geometric way. The guard can be considered by defining $R_l(X) \cap Guard(e) : \mathcal{L}_{R_l(X)} \cup \mathcal{L}_{Guard(e)}$. The polytope $R_e(R_l(X))$ can be defined by $\mathcal{L}_e \cup \mathcal{L}_{Reset(e)}$ where $\mathcal{L}_e$ is the set of the constraints computed from $\mathcal{L}_{R_l(X)} \cup \mathcal{L}_{Guard(e)}$ by eliminating all reset variables by Fourier-Motzkin elimination [12], and $\mathcal{L}_{Reset(e)}$ defines the box $Reset(e)$.

Example 4. We show an example in Figure 7, where $R_l(X) \cap Guard(e)$ is given by the polytope $P : -2x_1 + x_2 \leq 0 \wedge x_1 - 2x_2 \leq 0 \wedge x_1 \leq 2 \wedge -x_1 \leq -1$. The reset box is $Reset(e) : x_1 \leq 3 \wedge -x_1 \leq -1$, and $Inv(l')$ is the box $[0,5] \times [0,5]$. Firstly, we compute the maximum and minimum value of the variable x_1, and we obtain $x_1 \leq 2$ and $-x_1 \leq -1$. By using the constraint $x_1 \leq 2$, we eliminate the variable x_1 from $-2x_1 + x_2 \leq 0$ and obtain a new constraint $x_2 \leq 4$. Similarly, we use $-x_1 \leq -1$ to eliminate the variable x_1 from $x_1 - 2x_2 \leq 0$ and get $-x_2 \leq -0.5$. At last, the set $R_e(R_l(X))$ is the polytope

$$R_e(R_l(X)) : x_2 \leq 4 \wedge -x_2 \leq -0.5 \wedge x_1 \leq 3 \wedge -x_1 \leq -1$$

Algorithm 2 shows the computation of the reachable set after a jump. Although the Fourier-Motzkin elimination is double-exponential in general, in the next section we show that it is efficient on the reachable sets.

3.4 Complexity of the Reachability Computation

The reachable set of a rectangular automaton $\mathcal{A}$ can be computed by Algorithm 3. Any reachable set $R_l(X)$ in Algorithm 3 is computed by a sequence

$$X_0 \rightarrow R_{l_0}(X_0) \rightarrow X_1 \rightarrow R_{l_1}(X_1) \rightarrow \cdots \rightarrow X_k \rightarrow R_{l_k}(X_k)$$

where $X_j = R_{e_j}(R_{l_{j-1}}(X_{j-1}))$ for $1 \leq j \leq k$, and $X_0 = Init(l_0)$. Although the termination of Algorithm 3 is not guaranteed, if we lay an upper bound $\bar{k}$ on

---

Algorithm 2. Algorithm to compute the constraints of $R_e(R_l(X))$

Input: The jump $e = (l, l')$, the constraints of $R_l(X) : \mathcal{L}$
Output: The constraints of $R_e(R_l(X))$
1: Compute the constraint set $\mathcal{L}_P$ of $P = R_l(X) \cap Guard(e)$; $S \leftarrow \mathcal{L}_P$;
2: **for all** $x_i \in ResetVar(e)$ **do**
3:    Eliminate x_i from the constraints in S by Fourier-Motzkin elimination;
4: **end for**
5: **return** $S \cup \mathcal{L}_{Reset(e)}$

Algorithm 3. Reachability computation for a rectangular automaton

---

Input: A rectangular hybrid automaton $\mathcal{A}$
Output: The reachable set of $\mathcal{A}$
1: $R_{\mathcal{A}} \leftarrow \{(l, Init(l)) \mid l \in Loc\}$;
2: Define a queue Q with elements $(l, X) \in R_{\mathcal{A}}$;
3: **while** Q is not empty **do**
4:    Get (l, X) from Q; $Y \leftarrow R_l(X)$; $R_{\mathcal{A}} \leftarrow R_{\mathcal{A}} \cup \{(l, Y)\}$;
5:    **for all** $e = (l, l') \in Jump$ **do**
6:       $Z \leftarrow R_e(Y)$;
7:       **if** $(l', Z) \notin R_{\mathcal{A}}$ **then**
8:          Insert (l', Z) into Q; $R_{\mathcal{A}} \leftarrow R_{\mathcal{A}} \cup \{(l', Z)\}$;
9:       **end if**
10:    **end for**
11: **end while**
12: **return** $R_{\mathcal{A}}$

---

k then it always stops. We prove that if k is viewed as a constant, then the computation is polynomial in the number of the variables of $\mathcal{A}$.

We prove it by showing that an irredundant constraint set of X_j can be computed from an irredundant constraint set of X_{j-1} in PTIME. Notice that this property is not possessed by any of the methods proposed in the past.

Lemma 5. *For $1 \leq j \leq k$, both $NF(X_j)$ and $NF(X_{j-1} \oplus B_{j-1})$ are polynomial in $NF(X_{j-1})$.*

Proof. By Lemma 1, the size of the irredundant constraint set of X_j is proportional to $NF(X_j)$, then we consider the facets of X_j. We define $G_j = Inv(l_j) \cap Guard(e_{j+1})$ and $B_j = Flow(l_j)$. If the whole space is $\mathbb{R}^d$, in order to maximize the number of X_j's facets, we assume B_i, G_i for $0 \leq i \leq j-1$ are full-dimensional boxes, and X_j is also full-dimensional. Since X_j can be expressed by

$$\bigcup_{a_{j-1} \leq \lambda_{j-1} \leq b_{j-1}} \cdots \bigcup_{a_0 \leq \lambda_0 \leq b_0} R_{e_j}((\cdots R_{e_1}((X_0 \oplus \lambda_0 B_0) \cap G_0) \cdots \oplus \lambda_{j-1} B_{j-1}) \cap G_{j-1})$$

a facet F_{X_j} of it can be uniquely expressed by

$$\bigcup_{a'_{j-1} \leq \lambda_{j-1} \leq b'_{j-1}} \cdots \bigcup_{a'_0 \leq \lambda_0 \leq b'_0} F(\lambda_0, \ldots, \lambda_{j-1}) \qquad (4)$$

where $a_i \leq a_i'$ and $b_i' \leq b_i$ for $0 \leq i \leq j-1$, such that

(i) $F(\lambda_0, \ldots, \lambda_{j-1})$ is a face of the box $\Phi(\lambda_0, \ldots, \lambda_{j-1}) = R_{e_j}((\cdots R_{e_1}((X_0 \oplus \lambda_0 B_0) \cap G_0) \cdots \oplus \lambda_{j-1} B_{j-1}) \cap G_{j-1})$ and there is no higher dimensional face of $\Phi(\lambda_0, \ldots, \lambda_{j-1})$ can be used to express F_{X_j};
(ii) if the maximum dimension of all those faces $F(\lambda_0, \ldots, \lambda_{j-1})$ is d' where $d - d' - 1 \leq j$, then there are exactly $(d - d' - 1)$ many λ_i where $0 \leq i \leq j-1$ such that these parameters help to determine $\mathcal{N}(F_{X_j}, X_j)$;
(iii) for any $0 \leq i \leq j-1$, if λ_i helps to determine $\mathcal{N}(F_{X_j}, X_j)$, then the box G_i could also help to determine $\mathcal{N}(F_{X_j}, X_j)$;
(iv) for any $0 \leq i \leq j-1$, any γ_i, γ_i' where

$$\begin{cases} a_i' < \gamma_i, \gamma_i' < b_i', & \text{if } a_i' < b_i' \\ \gamma_i = \gamma_i' = a_i', & \text{otherwise} \end{cases}$$

we have that $F(\gamma_0, \ldots, \gamma_{j-1}), F(\gamma_0', \ldots, \gamma_{j-1}')$ have the maximum dimension among all the faces $F(\lambda_0, \ldots, \lambda_{j-1})$, and $\mathcal{N}(F(\gamma_0, \ldots, \gamma_{j-1}), \Phi(\gamma_0, \ldots, \gamma_{j-1}))$ $= \mathcal{N}(F(\gamma_0', \ldots, \gamma_{j-1}'), \Phi(\gamma_0', \ldots, \gamma_{j-1}'))$.

In brief, the above properties tell that $\mathcal{N}(F_{X_j}, X_j)$ depends on (a) the set $\mathcal{N}(F(\gamma_0, \ldots, \gamma_{j-1}), \Phi(\gamma_0, \ldots, \gamma_{j-1}))$ in the property (iv), i.e., the outer normals of a bounded box face (we call those faces related), (b) the $(d - d' - 1)$ parameters in the property (ii), and (c) the dependence of $\mathcal{N}(F_{X_j}, X_j)$ and G_i for every $0 \leq i \leq j-1$ such that λ_i helps to determine $\mathcal{N}(F_{X_j}, X_j)$. Thereby if F_B is a d'-face of a bounded box, it has at most $2^{d-d'-1}\binom{j}{d-d'-1}$ related facets in X_j.

Given a dimension d' where $d - d' - 1 \leq j$, as we said, if a d'-face F_B of a bounded box B is related to some facet F_{X_j} then there are exactly $(d - d' - 1)$ many λ_i's help to determine the outer normals of F_{X_j}. Thus there are $(d - d' - 1)$ steps to determine $\mathcal{N}(F_{X_j}, X_j)$. We define $\mathcal{P}_i$ as the set of the d'-faces in B which possibly have related facets in X_j after the ith step. Obviously, $\mathcal{P}_0$ contains all the d'-faces in B. In every $(i+1)$th step, at least half of the faces in $\mathcal{P}_i$ lose the possibility to have related facets in X_j since X_i is a union of boxes and every box is centrally symmetric. Hence, there are at most

$$2^{-(d-d'-1)} \mathcal{F}_{d'}^d = 2^{-(d-d'-1)} \left(2^{d-d'} \binom{d}{d'} \right) = 2 \binom{d}{d'}$$

d'-faces of B could have related facets in X_j, where $\mathcal{F}_{d'}^d$ is the number of the d'-faces in B. Therefore, there are at most $2^{d-d'} \binom{j}{d-d'-1} \binom{d}{d'}$ facets in X_j which are related to some d'-faces of B. By considering all $\max(d - j - 1, 0) \leq d' \leq d-1$, we can conclude that $NF(X_j)$ is polynomial in $NF(X_{j-1})$ for $j \geq 1$. Similarly, we can also prove that $NF(X_{j-1} \oplus B_{j-1})$ is polynomial in $NF(X_{j-1})$ for $j \geq 1$. □

Now we give our method to compute X_j from X_{j-1}. The most expensive part in the computation is computing $X_{j-1} \oplus B_{j-1}$. We decompose B_{j-1} by $B_{j-1} = [\underline{a}_1, \overline{a}_1]_1 \oplus [\underline{a}_2, \overline{a}_2]_2 \oplus \cdots \oplus [\underline{a}_d, \overline{a}_d]_d$, such that for $1 \leq i \leq d$, $x[i] \leq \overline{a}_i$, $-x[i] \leq -\underline{a}_i$

are irredundant constraints for B_{j-1} and $[\underline{a}_i, \overline{a}_i]_i$ is a line segment (1-dimensional box) defined by the following constraint set:

$$\{x[i] \leq \overline{a}_i, -x[i] \leq -\underline{a}_i\} \cup \{x[i'] \leq 0 \mid i' \neq i\} \cup \{-x[i'] \leq 0 \mid i' \neq i\}$$

We denote the polytope resulting from adding the first m line segments onto X_{j-1} by X_{j-1}^m, then for all $1 \leq m \leq d$, $NF(X_{j-1}^m)$ is polynomial in $NF(X_{j-1})$. Since an irredundant constraint set for X_{j-1}^m can be computed in PTIME based on an irredundant constraint set of X_{j-1}^{m-1}, we conclude that an irredundant constraint set which defines $X_{j-1} \oplus B_{j-1}$ can be computed in a time polynomial in d if j is viewed as a constant.

Next we consider the complexity of the Fourier-Motzkin elimination on the set $R_{l_j}(X_j)$. Since $NF(X_{j+1})$ is polynomial in $NF(X_j)$, the polyhedron resulting from the elimination of each reset variable has a number of facets which is polynomial in $NF(X_j)$. Since eliminating one variable is PTIME, we conclude that the Fourier-Motzkin elimination on $R_{l_j}(X_j)$ is polynomial in d if j is viewed as a constant. If we use *interior point methods* [14] to solve linear programs then the bounded reachability computation is polynomial in d.

Theorem 2. *The computational complexity of $R_{l_j}(X_j)$ is polynomial in d if j is viewed as a constant.*

Theorem 3. *The computational complexity of the reachable set with a bounded number of jumps is polynomial in d if the bound is viewed as a constant.*

Unfortunately, the worst-case complexity is exponential in j. However, it only happens in extreme cases. The exact complexity of our approach mainly depends on the complexity of solving linear programs.

4   Experimental Results

We implemented our method in MATLAB using the CDD tool [18] for linear programming. We compared our implementation with PHAVer (embedded in SpaceEx [19]) on a scalable artificial example. Since there are rare high dimensional examples published, we design a scalable example which is given in Figure 8, where d is the (user-defined) dimension of the automaton and i denotes all the integers from 1 to d. The automaton $\mathcal{A}_d$ helps to generate reachable sets with large numbers of vertices and facets, and for each jump, nearly half of the variables are reset.

The experiments were run on a computer with a 2.8 GHz CPU and 4GB memory, the operating system is Linux. The experimental results are given by Table 1. Since MATLAB does not provide a build-in function to monitor the memory usage of its programs on Linux, the listed memory usage is the total memory usage minus MATLAB memory usage before the experiment. Our method can handle $\mathcal{A}_{10}$ efficiently, however PHAVer stops at $\mathcal{A}_7$. Our implementation is a prototype and the running times can even be improved by a C++ implementation and a faster LP solver.

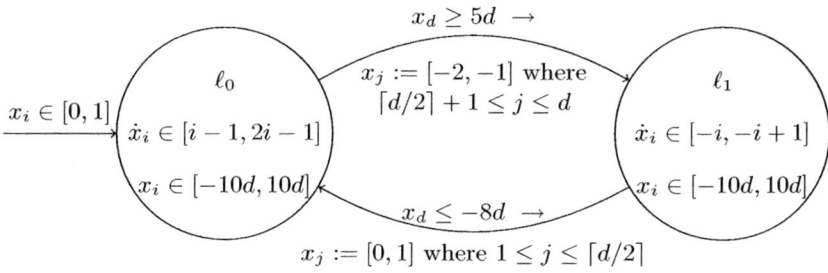

Fig. 8. Rectangular automaton $\mathcal{A}_d$

Table 1. Experimental results. Time is in seconds, memory in MBs. "MaxJmp" is the bound on the number of jumps, "ToLP" is the total linear programming time, "LPs" is the number of linear programs solved (including the detection of redundant constraints), "Constraints" is the number of irredundant constraints computed, "n.a." means not available, "t.o." means that the running time was greater than one hour.

Dimension	MaxJmp	PHAVer		Our method				
		Memory	Time	Memory	Time	ToLP	LPs	Constraints
5	2	9.9	0.81	< 10	2.36	2.20	1837	81
6	2	48.1	21.69	< 10	4.96	4.68	3127	112
7	2	235.7	529.01	< 10	15.95	15.28	7214	163
8	2	n.a.	t.o.	< 10	27.42	26.48	10517	209
9	2	n.a.	t.o.	< 10	107.99	105.59	23639	287
10	2	n.a.	t.o.	< 10	218.66	215.45	32252	354
5	4	10.2	1.51	< 10	4.82	4.50	3734	167
6	4	51.1	35.52	< 10	11.25	10.64	7307	240
7	4	248.1	1191.64	< 10	32.93	31.60	16101	352
8	4	n.a.	t.o.	< 10	72.04	69.81	27375	466
9	4	n.a.	t.o.	< 10	240.51	235.61	64863	641
10	4	n.a.	t.o.	< 10	543.05	535.77	86633	816

5   Conclusion

We introduced our efficient approach for the bounded reachability computation of rectangular automata. However, the method of computing the reachable set under a flow transition can also be applied to linear hybrid automata. With some more effort this approach can also be adapted for the approximative analysis of hybrid systems with nonlinear behavior.

References

1. Henzinger, T.A., Kopke, P.W., Puri, A., Varaiya, P.: What's decidable about hybrid automata? J. Comput. Syst. Sci. 57(1), 94–124 (1998)
2. Henzinger, T.A., Ho, P., Wong-Toi, H.: Algorithmic analysis of nonlinear hybrid systems. IEEE Transactions on Automatic Control 43(4), 540–554 (1998)

3. Preußig, J., Kowalewski, S., Wong-Toi, H., Henzinger, T.A.: An algorithm for the approximative analysis of rectangular automata. In: Ravn, A.P., Rischel, H. (eds.) FTRTFT 1998. LNCS, vol. 1486, pp. 228–240. Springer, Heidelberg (1998)
4. Wong-Toi, H., Preußig, J.: A procedure for reachability analysis of rectangular automata. In: Proc. of American Control Conference, vol. 3, pp. 1674–1678 (2000)
5. Doyen, L., Henzinger, T.A., Raskin, J.: Automatic rectangular refinement of affine hybrid systems. In: Pettersson, P., Yi, W. (eds.) FORMATS 2005. LNCS, vol. 3829, pp. 144–161. Springer, Heidelberg (2005)
6. Alur, R., Dill, D.L.: A theory of timed automata. Theor. Comput. Sci. 126(2), 183–235 (1994)
7. Chutinan, A., Krogh, B.H.: Computing polyhedral approximations to flow pipes for dynamic systems. In: Proc. of CDC 1998. IEEE Press, Los Alamitos (1998)
8. Stursberg, O., Krogh, B.H.: Efficient representation and computation of reachable sets for hybrid systems. In: Maler, O., Pnueli, A. (eds.) HSCC 2003. LNCS, vol. 2623, pp. 482–497. Springer, Heidelberg (2003)
9. Girard, A.: Reachability of uncertain linear systems using zonotopes. In: Morari, M., Thiele, L. (eds.) HSCC 2005. LNCS, vol. 3414, pp. 291–305. Springer, Heidelberg (2005)
10. Henzinger, T.A., Ho, P.-H., Wong-Toi, H.: Hytech: A model checker for hybrid systems. Software Tools for Technology Transfer (1), 110–122 (1997)
11. Frehse, G.: Phaver: Algorithmic verification of hybrid systems past hytech. In: Morari, M., Thiele, L. (eds.) HSCC 2005. LNCS, vol. 3414, pp. 258–273. Springer, Heidelberg (2005)
12. Ziegler, G.M.: Lectures on Polytopes. Graduate Texts in Mathematics, vol. 152. Springer, Heidelberg (1995)
13. Chen, X., Abraham, E., Frehse, G.: Efficient bounded reachability computation for rectangular automata. Technical report, RWTH Aachen University (2011), http://www-i2.informatik.rwth-aachen.de/i2/hybrid_research_pub0/
14. Boyd, S., Vandenberghe, L.: Convex Optimization. Cambridge University Press, Cambridge (2004)
15. Fukuda, K.: From the zonotope construction to the minkowski addition of convex polytopes. J. Symb. Comput. 38(4), 1261–1272 (2004)
16. Weibel, C., Fukuda, K.: Computing faces up to k dimensions of a minkowski sum of polytopes. In: Proc. of CCCG 2005, pp. 256–259 (2005)
17. Henzinger, T.A.: The theory of hybrid automata. In: Proc. of LICS 1996, pp. 278–292 (1996)
18. Fukuda, K.: cdd, cddplus and cddlib homepage, http://www.ifor.math.ethz.ch/~fukuda/cdd_home/
19. Frehse, G., Le Guernic, C., Donzé, A., Ray, R., Lebeltel, O., Ripado, R., Girard, A., Dang, T., Maler, O.: Spaceex: Scalable verification of hybrid systems. In: Gopalakrishnan, G., Qadeer, S. (eds.) CAV 2011. LNCS, vol. 6806, pp. 379–395. Springer, Heidelberg (2011)

Reachability and Deadlocking Problems in Multi-stage Scheduling

Christian E.J. Eggermont and Gerhard J. Woeginger

Department of Mathematics and Computer Science, TU Eindhoven, Netherlands

Abstract. We study reachability and deadlock detection questions in multi-stage scheduling systems. The jobs have partially ordered processing plans that dictate the order in which the job passes through the machines. Our results draw a sharp borderline between tractable and intractable cases of these questions: certain types of processing plans (that we call unconstrained and source-constrained) lead to algorithmically tractable problems, whereas all remaining processing plans lead to NP-hard problems.

We give conditions under which safe system states can be recognized in polynomial time, and we prove that without these conditions the recognition of safe system states is NP-hard. We show that deciding reachability of a given state is essentially equivalent to deciding safety. Finally, we establish NP-hardness of deciding whether the system can ever fall into a deadlock state.

Keywords: Scheduling, resource allocation, deadlock, computational complexity.

1   Introduction

A robotic cell consists of three machines that install colored nibbles onto dingbats; the first machine installs yellow nibbles, the second machine blue nibbles, and the third machine installs red nibbles. Currently two dingbats are moving through the cell and want to visit all three machines: dingbat A needs the blue nibble mounted before the red nibble, and dingbat B needs the blue nibble mounted before the yellow nibble. Dingbat A receives its yellow nibble on the first machine, and then its blue nibble on the second machine. In the meaintime dingbat B receives its red nibble on the red machine. Now a catastrophe has happened!! The robotic cell is stuck! Dingbat A blocks the second machine, and wants to move to the third machine. The third machine is blocked by dingbat B, which wants to move to the second machine. One cannot help but wonder: would there have been a better way of handling these dingbats?

More generally, we consider real-time multi-stage scheduling systems with m machines $M_1, \ldots, M_m$ and n jobs $J_1, \ldots, J_n$. Each machine M_i has a corresponding *capacity* $\mathrm{cap}(M_i)$, which means that at any moment in time it can simultaneously hold and process up to $\mathrm{cap}(M_i)$ jobs. Each job J_j has a processing plan: it requests processing on a certain subset $\mathcal{M}(J_j)$ of the machines,

G. Delzanno and I. Potapov (Eds.): RP 2011, LNCS 6945, pp. 153–164, 2011.

and the way in which the job passes through these machines is constrained by a partial order $\prec_j$. Whenever $M_a \prec_j M_b$ holds for two machines in $\mathcal{M}(J_j)$, then job J_j must complete its processing on machine M_a before it can visit M_b.

The various jobs move through the system in an unsynchronized fashion and hop from machine to machine. In the beginning a job is asleep and waiting outside the system. For technical reasons, we assume that the job occupies an artificial machine M_0 of unbounded capacity. After some time the job wakes up and starts looking for an available machine on which it can be processed next in agreement with its processing plan. If no such machine is available, then the job falls asleep again. If the job finds such a machine and if this machine has free capacity, then the job moves there, receives its processing, and falls asleep for some time; then it wakes up again, and the game repeats. As soon as the processing of the job on all relevant machines is completed, the job leaves the system; we assume that the job then moves to an artificial final machine M_{m+1} (with unbounded capacity), and disappears.

The following example puts the dingbat and nibble robotic cell (as described in the first paragraph of this paper) into our formal framework.

Example 1. A scheduling system has three machines M_1, M_2, M_3 of capacity 1. There are two jobs J_1 and J_2 that both require processing on all three machines, with $M_2 \prec_1 M_3$ and $M_2 \prec_2 M_1$ and no further constraints imposed by their processing plans.

Suppose the following happens. The first job moves to machine M_1. The first job moves to machine M_2. The second job moves to machine M_3. Once the two jobs have completed their processing on these machines, they keep blocking their machines and simultaneously keep waiting for the other machine to become idle. The processing never terminates, and the system is in deadlock.

The described framework is closely related to shop models from classical scheduling theory. In an open shop system, all partial orders $\prec_j$ are empty. In a flow shop system, all partial orders $\prec_j$ are total orders and all machine sets $\mathcal{M}(J_j)$ contain all available machines. In a job shop system, all partial orders $\prec_j$ are total orders and the machine sets $\mathcal{M}(J_j)$ depend on the job. For more information on multi-stage scheduling systems, the reader is referred to the survey [4].

Summary of Considered Problems and Derived Results

We discuss the computational complexity of several fundamental problems in multi-stage scheduling systems that are centered around the reachability of certain desired and certain undesired system states. We are interested in the borderline between easy and hard special cases. A particularly simple special case arises, if all jobs have *isomorphic processing plans*; this means that all partial orders $\prec_j$ are isomorphic to each other. One example for such a system are the flow shops, and another example is described in Example 1. Most of our complexity results hold for systems with such isomorphic processing plans, and we precisely separate the processing plans with good behavior (polynomially solvable problem variants) from the processing plans with bad behavior (NP-hard problem variants). Section 2 classifies some typical processing plans that arise in our theorems.

The most fundamental question is to distinguish the unsafe system states (for which an eventual deadlock is unavoidable) from the safe ones (for which deadlocks can permanently be avoided):

PROBLEM: SAFE STATE RECOGNITION

INSTANCE: A scheduling system. A system state s.

QUESTION: Is state s safe?

Lawley & Reveliotis [5] established NP-hardness of SAFE STATE RECOGNITION, even for the special job shop case where every processing plan is totally ordered. The construction in [5] crucially builds on jobs with very long processing plans.

We prove a substantially stronger result: SAFE STATE RECOGNITION is NP-hard, even in the extremely primitive special case every processing plan is a total order of **only three** machines. More generally, in Section 5 we will establish NP-hardness for all systems with isomorphic processing plans with only two exceptions. The first exception concerns processing plans where the partial order is empty. The second exception concerns processing plans where the partial order is *source-constrained*. This means that the partial order points out a machine to which the job must go first (and no further constraints); see Figure 1(c) for an illustration. Section 4 shows that for these two exceptional cases SAFE STATE RECOGNITION in fact is polynomially solvable.

Another fundamental question is to characterize those system states that actually can be reached while the system is running:

PROBLEM: REACHABLE STATE RECOGNITION

INSTANCE: A scheduling system. A system state s.

QUESTION: Can the system reach state s when starting from the initial situation where all machines are empty?

In Section 6 we show that deciding reachability essentially is equivalent to deciding safety in a closely related system. The (simple) idea is to reverse the time axis, and to make the system run backward.

A third class of problems is centered around the question whether a scheduling system can ever fall into a deadlock state. In case it can not, there are no reachable unsafe states: the system is fool-proof and will run smoothly without supervision.

PROBLEM: REACHABLE DEADLOCK

INSTANCE: A scheduling system.

QUESTION: Can the system ever reach a deadlock state when starting from the initial situation?

In Section 7 we will prove NP-hardness of REACHABLE DEADLOCK in several highly restricted situations. Our main result states that the problem is NP-hard for all systems with isomorphic processing plans with only two exceptions

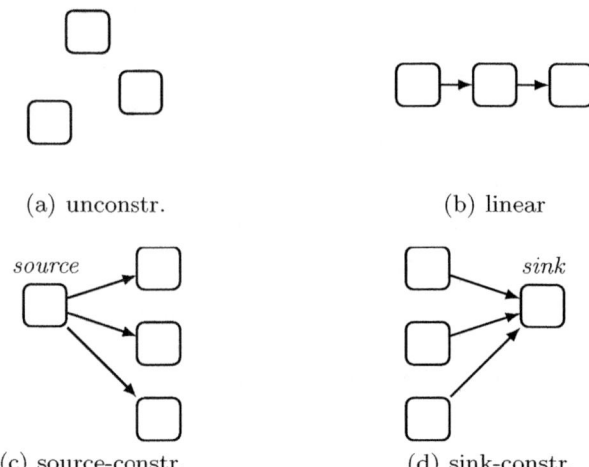

(a) unconstr.   (b) linear

(c) source-constr.   (d) sink-constr.

Fig. 1. Some specially structured job processing plans

(which are closely related to the two exceptions that we got for SAFE STATE RECOGNITION). The first exception concerns processing plans where the partial order is empty and uses at most three machines. The second exception concerns processing plans where the partial order is source-constrained and uses at most four machines. The computational complexity of these two exceptional cases remains unclear.

Due to lack of space, many proofs and details are deferred to the journal version of this paper.

2   A Taxonomy of Job Processing Plans

Consider a job J_j that requests processing on some subset $\mathcal{M}(J_j)$ of machines. The way in which job J_j passes through the machines in $\mathcal{M}(J_j)$ is governed by a strict order $\prec_j$ on the set $\mathcal{M}(J_j)$. Whenever $M_a \prec_j M_b$ holds for two machines $M_a, M_b \in \mathcal{M}(J_j)$ in this strict order, job J_j must complete its processing on machine M_a before it can visit machine M_b. The machine set $\mathcal{M}(J_j)$ together with the strict order $\prec_j$ is called the *processing plan* of job J_j. There is a number of specially structured processing plans that play a crucial role in our investigations:

Unconstrained. If the strict order $\prec_j$ is empty, then the ordering in which job J_j passes through the machines in $\mathcal{M}(J_j)$ is immaterial and can be chosen arbitrarily. We call such a processing plan *unconstrained*. Note that in classical multi-stage scheduling unconstrained processing plans occur in so-called open shops.

Linear. If the strict order $\prec_j$ is a total order on $\mathcal{M}(J_j)$, then we say that the processing plan is *linear*. In classical multi-stage scheduling linear processing plans occur in so-called job shops.

Source-constrained. Assume that there exists a machine M^{source} in $\mathcal{M}(J_j)$, such that $M^{source} \prec_j M$ holds for all machines $M \in \mathcal{M}(J_j) - \{M^{source}\}$ and such that the strict order imposes no further constraints. Such a processing plan is called *source-constrained*, and M^{source} is called the *source* of the processing plan.

Sink-constrained. In a symmetric fashion to source-constrained processing plans, we define *sink-constrained* processing plans: there exists a machine M^{sink} in $\mathcal{M}(J_j)$, such that $M \prec_j M^{sink}$ for all machines $M \in \mathcal{M}(J_j) - \{M^{sink}\}$ and there are no further constraints. The machine M^{sink} is called the *sink* of the processing plan.

Definition 1. *A processing plan has an $\alpha\beta\gamma\delta$-decomposition if there are two machines $D^\alpha \neq D^\beta$ and if there is a partition of the remaining machines in $\mathcal{M}(J_j) - \{D^\alpha, D^\beta\}$ into two sets $\mathcal{D}^\gamma$ and $\mathcal{D}^\delta$ with the following properties:*

(i) $D^\beta \not\prec_j D^\alpha$
(ii) $\mathcal{D}^\gamma := succ(D^\beta)$ forms a non-empty antichain
(iii) $pred(\mathcal{D}^\delta) \subseteq \mathcal{D}^\delta$

Lemma 1. *(Decomposition lemma). If a processing plan is neither unconstrained nor source-constrained, then it has an $\alpha\beta\gamma\delta$-decomposition.*

Proof. If $\mathcal{M}(J_j)$ and $\prec_j$ form a processing plan that is neither unconstrained nor source-constrained, then there exists an element that has at least one successor and at least one non-successor. Among all such elements, pick element D^β such that $succ(D^\beta)$ has minimum cardinality. Choose $D^\alpha \neq D^\beta$ as a sink on the ordered set induced by the non-successors of D^β, let $\mathcal{D}^\gamma := succ(D^\beta)$, and let $\mathcal{D}^\delta$ contain all remaining elements (which are non-successors of D^β).

Properties (i) and (iii) hold by construction. Furthermore $\mathcal{D}^\gamma$ is non-empty, and the existence of two elements M, M' in $\mathcal{D}^\gamma$ with $M \prec_j M'$ would contradict the choice of D^β (as M then would have fewer successors than D^β). Hence also (ii) is satisfied. $\qquad\square$

The simplest example of a processing plan that is neither unconstrained nor source-constrained consists of three machines M_a, M_b, M_c with constraints $M_a \prec_j M_b \prec_j M_c$. The corresponding $\alpha\beta\gamma\delta$-decomposition is unique and given by $D^\alpha = M_a$, $D^\beta = M_b$, $\mathcal{D}^\gamma = \{M_c\}$, and $\mathcal{D}^\delta = \emptyset$. The following example illustrates that $\alpha\beta\gamma\delta$-decompositions need not be unique.

Example 2. Consider a job J with $\mathcal{M}(J) = \{M_1, M_2, M_3, M_4\}$ and $M_1 \prec M_3$, $M_1 \prec M_4$, $M_2 \prec M_4$. There are two $\alpha\beta\gamma\delta$-decompositions:

- $D^\alpha = M_3$, $D^\beta = M_2$, $\mathcal{D}^\gamma = \{M_4\}$ and $\mathcal{D}^\delta = \{M_1\}$,
- $D^\alpha = M_2$, $D^\beta = M_1$, $\mathcal{D}^\gamma = \{M_3, M_4\}$ and $\mathcal{D}^\delta = \emptyset$.

These decompositions are illustrated in Figure 2.

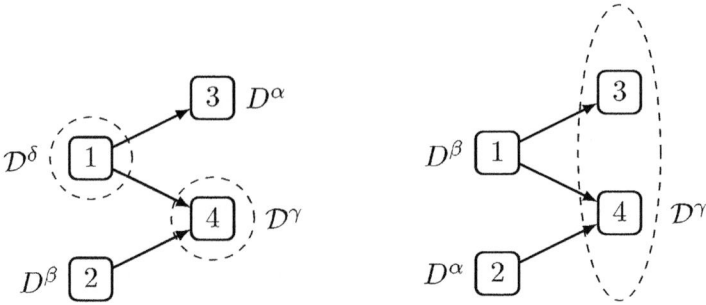

Fig. 2. Two $\alpha\beta\gamma\delta$-decompositions of the partial order of Example 2

Definition 2. *A processing plan is a* pitchfork, *if it possesses an $\alpha\beta\gamma\delta$-decomposition with $\mathcal{D}^\delta = \emptyset$. The cardinality of $\mathcal{D}^\gamma$ in such a decomposition is called its* prong-number.

The $\alpha\beta\gamma\delta$-decomposition in the righhand side of Figure 2 demonstrates that job J in Example 2 has a pitchfork processing plan. Also Example 1 is a scheduling system in which all processing plans are pitchforks. Some further typical examples of pitchforks are given in Figure 3.

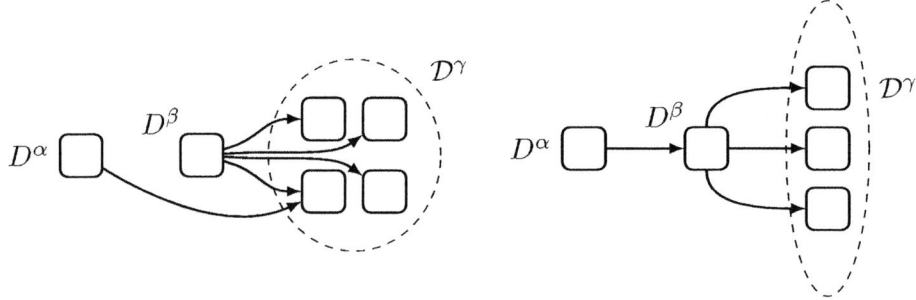

Fig. 3. Two typical pitchforks with $\alpha\beta\gamma\delta$-decomposition ($\mathcal{D}^\delta = \emptyset$)

3   System States: Safe, Unsafe, and Deadlocks

A *state* of a scheduling system is a snapshot of a situation that might potentially occur while the system is running. A state s specifies for every job J_j

- the machine $M^s(J_j)$ on which this job is currently waiting or currently being processed,
- the set $\mathcal{M}^s(J_j) \subseteq \mathcal{M}(J_j) - \{M^s(J_j)\}$ of machines on which the job still needs future processing.

The machines $M^s(J_j)$ implicitly determine

- the set $\mathcal{J}^s(M_i) \subseteq \{J_1, \ldots, J_n\}$ of jobs currently handled by machine M_i.

The *initial state* 0 is the state where all jobs are still waiting for their first processing; in other words in the initial state all jobs J_j satisfy $M^0(J_j) = M_0$ and $\mathcal{M}^0(J_j) = \mathcal{M}(J_j)$. The *final state* f is the state where all jobs have been completed; in other words in the final state all jobs J_j satisfy $M^f(J_j) = M_{m+1}$ and $\mathcal{M}^f(J_j) = \emptyset$.

A state t is called a *successor* of a state s, if it results from s by moving a single job J_j from its current machine $M^s(J_j)$ to some machine M that is a source in set $\mathcal{M}^s(J_j)$, or by moving a job J_j with $\mathcal{M}^s(J_j) = \emptyset$ from its current machine to M_{m+1}. In this case we also say that the entire system *moves* from s to t. This successor relation is denoted $s \to t$. A state t is said to be *reachable* from state s, if there exists a finite sequence $s = s_0, s_1, \ldots, s_k = t$ of states (with $k \geq 0$) such that $s_{i-1} \to s_i$ holds for $i = 1, \ldots, k$. A state s is called *reachable*, if it is reachable from the initial state 0. A state is called *safe*, if the final state f is reachable from it; otherwise the state is called *unsafe*. A state is a *deadlock*, if it has no successor states and if it is not the final state f. A state is called *super-safe*, if no deadlock states are reachable from it.

Example 3 (continuation of Example 1). We show several potential successor states of state 0 illustrating the terms mentioned. Unless otherwise noted in the description of a state each job still needs to visit all machines other than its current location:

- state u: $M^u(J_1) = M_1$ and $M^u(J_2) = M_3$,
- state s: $M^s(J_1) = M_2$ and $M^s(J_2) = M_3$,
- state d: $M^d(J_1) = M_2$, $\mathcal{M}^d(J_1) = \{M_3\}$ and $M^d(J_2) = M_3$,
- state p: $M^p(J_1) = M_2$, $\mathcal{M}^p(J_1) = \{M_3\}$ and $M^p(J_2) = M_1$, $\mathcal{M}^p(J_2) = \{M_3\}$,

Out of these four states, only p is not reachable (explained in Example 4), and only states s and p are safe. Furthermore, state u is unsafe but not deadlock. Finally note that state s is super-safe, whereas the initial state 0 is not super-safe.

4 Deciding Safety: Easy Cases

This section discusses special cases for which the safety of states can be decided in polynomial time. The following theorem unifies and extends several results from the literature; see Sulistyono & Lawley [6, Th.3, p.825] and Lawley & Reveliotis [5, p.398] (which also contains further references). The proof method is standard for the area, but the theorem in its full generality appears to be new.

Theorem 1. SAFE STATE RECOGNITION *can be decided in polynomial time, if machines of unit capacity only occur in unconstrained and source-constrained processing plans.*

If a job requires processing on only one or two machines, then its processing plan is either unconstrained or source-constrained. This yields the following trivial corollary to Theorem 1.

Corollary 1. SAFE STATE RECOGNITION *can be decided in polynomial time, if every job requires processing on at most two machines.*

The rest of this section presents the proof of Theorem 1. We will show that in the setting of this theorem, the existence of an unsafe state is equivalent to the existence of a so-called blocking set (Lemma 4). Since the existence of a blocking set is decidable in polynomial time (Lemma 2), the theorem follows.

A machine M is called *full* in state s, if it is handling exactly $\mathrm{cap}(M)$ many jobs. For a subset $\mathcal{J}$ of jobs we let $\mathrm{Next}^s(\mathcal{J})$ denote the set of machines M for which there is a job $J_j \in \mathcal{J}$ with $M \in \mathcal{M}^s(J_j)$ and for which there is no $N \in \mathcal{M}^s(J_j)$ with $N \prec_j M$; in other words, such a machine M is a potential candidate for being requested by job J_j in its next processing step. A non-empty subset $\mathcal{B}$ of the machines is called *blocking* for state s,

- if every machine in $\mathcal{B}$ is full, and
- if every job J_j that is on some machine in $\mathcal{B}$ satisfies $\emptyset \neq \mathrm{Next}^s(J_j) \subseteq \mathcal{B}$.

The machines in a blocking set $\mathcal{B}$ all operate at full capacity on jobs that in the immediate future only want to move to other machines in $\mathcal{B}$; in other words $\mathrm{Next}^s(\mathcal{J}^s(\mathcal{B})) \subseteq \mathcal{B}$. Since these jobs are permanently blocked from moving, the state s must eventually lead to a deadlock and hence is unsafe.

Lemma 2. *For a given state s, it can be decided in polynomial time whether s has a blocking set of machines.*

Proof. Create an auxiliary directed graph that corresponds to state s: the vertices are the machines $M_1, \ldots, M_m$. Whenever some job J_j occupies a machine M_i, the directed graph contains an arc from M_i to every machine in $\mathrm{Next}^s(J_j)$. Obviously state s has a blocking set of machines if and only if the auxiliary directed graph contains a strongly connected component with the following two properties: (i) All vertices in the component are full. (ii) There are no arcs leaving the component. Since the strongly connected components of a directed graph can be determined and analyzed in linear time, the statement follows. □

Here is a simple procedure that determines whether a given machine M_i is part of a blocking set in state s: Let $\mathcal{B}_0 = \{M_i\}$. For $k \geq 1$ let $\mathcal{B}_k = \mathrm{Next}^s(\mathcal{J}^s(\mathcal{B}_{k-1})) \cup \mathcal{B}_{k-1}$. Clearly $\mathcal{B}_0 \subseteq \mathcal{B}_1 \subseteq \cdots \subseteq \mathcal{B}_{m-1} = \mathcal{B}_m$. Furthermore machine M_i belongs to a blocking set, if and only if $\mathcal{B}_m$ is a blocking set, if and only if all machines in $\mathcal{B}_m$ are full. In case $\mathcal{B}_m$ is a blocking set, we denote it by $\mathcal{B}_{\min}^s(M_i)$ and call it the *canonical* blocking set for machine M_i in state s. The same procedure can be used starting from any subset $\mathcal{B}_0$ of machines, and the resulting set of machines $\mathcal{B}_m$ will be denoted by $\mathcal{R}^s(\mathcal{B}_0)$. In particular $\mathcal{B}_{\min}^s(M_i) = \mathcal{R}^s(\{M_i\})$. The canonical blocking set is the smallest blocking set containing M_i:

Lemma 3. *If M_i belongs to a blocking set $\mathcal{B}$ in state s, then $\mathcal{B}_{\min}^s(M_i) \subseteq \mathcal{B}$.*

In the case of open shop it is known (Eggermont, Schrijver & Woeginger [2]) that *every* unsafe state is caused by blocking sets. The following shows a generalization.

Lemma 4. *Consider a scheduling system where machines of unit capacity only occur in unconstrained and source-constrained processing plans. Then a state s is unsafe if and only if state s contains a blocking set of machines.*

Proof. Without restriction of generality, we can assume all jobs have entered the system and require further processing. For deadlock states the statement is obvious. Clearly a state containing a blocking set is unsafe. For the other direction, assume there is an unsafe state s without a blocking set where any possible move will result in a state containing a blocking set. Let E be a machine contained in $\text{Next}^s(J_j)$ for some job J_j, that is not full in s. Without loss of generality, we assume that among all successors of state s resulting from a job moving to E, state t has smallest cardinality canonical blocking set for machine E. Clearly E has to be in every blocking set of t.

Case where $\text{cap}(E) \geq 2$: Since there have to be jobs on E in state s (otherwise E would not be full in t), we can look at the set $\mathcal{B} = \mathcal{R}^s(\text{Next}^s(\mathcal{J}^s(E)))$. The jobs on E in state s are part of a blocking set in t, so $\text{Next}^s(\mathcal{J}^s(E)) \neq \emptyset$. Clearly $\mathcal{B}$ is contained in every minimal blocking set in every successor of state s which is the result of a move of some job to E. If $E \notin \mathcal{B}$, then $\mathcal{B}$ is a blocking set in s, which we assumed was not the case. So $E \in \mathcal{B}$ and let J' be on $M' \in \mathcal{B}$ for which $E \in \text{Next}^s(J')$. Let state t' be the successor of state s resulting from moving J' (from M') to E. As noted before, the minimal blocking set $\mathcal{B}^{t'}_{\min}(E)$ contains E and hence also $\mathcal{B}$. However $M' \in \mathcal{B}$ is not full in t' since J' just left M', so this gives the required contradiction.

Case where $\text{cap}(E) = 1$: There is a job J' on a machine in $\mathcal{B}^t_{min}(E)$ with $E \in \text{Next}^t(J') = \text{Next}^s(J')$, as otherwise $\mathcal{B}^t_{min}(E) - \{E\}$ would be a blocking set in state s. Since J' had already entered the system in state s and requires processing on E of unit capacity, by assumption the processing plan of J' in state s must be unconstrained, and thus $\mathcal{B} \subseteq \mathcal{B}^t_{min}(E)$. Let t' be the successor of state s resulting from moving J' to E and denote by $\mathcal{B}$ the set $\mathcal{R}^s(\mathcal{M}^s(J'))$. Since J' was on a machine in $\mathcal{B}^t_{min}(E)$ we have $\mathcal{B} \subseteq \mathcal{B}^t_{min}(E)$. Similarly $\mathcal{B} \subseteq \mathcal{B}^{t'}_{min}(E)$. Finally $E \in \mathcal{M}^s(J')$ implies that

$$\mathcal{B}^{t'}_{\min}(E) = \mathcal{R}^{t'}(\{E\}) = \mathcal{R}^s(\mathcal{M}^s(J')) = \mathcal{B}.$$

Since the machine $M^s(J') = M^t(J') \in \mathcal{B}^t_{\min}(E)$ is no longer full in t' so we have that $\mathcal{B}^{t'}_{\min}(E) \subsetneq \mathcal{B}^t_{\min}(E)$, resulting in a contradiction with the minimality of the cardinality of $\mathcal{B}^t_{\min}(E)$ taken over all successors of s resulting from a job moving to E. $\qquad\square$

5   Deciding Safety: Hard Cases

In this section we show that SAFE STATE RECOGNITION is NP-hard in all cases that are not covered by Theorem 1.

Theorem 2. *Let D be a fixed directed graph that is neither unconstrained nor source-constrained. SAFE STATE RECOGNITION is NP-hard, even if all processing* *plans are isomorphic to D. Furthermore this problem variant (with all processing plans isomorphic to D) remains NP-hard, even if the system contains only a single machine of unit capacity.*

Theorem 2 is proved by a reduction from the satisfiability problem; see Garey & Johnson [3].

PROBLEM: THREE-SATISFIABILITY

INPUT: A set $X = \{x_1, \ldots, x_n\}$ of n logical variables; a set $C = \{c_1, \ldots, c_m\}$ of m clauses over X that each contain three literals.

QUESTION: Is there a truth assignment for X that satisfies all the clauses in C?

Given an instance (X, C) of the problem THREE-SATISFIABILITY, we construct a scheduling system with a state s. We will show that state s is safe if there is a truth assignment for X that satisfies all clauses in C, and unsafe otherwise.

The (technical and somewhat lengthy) proof can be found in the full version of the paper.

6   Reachability

In this section we show that deciding reachability is essentially equivalent to deciding safeness.

Consider a scheduling system and some fixed system state s. We define a new (artificial) scheduling system and a new state $\rho(s)$ where $M^{\rho(s)}(J_j) := M^s(J_j)$, and $\mathcal{M}^{\rho(s)}(J_j) := \mathcal{M}(J_j) - M^s(J_j) - \{M^s(J_j)\}$. Furthermore for all jobs J_j, $\prec_j$ in $\rho(s)$ coincides with $\succ_j$ in s. Note that in both states s and $\rho(s)$ every job is sitting on the same machine, but the work that has already been performed in state s is exactly the work that still needs to be done in state $\rho(s)$.

Lemma 5. *State s is reachable in the old system, if and only if state $\rho(s)$ is safe in the new system.*

Proof. First assume that s is reachable, and let $0 = s_0 \rightarrow s_1 \rightarrow \cdots \rightarrow s_k = s$ denote a corresponding witness sequence of moves. Define a new sequence $\rho(s) = t_k \rightarrow t_{k-1} \rightarrow \cdots \rightarrow t_0 = f$ of moves: whenever the move $s_\ell \rightarrow s_{\ell+1}$ ($0 \le \ell \le k - 1$) results from moving job J_j from machine M_a to machine M_b, then the move $t_{\ell+1} \rightarrow t_\ell$ results from moving job J_j from machine M_b to machine M_a. Hence $\rho(s)$ is safe. A symmetric argument shows that if $\rho(s)$ is safe then s is reachable. □

Example 4. (Continuation of Example 3). State p is not reachable, since $\rho(p)$ is a deadlock: $M^{\rho(p)}(J_1) = M_2$, $\mathcal{M}^{\rho(p)}(J_1) = \{M_1\}$, and $M^{\rho(p)}(J_2) = M_1$, $\mathcal{M}^{\rho(p)}(J_2) = \{M_2\}$.

As the construction of the new scheduling system and state $\rho(s)$ can be done in polynomial time, deciding reachability is algorithmically equivalent to deciding safeness. Hence Theorems 1 and 2 yield the following:

Theorem 3. *Problem* REACHABLE STATE RECOGNITION *is decidable in polynomial time if machines of unit capacity only occur in unconstrained and sink-constrained processing plans, and NP-hard otherwise.*

The following lemma is used in the proof of Theorem 6.

Lemma 6. *Let s be a state, and let $\mathcal{K}$ be a subset of machines such that every job J_j that still needs further processing in s satisfies $M^s(J_j) \in \mathcal{K}$, and, $K \nprec_j N$ holds for each $K \in \mathcal{K}$ and $N \in \mathcal{M}(J_j) \backslash \mathcal{K}$, and finally*

$$\mathcal{M}^s(J_j) \cup \{M^s(J_j)\} = \mathcal{K} \cap \mathcal{M}(J_j).$$

Then s is a reachable system state.   □

7   Reachable Deadlock

In this section we establish NP-hardness of REACHABLE DEADLOCK, even for some highly restricted special cases. We first recall a hardness result from [2]:

Theorem 4. *(Eggermont, Schrijver & Woeginger [2])*
REACHABLE DEADLOCK *is NP-hard for jobs with unconstrained processing plans, even if every job requires processing on at most four machines and if all machine capacities are at most three.*   □

This theorem easily implies the following.

Theorem 5. REACHABLE DEADLOCK *is NP-hard for jobs with source-constrained processing plans, even if every job requires processing on at most five machines and if all machine capacities are at most three.*

Proof. We modify a scheduling system with unconstrained processing plans into another scheduling system with source-constrained processing plans: for each job J we create a new machine $M^{src}(J)$ of unit capacity and add this machine as new source to J's processing plan. It is straightforward to see that the old system has a reachable deadlock if and only if the new system has a reachable deadlock.   □

The main result of this section is the following.

Theorem 6. *Let D be a fixed directed graph that is neither unconstrained nor source-constrained.* REACHABLE DEADLOCK *is NP-hard, even if all processing plans are isomorphic to D and if all machines have unit capacity.*

The proof of Theorem 6 can be found in the full version of the paper. The following observation will be useful.

Lemma 7. *If there exists a reachable deadlock, then there also exists a reachable deadlock where the set of all occupied machines forms a minimal blocking set.*

References

1. Aho, A.V., Garey, M.R., Ullman, J.D.: The transitive reduction of a directed graph. SIAM Journal on Computing 1, 131–137 (1972)
2. Eggermont, C.E.J., Schrijver, A., Woeginger, G.J.: Analysis of multi-state open shop processing systems. In: Proceedings of 28th International Symposium on Theoretical Aspects of Computer Science, LIPIcs, vol. 9, pp. 484–494 (2011)
3. Garey, M.R., Johnson, D.S.: Computers and Intractability: A Guide to the Theory of NP-Completeness. Freeman, San Francisco (1979)
4. Lawler, E.L., Lenstra, J.K., Rinnooy Kan, A.H.G., Shmoys, D.B.: Sequencing and scheduling: Algorithms and complexity. In: Handbooks in Operations Research and Management Science, vol. 4, pp. 445–522. North Holland, Amsterdam (1993)
5. Lawley, M., Reveliotis, S.: Deadlock avoidance for sequential resource allocation systems: hard and easy cases. The International Journal of Flexible Manufacturing Systems 13, 385–404 (2001)
6. Sulistyono, W., Lawley, M.A.: Deadlock avoidance for manufacturing systems with partially ordered process plans. IEEE Transactions on Robotics and Automation 17, 819–832 (2001)

Improving Reachability Analysis of Infinite State Systems by Specialization

Fabio Fioravanti[1], Alberto Pettorossi[2],
Maurizio Proietti[3], and Valerio Senni[2,4]

[1] Dipartimento di Scienze, University 'G. D'Annunzio', Pescara, Italy
fioravanti@sci.unich.it
[2] DISP, University of Rome Tor Vergata, Rome, Italy
{pettorossi,senni}@disp.uniroma2.it
[3] CNR-IASI, Rome, Italy
maurizio.proietti@iasi.cnr.it
[4] LORIA-INRIA, Villers-les-Nancy, France
valerio.senni@loria.fr

Abstract. We consider infinite state reactive systems specified by using linear constraints over the integers, and we address the problem of verifying safety properties of these systems by applying reachability analysis techniques. We propose a method based on program specialization, which improves the effectiveness of the backward and forward reachability analyses. For backward reachability our method consists in: (i) specializing the reactive system with respect to the initial states, and then (ii) applying to the specialized system a reachability analysis that works backwards from the unsafe states. For forward reachability our method works as for backward reachability, except that the role of the initial states and the unsafe states are interchanged. We have implemented our method using the MAP transformation system and the ALV verification system. Through various experiments performed on several infinite state systems, we have shown that our specialization-based verification technique considerably increases the number of successful verifications without significantly degrading the time performance.

1 Introduction

One of the present challenges in the field of automatic verification of reactive systems is the extension of the model checking techniques [5] to systems with an infinite number of states. For these systems exhaustive state exploration is impossible and, even for restricted classes, simple properties such as *safety* (or *reachability*) properties are undecidable (see [10] for a survey of relevant results).

In order to overcome this limitation, several authors have advocated the use of *constraints* over the integers (or the reals) to represent infinite sets of states [4,8,9,15,17]. By manipulating constraint-based representations of sets of states, one can verify a safety property φ of an infinite state system by one of the following two strategies:

G. Delzanno and I. Potapov (Eds.): RP 2011, LNCS 6945, pp. 165–179, 2011.

(i) Backward Strategy: one applies a *backward reachability* algorithm, thereby computing the set BR of states from which it is possible to reach an *unsafe* state (that is, a state where $\neg\varphi$ holds), and then one checks whether or not BR has an empty intersection with the set I of the initial states;

(ii) Forward Strategy: one applies a *forward reachability* algorithm, thereby computing the set FR of states reachable from an initial state, and then one checks whether or not FR has an empty intersection with the set U of the unsafe states.

Variants of these two strategies have been proposed and implemented in various automatic verification tools [2,3,14,20,25]. Some of them also use techniques borrowed from the field of *abstract interpretation* [6], whereby in order to check whether or not a safety property φ holds for all states which are reachable from the initial states, an *upper approximation* $\overline{BR}$ (or $\overline{FR}$) of the set BR (or FR) is computed. These techniques improve the termination of the verification tools at the expense of a possible loss in precision. Indeed, whenever $\overline{BR} \cap I \neq \emptyset$ (or $\overline{FR} \cap U \neq \emptyset$), one cannot conclude that, for some state, φ does not hold.

One weakness of the Backward Strategy is that, when computing BR, it does not take into account the properties holding on the initial states. This may lead, even if the formula φ does hold, to a failure of the verification process, because either the computation of BR does not terminate or one gets an overly approximated $\overline{BR}$ with a non-empty intersection with the set I. A similar weakness is also present in the Forward Strategy as it does not take into account the properties holding on the unsafe states when computing FR or $\overline{FR}$.

In this paper we present a method, based on *program specialization* [19], for overcoming these weaknesses. Program specialization is a program transformation technique that, given a program and a specific context of use, derives a specialized program that is more effective in the given context. Our specialization method is applied before computing BR (or FR). Its objective is to transform the constraint-based specification of a reactive system into a new specification that, when used for computing BR (or FR), takes into consideration also the properties holding on the initial states (or the unsafe states, respectively).

Our method consists of the following three steps: (1) the translation of a reactive system specification into a *constraint logic program* (CLP) [18] that implements backward (or forward) reachability; (2) the specialization of the CLP program with respect to the initial states (or the unsafe states, respectively), and (3) the reverse translation of the specialized CLP program into a specialized reactive system. We prove that our specialization method is correct, that is, it transforms a given specification into one which satisfies the same safety properties.

We have implemented our specialization method on the MAP transformation system for CLP programs [22] and we have performed experiments on several infinite state systems by using the *Action Language Verifier* (ALV) [25]. These experiments show that specialization determines a relevant increase of the number of successful verifications, in the case of both backward and forward reachability analysis, without a significant degradation of the time performance.

2   Specifying Reactive Systems

In order to specify reactive systems and their safety properties, we use a simplified version of the languages considered in [2,3,20,25]. Our language allows us to specify systems and properties by using constraints over the set $\mathbb{Z}$ of the integers.

A *system* is a triple $\langle Var, Init, Trans \rangle$, where: (i) *Var* is a *variable declaration*, (ii) *Init* is a formula denoting the set of *initial states*, and (iii) *Trans* is a formula denoting a *transition relation* between states.

Now we formally define these notions. A variable declaration *Var* is a sequence of declarations of (distinct) variables each of which may be either: (i) an *enumerated* variable, or (ii) an *integer* variable. (i) An enumerated variable x is declared by the statement: **enumerated** x D, meaning that x ranges over a finite set D of constants. The set D is said to be the *type* of x and it is also said to be the type of every constant in D. (ii) An integer variable x is declared by the statement: **integer** x, meaning that x is a variable ranging over the set $\mathbb{Z}$ of the integers. By $\mathcal{X}$ we denote the set of variables declared in *Var*, and by $\mathcal{X}'$ we denote the set $\{x' \mid x \in \mathcal{X}\}$ of primed variables.

Constraints are defined as follows. If e_1 and e_2 are enumerated variables or constants of the same type, then $e_1 = e_2$ and $e_1 \neq e_2$ are *atomic constraints*. If p_1 and p_2 are linear polynomials with integer coefficients, then $p_1 = p_2$, $p_1 \geq p_2$, and $p_1 > p_2$ are *atomic constraints*. A *constraint* is either *true*, or *false*, or an atomic constraint, or a *conjunction* of constraints. *Init* is a disjunction of constraints on the variables in $\mathcal{X}$. *Trans* is a disjunction of constraints on the variables in $\mathcal{X} \cup \mathcal{X}'$.

A *specification* is a pair $\langle Sys, Safe \rangle$, where *Sys* is a system and *Safe* is a formula of the form $\neg \mathsf{EF}\, Unsafe$, specifying a *safety property* of the system, and *Unsafe* is a disjunction of constraints on the variables in $\mathcal{X}$.

Example 1. Here we show a reactive system (1.1) and its specification (1.2) in our language.

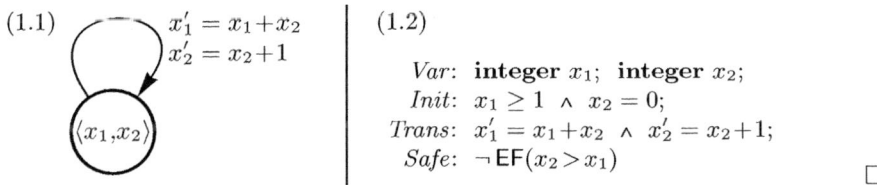

(1.1) $x_1' = x_1 + x_2$ $x_2' = x_2 + 1$ $\langle x_1, x_2 \rangle$

(1.2)

Var: **integer** x_1; **integer** x_2;
Init: $x_1 \geq 1 \ \wedge \ x_2 = 0$;
Trans: $x_1' = x_1 + x_2 \ \wedge \ x_2' = x_2 + 1$;
Safe: $\neg\,\mathsf{EF}(x_2 > x_1)$

$\square$

Now we define the semantics of a specification. Let D_i be a finite set of constants, for $i = 1, \ldots, k$. Let $X = \langle x_1, \ldots, x_k, x_{k+1}, \ldots, x_n \rangle$ be a listing of the variables in $\mathcal{X}$, where: (i) for $i = 1, \ldots, k$, x_i is an enumerated variable of type D_i, and (ii) for $i = k+1, \ldots, n$, x_i is an integer variable. Let X' be a listing $\langle x_1', \ldots, x_k', x_{k+1}', \ldots, x_n' \rangle$ of the variables in $\mathcal{X}'$. A *state* is an n-tuple $\langle r_1, \ldots, r_k, z_{k+1}, \ldots, z_n \rangle$ of constants in $D_1 \times \ldots \times D_k \times \mathbb{Z}^{n-k}$.

A state s of the form $\langle r_1, \ldots, r_k, z_{k+1}, \ldots, z_n \rangle$ *satisfies* a disjunction d of constraints on $\mathcal{X}$, denoted $s \models d$, if the formula $d[s/X]$ holds, where $[s/X]$ denotes the substitution $[r_1/x_1, \ldots, r_k/x_k, z_{k+1}/x_{k+1}, \ldots, z_n/x_n]$. A state satisfying *Init* (resp., *Unsafe*) will be called an *initial* (resp., *unsafe*) state.

A pair of states $\langle s, s' \rangle$ *satisfies* a constraint c on the variables in $\mathcal{X} \cup \mathcal{X}'$, denoted $\langle s, s' \rangle \models c$, if the constraint $c[s/X, s'/X']$ holds. A *computation sequence* is a sequence of states $s_0, \ldots, s_m$, with $m \geq 0$, such that, for $i = 0, \ldots, m-1$, $\langle s_i, s_{i+1} \rangle \models c$, for some constraint c in $\{c_j \mid j \in J\}$, where $\mathit{Trans} = \bigvee_{j \in J} c_j$. State s_m is *reachable* from state s_0 if there exists a computation sequence $s_0, \ldots, s_m$. The system Sys *satisfies* the safety property, called Safe, of the form $\neg \mathsf{EF}\, \mathit{Unsafe}$, if there is no state s which is reachable from an initial state and $s \models \mathit{Unsafe}$.

A specification $\langle \mathit{Sys}_1, \mathit{Safe}_1 \rangle$ is *equivalent* to a specification $\langle \mathit{Sys}_2, \mathit{Safe}_2 \rangle$ if Sys_1 satisfies Safe_1 if and only if Sys_2 satisfies Safe_2.

3 Constraint-Based Specialization of Reactive Systems

Now we present a method for transforming a specification $\langle \mathit{Sys}, \mathit{Safe} \rangle$ into an equivalent specification whose safety property is easier to verify. This method has two variants, called *Bw-Specialization* and *Fw-Specialization*. Bw-Specialization specializes the given system with respect to the disjunction *Init* of constraints that characterize the initial states. Thus, backward reachability analysis of the specialized system may be more effective because it takes into account the information about the initial states. A symmetric situation occurs in the case of Fw-Specialization where the given system is specialized with respect to the disjunction *Unsafe* of constraints that characterize the unsafe states.

Here we present the Bw-Specialization method only. (The Fw-Specialization method is similar and it is described in Appendix.) Bw-Specialization transforms the specification $\langle \mathit{Sys}, \mathit{Safe} \rangle$ into an equivalent specification $\langle \mathit{SpSys}, \mathit{SpSafe} \rangle$ according to the following three steps.

Step (1). *Translation*: The specification $\langle \mathit{Sys}, \mathit{Safe} \rangle$ is translated into a CLP program, called *Bw*, that implements the backward reachability algorithm.

Step (2). *Specialization*: The CLP program *Bw* is specialized into a program *SpBw* by taking into account the disjunction *Init* of constraints.

Step (3). *Reverse Translation*: The specialized CLP program *SpBw* is translated back into a new, specialized specification $\langle \mathit{SpSys}, \mathit{SpSafe} \rangle$, which is equivalent to $\langle \mathit{Sys}, \mathit{Safe} \rangle$.

The specialized specification $\langle \mathit{SpSys}, \mathit{SpSafe} \rangle$ contains new constraints that are derived by propagating through the transition relation of the system Sys the constraints *Init* holding in the initial states. Thus, the backward reachability analysis that uses the transition relation of the specialized system SpSys, takes into account the information about the initial states and, for this reason, it is often more effective (see Section 4 for an experimental validation of this fact).

Let us now describe Steps (1), (2), and (3) in more detail.

Step (1). Translation. Let us consider the system $\mathit{Sys} = \langle \mathit{Var}, \mathit{Init}, \mathit{Trans} \rangle$ and the property Safe. Suppose that:

(1) X and X' are listings of the variables in the sets $\mathcal{X}$ and $\mathcal{X}'$, respectively,
(2) *Init* is a disjunction $\mathit{init}_1(X) \vee \ldots \vee \mathit{init}_k(X)$ of constraints,

(3) *Trans* is a disjunction $t_1(X, X') \vee \ldots \vee t_m(X, X')$ of constraints,

(4) *Safe* is the formula $\neg\, \mathsf{EF}\, \mathit{Unsafe}$, where *Unsafe* is a disjunction $u_1(X) \vee \ldots \vee u_n(X)$ of constraints.

Then, program *Bw* consists of the following clauses:

I_1: *unsafe* $\leftarrow$ $\mathit{init}_1(X) \wedge \mathit{bwReach}(X)$

    $\ldots$

I_k: *unsafe* $\leftarrow$ $\mathit{init}_k(X) \wedge \mathit{bwReach}(X)$

T_1: *bwReach*$(X) \leftarrow t_1(X, X') \wedge \mathit{bwReach}(X')$

    $\ldots$

T_m: *bwReach*$(X) \leftarrow t_m(X, X') \wedge \mathit{bwReach}(X')$

U_1: *bwReach*$(X) \leftarrow u_1(X)$

    $\ldots$

U_n: *bwReach*$(X) \leftarrow u_n(X)$

The meaning of the predicates defined in the program *Bw* is as follows:
(i) *bwReach*(X) holds iff an unsafe state can be reached from the state X in zero or more applications of the transition relation, and (ii) *unsafe* holds iff there exists an initial state X such that *bwReach*(X) holds.

Example 2. For the system of Example 1 we get the following CLP program:

I_1: *unsafe* $\leftarrow x_1 \geq 1 \wedge x_2 = 0 \wedge \mathit{bwReach}(x_1, x_2)$

T_1: *bwReach*$(x_1, x_2) \leftarrow x_1' = x_1 + x_2 \wedge x_2' = x_2 + 1 \wedge \mathit{bwReach}(x_1', x_2')$

U_1: *bwReach*$(x_1, x_2) \leftarrow x_2 > x_1$ $\square$

The semantics of program *Bw* is given by the *least* $\mathbb{Z}$-*model*, denoted $M(\mathit{Bw})$, that is, the set of ground atoms derived by using: (i) the theory of linear equations and inequations over the integers $\mathbb{Z}$ for the evaluation of the constraints, and (ii) the usual least model construction (see [18] for more details).

The translation of the specification $\langle \mathit{Sys}, \mathit{Safe} \rangle$ performed during Step (1) is correct in the sense stated by Theorem 1. The proof of this theorem is based on the fact that the definition of the predicate *bwReach* in the program *Bw* is a recursive definition of the reachability relation defined in Section 2.

Theorem 1 (Correctness of Translation). *The system Sys satisfies the formula Safe iff unsafe* $\notin M(\mathit{Bw})$.

Step (2). Specialization. Program *Bw* is transformed into a specialized program *SpBw* such that *unsafe* $\in M(\mathit{Bw})$ iff *unsafe* $\in M(\mathit{SpBw})$ by applying the specialization algorithm shown in Figure 1.

This algorithm modifies the initial program *Bw* by propagating the information about the initial states *Init* and it does so by using the *definition introduction, unfolding, clause removal,* and *folding* rules for transforming constraint logic programs (see, for instance, [11]). In particular, our specialization algorithm: (i) introduces new predicates defined by clauses of the form $newp(X) \leftarrow c(X) \wedge \mathit{bwReach}(X)$, corresponding to specialized versions of the *bwReach* predicate, and (ii) derives mutually recursive definitions of these new predicates by applying the unfolding, clause removal, and folding rules.

Input: Program *Bw*.
Output: Program *SpBw* such that *unsafe* $\in M(Bw)$ iff *unsafe* $\in M(SpBw)$.

INITIALIZATION:
$SpBw := \{J_1, \ldots, J_k\}$, where J_1: *unsafe* $\leftarrow init_1(X) \wedge newu_1(X)$
$$\cdots$$
J_k: *unsafe* $\leftarrow init_k(X) \wedge newu_k(X)$;

$InDefs := \{I'_1, \ldots, I'_k\}$, where I'_1: $newu_1(X) \leftarrow init_1(X) \wedge bwReach(X)$
$$\cdots$$
I'_k: $newu_k(X) \leftarrow init_k(X) \wedge bwReach(X)$;

Defs := *InDefs*;
while there exists a clause C: $newp(X) \leftarrow c(X) \wedge bwReach(X)$ in *InDefs* do

UNFOLDING: $SpC := \{\, newp(X) \leftarrow c(X) \wedge t_1(X, X') \wedge bwReach(X'),$
$$\cdots$$
$newp(X) \leftarrow c(X) \wedge t_m(X, X') \wedge bwReach(X'),$
$newp(X) \leftarrow c(X) \wedge u_1(X),$
$$\cdots$$
$newp(X) \leftarrow c(X) \wedge u_n(X) \,\}$;

CLAUSE REMOVAL:
while in *SpC* there exist two distinct clauses E and F such that E $\mathbb{R}$-subsumes F or
there exists a clause F whose body has a constraint which is not $\mathbb{R}$-satisfiable
do $SpC := SpC - \{F\}$ *end-while*;

DEFINITION-INTRODUCTION & FOLDING:
while in *SpC* there is a clause E of the form: $newp(X) \leftarrow e(X, X') \wedge bwReach(X')$
do

if in *Defs* there is a clause D of the form: $newq(X) \leftarrow d(X) \wedge bwReach(X)$ such
that $e(X, X') \sqsubseteq_{\mathbb{R}} d(X')$, where $d(X')$ is $d(X)$ with X replaced by X'
then $SpC := (SpC - \{E\}) \cup \{newp(X) \leftarrow e(X, X') \wedge newq(X')\}$;
else let $Gen(E, Defs)$ be the clause $newr(X) \leftarrow g(X) \wedge bwReach(X)$ where:
(i) *newr* is a predicate symbol not in *Defs* and (ii) $e(X, X') \sqsubseteq_{\mathbb{R}} g(X')$;
$Defs := Defs \cup \{Gen(E, Defs)\}$; $InDefs := InDefs \cup \{Gen(E, Defs)\}$;
$SpC := (SpC - \{E\}) \cup \{newp(X) \leftarrow e(X, X') \wedge newr(X')\}$;
end-while;

$SpBw := SpBw \cup SpC$;
end-while

Fig. 1. The specialization algorithm

An important feature of our specialization algorithm is that the applicability
conditions of the transformation rules used by the algorithm are expressed in
terms of the unsatisfiability (or entailment) of constraints on the domain $\mathbb{R}$ of the
real numbers, instead of the domain $\mathbb{Z}$ of the integer numbers, thereby allowing
us to use more efficient constraint solvers (according to the present state-of-the-
art solvers). Note that, despite this domain change from $\mathbb{Z}$ to $\mathbb{R}$, the specialized
reachability program *SpBw* is *equivalent* to the initial program *Bw* w.r.t. the least
$\mathbb{Z}$-model semantics (see Theorem 4 below). This result is based on the correctness
of the transformation rules [11] and on the fact that the unsatisfiability (or
entailment) of constraints on $\mathbb{R}$ implies the unsatisfiability (or entailment) of
those constraints on $\mathbb{Z}$. For instance, let us consider the rule that removes a
clause of the form $H \leftarrow c \wedge B$ if the constraint c is unsatisfiable on the integers.

Our specialization algorithm removes the clause if c is unsatisfiable on the reals. Clearly, we may miss the opportunity of removing a clause whose constraint is satisfiable on the reals and unsatisfiable on the integers, thereby deriving a specialized program with redundant satisfiability tests. More in general, the use of constraint solvers on the reals may reduce the specialization time, but may leave in the specialized programs residual satisfiability tests on the integers that should be performed at verification time on the specialized system.

Let us define the notions of $\mathbb{R}$-satisfiability, $\mathbb{R}$-entailment, and $\mathbb{R}$-subsumption that we have used in the specialization algorithm. Let X and X' be n-tuples of variables as indicated in Section 2. The constraint $c(X)$ is $\mathbb{R}$-*satisfiable*, if there exists an n-tuple A in $D_1 \times \ldots \times D_k \times \mathbb{R}^{n-k}$ such that $c(A)$ holds. A constraint $c(X,X')$ $\mathbb{R}$-*entails* a constraint $d(X,X')$, denoted $c(X,X') \sqsubseteq_\mathbb{R} d(X,X')$, if for all A, A' in $D_1 \times \ldots \times D_k \times \mathbb{R}^{n-k}$, if $c(A,A')$ holds then $d(A,A')$ holds. (Note that the variables X or X' may be absent from $c(X,X')$ or $d(X,X')$.) Given two clauses of the forms $C\colon H \leftarrow c(X)$ and $D\colon H \leftarrow d(X) \wedge e(X,X') \wedge B$, where the constraint $e(X,X')$ and the atom B may be absent, we say that C $\mathbb{R}$-*subsumes* D, if $d(X) \wedge e(X,X') \sqsubseteq_\mathbb{R} c(X)$.

As usual when performing program specialization, our algorithm also makes use of a *generalization operator Gen* for introducing definitions of new predicates by generalizing constraints. Given a clause $E\colon newp(X) \leftarrow e(X,X') \wedge bwReach(X')$ and the set *Defs* of clauses that define the new predicates introduced so far by the specialization algorithm, $Gen(E, Defs)$ returns a clause G of the form $newr(X) \leftarrow g(X) \wedge bwReach(X)$ such that: (i) $newr$ is a fresh, new predicate symbol, and (ii) $e(X,X') \sqsubseteq_\mathbb{R} g(X')$ (where $g(X')$ is the constraint $g(X)$ with X replaced by X'). Then, clause E is folded by using clause G, thereby deriving $newp(X) \leftarrow e(X,X') \wedge newr(X')$. This transformation step preserves equivalence with respect to the least $\mathbb{Z}$-model semantics. Indeed, $newr(X')$ is equivalent to $g(X') \wedge bwReach(X')$ by definition and, as already mentioned, $e(X,X') \sqsubseteq_\mathbb{R} g(X')$ implies that $e(X,X')$ entails $g(X')$ in $\mathbb{Z}$.

The generalization operator we use in our experiments reported in Section 4, is defined in terms of relations and operators on constraints such as *widening* and *well-quasi orders* based on the coefficients of the polynomials occurring in the constraints. For lack of space we will not describe in detail the generalization operator we apply, and we refer to [13,23] for various operators which can be used for specializing constraint logic programs. It will be enough to say that the termination of the specialization algorithm is ensured by the fact that, similarly to the widening operator presented in [6], our generalization operator guarantees that during specialization only a finite number of new predicates is introduced.

Thus, we have the following result.

Theorem 2 (Termination and Correctness of Specialization). (i) *The specialization algorithm terminates.* (ii) *Let program SpBw be the output of the specialization algorithm. Then* $unsafe \in M(Bw)$ *iff* $unsafe \in M(SpBw)$.

Example 3. The following program is obtained as output of the specialization algorithm when it takes as input the CLP program of Example 2:

J_1: *unsafe* $\leftarrow$ $x_1 \geq 1 \wedge x_2 = 0 \wedge new1(x_1, x_2)$
S_1: $new1(x_1, x_2) \leftarrow x_1 \geq 1 \wedge x_2 = 0 \wedge x_1' = x_1 \wedge x_2' = 1 \wedge new2(x_1', x_2')$
S_2: $new2(x_1, x_2) \leftarrow x_1 \geq 1 \wedge x_2 = 1 \wedge x_1' = x_1 + 1 \wedge x_2' = 2 \wedge new3(x_1', x_2')$
S_3: $new3(x_1, x_2) \leftarrow x_1 \geq 1 \wedge x_2 \geq 1 \wedge x_1' = x_1 + x_2 \wedge x_2' = x_2 + 1 \wedge new3(x_1', x_2')$
V_1: $new3(x_1, x_2) \leftarrow x_1 \geq 1 \wedge x_2 > x_1$ $\square$

Step (3). Reverse Translation. The output of the specialization algorithm is a specialized program *SpBw* of the form:

J_1: *unsafe* $\leftarrow$ $init_1(X) \wedge newu_1(X)$

   ...

J_k: *unsafe* $\leftarrow$ $init_k(X) \wedge newu_k(X)$
S_1: $newp_1(X) \leftarrow s_1(X, X') \wedge newt_1(X')$

   ...

S_m: $newp_m(X) \leftarrow s_m(X, X') \wedge newt_m(X')$
V_1: $newq_1(X) \leftarrow v_1(X)$

   ...

V_n: $newq_n(X) \leftarrow v_n(X)$

where: (i) $s_1(X, X'), \ldots, s_m(X, X'), v_1(X), \ldots, v_m(X)$ are constraints, and (ii) the (possibly non-distinct) predicate symbols $newu_i$'s, $newp_i$'s, $newt_i$'s, and $newq_i$'s are the new predicate symbols introduced by the specialization algorithm. Let *NewPred* be the set of all of those new predicate symbols.

We derive a new specification $\langle SpSys, SpSafe \rangle$, where *SpSys* is a system of the form $\langle SpVar, SpInit, SpTrans \rangle$, as follows.

(1) Let x_p be a new enumerated variable ranging over the set *NewPred* of predicate symbols introduced by the specialization algorithm.
   Let the variable X occurring in the program *SpBw* denote the n-tuple of variables $\langle x_1, \ldots, x_k, x_{k+1}, \ldots, x_n \rangle$, where: (i) for $i = 1, \ldots, k$, x_i is an enumerated variable ranging over the finite set D_i, and (ii) for $i = k+1, \ldots, n$, x_i is an integer variable.
   We define *SpVar* to be the following sequence of declarations of variables:
      enumerated x_p *NewPred*;
      enumerated x_1 D_1; ...; **enumerated** x_k D_k;
      integer x_{k+1}; ...; **integer** x_n.
(2) From clauses $J_1, \ldots, J_k$ we get the disjunction *SpInit* of k constraints, each of which is of the form: $init_i(X) \wedge x_p = newu_i$.
(3) From clauses $S_1, \ldots, S_m$ we get the disjunction *SpTrans* of m constraints, each of which is of the form: $s_i(X, X') \wedge x_p = newp_i \wedge x_p' = newt_i$.
(4) From clauses $V_1, \ldots, V_n$ we get the disjunction *SpUnsafe* of n constraints, each of which is of the form: $v_i(X) \wedge x_p = newq_i$.
   SpSafe is the formula $\neg \mathrm{EF}\, SpUnsafe$.

The reverse translation of the program *SpBw* into the specification $\langle SpSys, SpSafe \rangle$ is correct in the sense stated by the following theorem.

Theorem 3 (Correctness of Reverse Translation). *The following equivalence holds: unsafe $\notin M(SpBw)$ iff SpSys satisfies SpSafe.*

Example 4. The following specialized specification is the result of the reverse translation of the specialized CLP program of Example 3:

$SpVar$: **enumerated** x_p {$new1, new2, new3$}; **integer** x_1; **integer** x_2;

$SpInit$: $x_1 \geq 1 \land x_2 = 0 \land x_p = new1$;

$SpTrans$: $(x_1 \geq 1 \land x_2 = 0 \land x_p = new1 \land x_1' = x_1 \land x_2' = 1 \land x_p' = new2) \lor$
$(x_1 \geq 1 \land x_2 = 1 \land x_p = new2 \land x_1' = x_1 + 1 \land x_2' = 2 \land x_p' = new3) \lor$
$(x_1 \geq 1 \land x_2 \geq 1 \land x_p = new3 \land x_1' = x_1 + x_2 \land x_2' = x_2 + 1 \land x_p' = new3)$

$SpSafe$: $\neg \mathsf{EF}(x_1 \geq 1 \land x_2 > x_1 \land x_p = new3)$

Note that the backward reachability algorithm implemented in the ALV tool [25] is *not* able to verify (within 600 seconds) the safety property of the *initial specification* (see Example 1). Basically, this is due to the fact that working backward from the unsafe states where $x_2 > x_1$ holds, ALV is not able to infer that, for all reachable states, $x_2 \geq 0$ holds. The Bw-Specialization method is able to derive, from the constraint characterizing the initial states, a new transition relation $SpTrans$ whose constraints imply $x_2 \geq 0$. By exploiting this constraint, ALV successfully verifies the safety property of the *specialized specification*.   □

The correctness of our Bw-Specialization method is stated by the following theorem, which is a straightforward consequence of Theorems 1, 2, and 3.

Theorem 4 (Correctness of Bw-Specialization). *Let $\langle SpSys, SpSafe \rangle$ be the specification derived by applying the Bw-Specialization method to the specification $\langle Sys, Safe \rangle$. Then, $\langle Sys, Safe \rangle$ is equivalent to $\langle SpSys, SpSafe \rangle$.*

4   Experimental Evaluation

In this section we present the results of the verification experiments we have performed on various infinite state systems taken from the literature [3,8,9,25].

We have run our experiments by using the ALV tool, which is based on a BDD-based symbolic manipulation for enumerated types and on a solver for linear constraints on integers [25]. ALV performs backward and forward reachability analysis by an approximate computation of the least fixpoint of the transition relation of the system. We have run ALV using the options: 'default' and 'A' (both for backward analysis), and the option 'F' (for forward analysis). The Bw-Specialization and the Fw-Specialization methods were implemented on MAP [22], a tool for transforming CLP programs which uses the SICStus Prolog `clpr` library to operate on constraints on the reals. All experiments were performed on an Intel Core 2 Duo E7300 2.66 GHz under Linux.

The results of our experiments are reported in Table 1, where we have indicated, for each system and for each ALV option used, the following times expressed in seconds: (i) the time taken by ALV for verifying the given system (columns *Sys*), and (ii) the total time taken by MAP for specializing the system and by ALV for verifying the specialized system (columns *SpSys*).

The experiments show that our specialization method always increases the *precision* of ALV, that is, for every ALV option used, the number of properties verified increases when considering the specialized systems (columns *SpSys*) instead of the given, non-specialized systems (columns *Sys*). There are also some examples (Consistency, Selection Sort, and Reset Petri Net) where ALV is not able to verify the property on the given reactive system (regardless of the option used), but it verifies the property on the corresponding specialized system.

Now, let us compare the verification times. The time in column *Sys* and the time in column *SpSys* are of the same order of magnitude in almost all cases. In two examples (Peterson and CSM, with the 'default' option) our method substantially reduces the total verification time. Finally, in the Bounded Buffer example (with options 'default' and 'A') our specialization method significantly increases the verification time. Thus, overall, the increase of precision due to the specialization method we have proposed, does not determine a significant degradation of the time performance.

The increase of the verification times in the Bounded Buffer example is due to the fact that the non-specialized system can easily be verified by a backward reachability analysis and, thus, our pre-processing based on specialization is unnecessary. Moreover, after specializing the Bounded Buffer system, we get a

Table 1. Verification times (in seconds) using ALV [25]. '$\perp$' means termination with the answer 'Unable to verify' and '∞' means 'No answer' within 10 minutes.

	default		A		F	
EXAMPLES	*Sys*	*SpSys*	*Sys*	*SpSys*	*Sys*	*SpSys*
1. Bakery2	0.03	0.05	0.03	0.05	0.06	0.04
2. Bakery3	0.70	0.25	0.69	0.25	∞	3.68
3. MutAst	1.46	0.37	1.00	0.37	0.22	0.59
4. Peterson	56.49	0.10	∞	0.10	∞	13.48
5. Ticket	∞	0.03	0.10	0.03	0.02	0.19
6. Berkeley RISC	0.01	0.04	$\perp$	0.04	0.01	0.02
7. DEC Firefly	0.01	0.02	$\perp$	0.03	0.01	0.07
8. IEEE Futurebus	0.26	0.68	$\perp$	$\perp$	∞	∞
9. Illinois University	0.01	0.03	$\perp$	0.03	∞	0.07
10. MESI	0.01	0.02	$\perp$	0.03	0.02	0.07
11. MOESI	0.01	0.06	$\perp$	0.05	0.02	0.08
12. Synapse N+1	0.01	0.02	$\perp$	0.02	0.01	0.01
13. Xerox PARC Dragon	0.01	0.05	$\perp$	0.06	0.02	0.10
14. Barber	0.62	0.21	$\perp$	0.21	∞	0.08
15. Bounded Buffer	0.01	3.10	0.01	3.16	∞	0.03
16. Unbounded Buffer	0.01	0.06	0.01	0.06	0.04	0.04
17. CSM	56.39	7.69	$\perp$	7.69	∞	125.32
18. Consistency	∞	0.11	$\perp$	0.11	∞	324.14
19. Insertion Sort	0.03	0.06	0.04	0.06	0.18	0.02
20. Selection Sort	∞	0.21	$\perp$	0.21	∞	0.33
21. Reset Petri Net	∞	0.02	$\perp$	$\perp$	∞	0.01
22. Train	42.24	59.21	$\perp$	$\perp$	∞	0.46
Number of verified properties	18	22	7	19	11	21

new system whose specification is quite large (because the MAP system generates a large number of clauses). We will return to this point in the next section.

5   Related Work and Conclusions

We have considered infinite state reactive systems specified by constraints over the integers and we have proposed a method, based on the specialization of CLP programs, for pre-processing the given systems and getting new, equivalent systems so that their backward (or forward) reachability analysis terminates with success more often (that is, precision is improved), without a significant increase of the verification time. The improvement of precision of the analysis is due to the fact that the backward (or forward) verification of the specialized systems takes into account the properties which are true on the initial states (or on the unsafe states, respectively).

The use of constraint logic programs in the area of system verification has been proposed by several authors (see [8,9], and [15] for a survey of early works). Also transformation techniques for constraint logic programs have been shown to be useful for the verification of infinite state systems [12,13,21,23,24]. In the approach presented in this paper, constraint logic programs provide as an intermediate representation of the systems to be verified so that one can easily specialize those systems. To these constraint logic programs we apply a variant of the specialization technique presented in [13]. However, unlike [12,13,21,23,24], the final result of our specialization is not a constraint logic program, but a new reactive system which can be analyzed by using *any* verification tool for reactive systems specified by linear constraints on the integers. In this paper we have used the ALV tool [25] to perform the verification task on the specialized systems (see Section 4), but we could have also used (with minor syntactic modifications) other verification tools, such as TReX [2], FAST [3], and LASH [20]. Thus, one can apply to the specialized systems any of the optimization techniques implemented in those verification tools, such as *fixpoint acceleration*. We leave it for future research to evaluate the combined use of our specialization technique with other available optimization techniques.

Our specialization method is also related to some techniques for abstract interpretation [6] and, in particular, to those proposed in the field of verification of infinite state systems [1,5,7,16]. For instance, program specialization makes use of *generalization* operators [13] which are similar to the widening operators used in abstract interpretation. The main difference between program specialization and abstract interpretation is that, when applied to a given system specification, the former produces an *equivalent* specification, while the latter produces a more abstract (possibly, finite state) model whose semantics is an approximation of the semantics of the given specification. Moreover, since our specialization method returns a new system specification which is written in the same language of the given specification, after performing specialization we may also apply abstract interpretation techniques for proving system properties. Finding combinations of program specialization and abstract interpretation techniques that are most suitable for the verification of infinite state systems is an interesting issue for future research.

A further relevant issue we would like to address in the future is the reduction of the size of the specification of the specialized systems. Indeed, in one of the examples considered in Section 4, the time performance of the verification was not quite good, because the (specification of the) specialized system had a large size, due to the introduction of a large number of new predicate definitions. This problem can be tackled by using techniques for controlling *polyvariance* (that is, for reducing the number of specialized versions of the same predicate), which is an important issue studied in the field of program specialization [19].

Finally, we plan to extend our specialization technique to specifications of other classes of reactive systems such as *linear hybrid systems* [14,17].

Acknowledgements. This work has been partially supported by PRIN-MIUR and by a joint project between CNR (Italy) and CNRS (France). The last author has been supported by an ERCIM grant during his stay at LORIA-INRIA. Thanks to Laurent Fribourg and John Gallagher for many stimulating conversations.

References

1. Abdulla, P.A., Delzanno, G., Ben Henda, N., Rezine, A.: Monotonic abstraction (On efficient verification of parameterized systems). Int. J. of Foundations of Computer Science 20(5), 779–801 (2009)
2. Annichini, A., Bouajjani, A., Sighireanu, M.: TReX: A tool for reachability analysis of complex systems. In: Berry, G., Comon, H., Finkel, A. (eds.) CAV 2001. LNCS, vol. 2102, pp. 368–372. Springer, Heidelberg (2001)
3. Bardin, S., Finkel, A., Leroux, J., Petrucci, L.: FAST: Acceleration from theory to practice. Int. J. on Software Tools for Technology Transfer 10(5), 401–424 (2008)
4. Bultan, T., Gerber, R., Pugh, W.: Model-checking concurrent systems with unbounded integer variables: symbolic representations, approximations, and experimental results. ACM TOPLAS 21(4), 747–789 (1999)
5. Clarke, E.M., Grumberg, O., Peled, D.: Model Checking. MIT Press, Cambridge (1999)
6. Cousot, P., Cousot, R.: Abstract interpretation: A unified lattice model for static analysis of programs by construction of approximation of fixpoints. In: Proc. POPL 1977, pp. 238–252. ACM Press, New York (1977)
7. Dams, D., Grumberg, O., Gerth, R.: Abstract interpretation of reactive systems. ACM TOPLAS 19(2), 253–291 (1997)
8. Delzanno, G.: Constraint-based verification of parameterized cache coherence protocols. Formal Methods in System Design 23(3), 257–301 (2003)
9. Delzanno, G., Podelski, A.: Constraint-based deductive model checking. Int. J. on Software Tools for Technology Transfer 3(3), 250–270 (2001)
10. Esparza, J.: Decidability of model checking for infinite-state concurrent systems. Acta Informatica 34(2), 85–107 (1997)
11. Etalle, S., Gabbrielli, M.: Transformations of CLP modules. Theoretical Computer Science 166, 101–146 (1996)

12. Fioravanti, F., Pettorossi, A., Proietti, M.: Verifying CTL properties of infinite state systems by specializing constraint logic programs. In: Proc. VCL 2001, Tech. Rep. DSSE-TR-2001-3, pp. 85–96. Univ. of Southampton, UK (2001)

13. Fioravanti, F., Pettorossi, A., Proietti, M., Senni, V.: Program specialization for verifying infinite state systems: An experimental evaluation. In: Alpuente, M. (ed.) LOPSTR 2010. LNCS, vol. 6564, pp. 164–183. Springer, Heidelberg (2011)

14. Frehse, G.: PHAVer: Algorithmic verification of hybrid systems past HYTECH. In: Morari, M., Thiele, L. (eds.) HSCC 2005. LNCS, vol. 3414, pp. 258–273. Springer, Heidelberg (2005)

15. Fribourg, L.: Constraint logic programming applied to model checking. In: Bossi, A. (ed.) LOPSTR 1999. LNCS, vol. 1817, pp. 31–42. Springer, Heidelberg (2000)

16. Godefroid, P., Huth, M., Jagadeesan, R.: Abstraction-based model checking using modal transition systems. In: Larsen, K.G., Nielsen, M. (eds.) CONCUR 2001. LNCS, vol. 2154, pp. 426–440. Springer, Heidelberg (2001)

17. Henzinger, T.A.: The theory of hybrid automata. In: Proc., LICS 1996, pp. 278–292 (1996)

18. Jaffar, J., Maher, M.: Constraint logic programming: A survey. J. of Logic Programming 19/20, 503–581 (1994)

19. Jones, N.D., Gomard, C.K., Sestoft, P.: Partial Evaluation and Automatic Program Generation. Prentice Hall, Englewood Cliffs (1993)

20. LASH homepage, http://www.montefiore.ulg.ac.be/~boigelot/research/lash

21. Leuschel, M., Massart, T.: Infinite state model checking by abstract interpretation and program specialization. In: Bossi, A. (ed.) LOPSTR 1999. LNCS, vol. 1817, pp. 63–82. Springer, Heidelberg (2000)

22. MAP homepage, http://www.iasi.cnr.it/~proietti/system.html

23. Peralta, J.C., Gallagher, J.P.: Convex hull abstractions in specialization of CLP programs. In: Leuschel, M. (ed.) LOPSTR 2002. LNCS, vol. 2664, pp. 90–108. Springer, Heidelberg (2003)

24. Roychoudhury, A., Narayan Kumar, K., Ramakrishnan, C.R., Ramakrishnan, I.V., Smolka, S.A.: Verification of parameterized systems using logic program transformations. In: Graf, S. (ed.) TACAS 2000. LNCS, vol. 1785, pp. 172–187. Springer, Heidelberg (2000)

25. Yavuz-Kahveci, T., Bultan, T.: Action Language Verifier: An infinite-state model checker for reactive software specifications. Formal Methods in System Design 35(3), 325–367 (2009)

Appendix. Specialization Method for Forward Reachability

Let us briefly describe the *Fw-Specialization* method to be applied as a pre-processing step before performing a forward reachability analysis.

Fw-Specialization consists of three Steps (1f), (2f), and (3f), analogous to Steps (1), (2), and (3) of the backward reachability case described in Section 3.

Step (1f). Translation. Consider the system $Sys = \langle Var, Init, Trans \rangle$ and the property *Safe* specified as indicated in Step (1) of Section 3. The specification $\langle Sys, Safe \rangle$ is translated into the following constraint logic program Fw that encodes the forward reachability algorithm.

$$G_1 \colon\ unsafe \leftarrow u_1(X) \wedge fwReach(X)$$
$$\cdots$$
$$G_n \colon\ unsafe \leftarrow u_n(X) \wedge fwReach(X)$$

$$R_1\colon fwReach(X') \leftarrow t_1(X, X') \wedge fwReach(X)$$
$$\cdots$$
$$R_m\colon fwReach(X') \leftarrow t_m(X, X') \wedge fwReach(X)$$
$$H_1\colon fwReach(X) \leftarrow init_1(X)$$
$$\cdots$$
$$H_k\colon fwReach(X) \leftarrow init_k(X)$$

Note that we have interchanged the roles of the initial and unsafe states (compare the clauses G_i's and H_i's of program Fw with clauses I_i's and U_i's of program Bw presented in Section 3), and we have reversed the direction of the derivation of new states from old ones (compare clauses R_i's of program Fw with clauses T_i's of program Bw).

Step (2f). Forward Specialization. Program Fw is transformed into an equivalent program $SpFw$ by applying a variant of the specialization algorithm described in Figure 1 to the input program Fw, instead of program Bw. This transformation consists in specializing Fw with respect to the disjunction $Unsafe$ of constraints that characterizes the unsafe states of the system Sys.

Step (3f). Reverse Translation. The output of the specialization algorithm is a program $SpFw$ of the form:

$$L_1\colon unsafe \leftarrow u_1(X) \wedge newu_1(X)$$
$$\cdots$$
$$L_n\colon unsafe \leftarrow u_n(X) \wedge newu_n(X)$$
$$P_1\colon newp_1(X') \leftarrow p_1(X, X') \wedge newd_1(X)$$
$$\cdots$$
$$P_r\colon newp_r(X') \leftarrow p_r(X, X') \wedge newd_r(X)$$
$$W_1\colon newq_1(X) \leftarrow w_1(X)$$
$$\cdots$$
$$W_s\colon newq_s(X) \leftarrow w_s(X)$$

where (i) $p_1(X, X'), \ldots, p_r(X, X')$, $w_1(X), \ldots, w_s(X)$ are constraints, and (ii) the (possibly non-distinct) predicate symbols $newu_i$'s, $newp_i$'s, $newd_i$'s, and $newq_i$'s are the new predicate symbols introduced by the specialization algorithm.

Now we translate the program $SpFw$ into a new specification $\langle SpSys, SpSafe \rangle$, where $SpSys = \langle SpVar, SpInit, SpTrans \rangle$. The translation is like the one presented in Step (3), the only difference being the interchange of the initial states and the unsafe states. In particular, (i) we derive a new variable declaration $SpVar$ by introducing a new enumerated variable ranging over the set of new predicate symbols, (ii) we extract the disjunction $SpInit$ of constraints characterizing the new initial states from the constrained facts W_i's, (iii) we extract the disjunction $SpTrans$ of constraints characterizing the new transition relation from the clauses P_i's, (iv) we extract the disjunction $SpUnsafe$ of constraints characterizing the new unsafe states from the clauses L_i's which define the $unsafe$ predicate, and finally, (v) we define $SpSafe$ as the formula $\neg\, \textsf{EF}\, SpUnsafe$.

Similarly to Section 3, we can prove the correctness of the transformation consisting of Steps (1f), (2f), and (3f).

Theorem 5 (Correctness of Fw-Specialization). *Let $\langle SpSys, SpSafe \rangle$ be the specification derived by applying the Fw-Specialization method to the specification $\langle Sys, Safe \rangle$. Then, $\langle Sys, Safe \rangle$ is equivalent to $\langle SpSys, SpSafe \rangle$.*

Starting from the specification of Example 1, by applying our Fw-Specialization method, we get the following specialized specification:

$SpVar$: **enumerated** x_p $\{new1, new2\}$; **integer** x_1; **integer** x_2;
$SpInit$: $x_1 \geq 1 \wedge x_2 = 0 \wedge x_p = new2$;
$SpTrans$: $(x_1 < 1 \wedge x_p = new2 \wedge x_1' = x_1 + x_2 \wedge x_2' = x_2 + 1 \wedge x_p' = new1) \vee$
$\quad\quad\quad (x_p = new2 \wedge x_1' = x_1 + x_2 \wedge x_2' = x_2 + 1 \wedge x_p' = new2)$
$SpSafe$: $\neg\, \mathsf{EF}(x_2 > x_1 \wedge x_p = new2)$

The forward reachability algorithm implemented in ALV successfully verifies the safety property of this specialized specification, while it is not able to verify (within 600 seconds) the safety property of the initial specification of Example 1.

Lower Bounds for the Length of Reset Words in Eulerian Automata*

Vladimir V. Gusev

Ural Federal University, Ekaterinburg, Russia
vl.gusev@gmail.com

Abstract. For each odd $n \geq 5$ we present a synchronizing Eulerian automaton with n states for which the minimum length of reset words is equal to $\frac{n^2 - 3n + 4}{2}$. We also discuss various connections between the reset threshold of a synchronizing automaton and a sequence of reachability properties in its underlying graph.

1 Background and Overview

A complete deterministic finite automaton $\mathscr{A}$ is called *synchronizing* if the action of some word w resets $\mathscr{A}$, that is, leaves the automaton in one particular state no matter at which state it is applied. Any such word w is said to be a *reset word* for the automaton. The minimum length of reset words for $\mathscr{A}$ is called the *reset threshold* of $\mathscr{A}$ and denoted by $\mathrm{rt}(\mathscr{A})$. Synchronizing automata constitute an interesting combinatorial object and naturally appear in many applications such as coding theory, robotics and testing of reactive systems. For a brief introduction to the theory of synchronizing automata we refer the reader to the recent surveys [10,14]. The interest to the field is also heated by the famous Černý conjecture.

In 1964 Jan Černý [2] constructed for each $n > 1$ a synchronizing automaton $\mathscr{C}_n$ with n states whose reset threshold is $(n-1)^2$. Soon after that he conjectured that these automata represent the worst possible case, that is, every synchronizing automaton with n states can be reset by a word of length $(n-1)^2$. Despite intensive research, the best upper bound on the reset threshold of synchronizing automata with n states achieved so far is $\frac{n^3 - n}{6}$, see [8], so it is much larger than the conjectured value. Though the Černý conjecture is open in general, it has been confirmed for various restricted classes of synchronizing automata, see, e.g., [5,3,6,13,15]. We recall here a result by Jarkko Kari from [6] as it has served as a departure point for the present paper.

Kari [6] has shown that every synchronizing Eulerian automaton with n states possesses a reset word of length at most $n^2 - 3n + 3$. Even though this result confirms the Černý conjecture for Eulerian automata, it does not close the synchronizability question for this class of automata since no matching lower bound

---

* Supported by the Russian Foundation for Basic Research, grant 10-01-00524, and by the Federal Education Agency of Russia, grant 2.1.1/13995.

G. Delzanno and I. Potapov (Eds.): RP 2011, LNCS 6945, pp. 180–190, 2011.

for the reset threshold of Eulerian automata has been found so far. In order to find such a matching bound, we need a series of Eulerian automata with large reset threshold, which is the main problem that we address in the present paper.

Our first attempt was following an approach from [1]. In that paper, several examples of slowly synchronizing automata, which had been discovered in the course of a massive computational experiment, have been related to known examples of primitive graphs with large exponent from [4] and then have been expanded to infinite series. The idea was to apply a similar analysis to Eulerian graphs with large exponent that have been characterized in [12]. However, it turns out that in this way we cannot achieve results close to what we can get by computational experiments. Thus, a refinement of the approach from [1] appears to be necessary. Here we suggest such a refinement, and this is the main novelty of the present paper. As a concrete demonstration of our modified approach, we exhibit a series of slowly synchronizing Eulerian automata whose reset threshold is twice as large as the reset threshold of automata that can be obtained by a direct application of techniques from [1]. We believe that the method suggested in this paper can find a number of other applications and its further development may shed a new light on the properties of synchronizing automata.

2   Preliminaries

A *complete deterministic finite automaton* (DFA) is a couple $\mathscr{A} = \langle Q, \Sigma \rangle$, where Q stands for the *state set* and Σ for the *input alphabet* whose letters act on Q by totally defined transformations. The action of Σ on Q extends in natural way to an action of the set Σ^* of all words over Σ. The result of the action of a word $w \in \Sigma^*$ on the state $q \in Q$ is denoted by $q \cdot w$. Triples of the form $(q, a, q \cdot a)$ where $q \in Q$ and $a \in \Sigma$ are called *transitions* of the DFA; q, a and $q \cdot a$ are referred to as, respectively, the *source*, the *label* and the *target* of the transition $(q, a, q \cdot a)$.

By a *graph* we mean a tuple of sets and maps: the set of *vertices* V, the set of *edges* E, a map $t : E \to V$ that maps every edge to its *tail* vertex, and a map $h : E \to V$ that maps every edge to its *head* vertex. Notice that in a graph, there may be several edges with the same tail and head. [1] We assume the reader's acquaintance with basic notions of the theory of graphs such as path, cycle, isomorphism etc.

Given a DFA $\mathscr{A} = \langle Q, \Sigma \rangle$, its *underlying graph* $D(\mathscr{A})$ has Q as the vertex set and has an edge e_τ with $t(e_\tau) = q$, $h(e_\tau) = q \cdot a$ for each transition $\tau = (q, a, q \cdot a)$ of $\mathscr{A}$. We stress that if two transitions have a common source and a common target (but different labels), then they give rise to different edges (with a common tail and a common head). It is easy to see that a graph D is isomorphic to the underlying graph of some DFA if and only if each vertex of D serves as the tail for the same number of edges (the number is called the *outdegree* of D). In the sequel, we always consider only graphs satisfying this property. Every DFA $\mathscr{A}$

---

[1] Our graphs are in fact directed multigraphs with loops. But we use a short name, since no other graph species will show up in this paper.

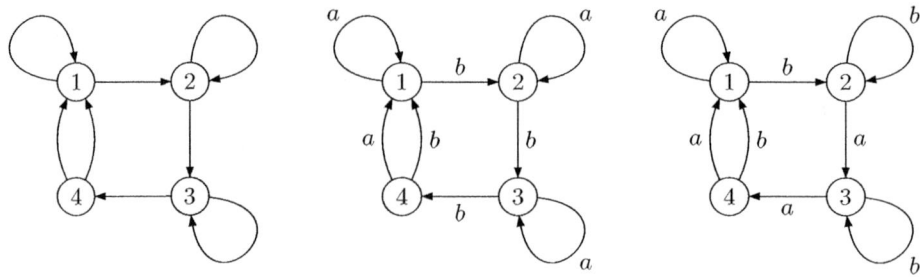

Fig. 1. A graph and two of its colorings

such that $D \cong D(\mathscr{A})$ is called a *coloring* of D. Thus, every coloring of D is labeling its edges of by letters from some alphabet whose cardinality is equal to the outdegree of D such that edges with a common tail get different colors. Fig. 1 shows a graph and two of its colorings by $\Sigma = \{a, b\}$.

A graph $D = \langle V, E \rangle$ is said to be *strongly connected* if for every pair $(v, v') \in V \times V$, there exists a path from v to v'. A graph is *Eulerian* if it is strongly connected and each of its vertices serves as the tail and as the head for the same number of edges. A DFA is said to be *Eulerian* if so is its underlying graph. More generally, we will freely transfer graph notions (such as path, cycle, etc) from graphs to automata they underlie.

A graph $D = \langle V, E \rangle$ is called *primitive* if there exists a positive integer t such that for every pair $(v, v') \in V \times V$, there exists a path from v to v' of length precisely t. The least t with this property is called the *exponent* of the digraph D and is denoted by $\exp(D)$.

Let w be a word over the alphabet $\Sigma = \{a_1, a_2, \ldots, a_k\}$. We say that a word $u \in \Sigma^*$ *occurs ℓ times as a factor of w* if there are exactly ℓ different words $x_1, \ldots, x_\ell \in \Sigma^*$ such that for each i, $1 \leq i \leq \ell$, there is a word $y_i \in \Sigma^*$ for which w decomposes as $w = x_i u y_i$. The number ℓ is called the *number of occurrences of u in w* and is denoted by $|w|_u$. The vector $(|w|_{a_1}, |w|_{a_2}, \ldots, |w|_{a_k}) \in \mathbb{N}_0^k$ is called the *Parikh vector* of the word w; here $\mathbb{N}_0$ stands for the set of non-negative integers.

Now suppose that $\mathscr{A} = \langle Q, \Sigma \rangle$ is a DFA and α is a path in $\mathscr{A}$ labelled by a word $w \in \Sigma^*$. If a vector $\mathsf{v} \in \mathbb{N}_0^k$ is equal to the Parikh vector of w, then we say that v is the *Parikh vector* of the path α. We refer to any path that has v as its Parikh vector as a v-*path*.

3   Main Results

We start with revisiting the technique used in [1] to obtain lower bounds for the reset threshold of certain synchronizing automata.

Consider an arbitrary synchronizing automaton $\mathscr{A} = \langle Q, \Sigma \rangle$. Let w be a reset word for $\mathscr{A}$ that leaves the automaton in some state $r \in Q$, that is, $p \cdot w = r$ for every $p \in Q$. Then, for every state $p \in Q$, the word w labels a path from

p to r. Therefore, for every state $p \in Q$ there is a path of length $|w|$ from p to r in the underlying graph $D(\mathscr{A})$. This leads us to the following notion. We say that a strongly connected graph $D = (V, E)$ is 0-*primitive* if there exists an integer $k > 0$ and a vertex $r \in V$ such that for every vertex $p \in V$ there is a path of length exactly k from p to r. The minimal integer k with this property (over all possible choices of r) is called the 0-*exponent* of D and is denoted by $\exp_0(D)$. We write $\exp_0(\mathscr{A})$ instead of $\exp_0(D(\mathscr{A}))$. Then we have that every synchronizing automaton $\mathscr{A}$ is 0-primitive and

$$\mathrm{rt}(\mathscr{A}) \geq \exp_0(\mathscr{A}). \tag{1}$$

It is not hard to see that the notions of 0-primitivity and primitivity are equivalent. Indeed, every primitive digraph is obviously 0-primitive. Conversely, let D be a 0-primitive digraph with n vertices. By the definition there are paths of length exactly $\exp_0(D)$ from every vertex to some fixed vertex r. Consider two arbitrary vertices p and q of D. Since D is strongly connected, there is a path α of length at most $n - 1$ from r to q. Now take any path β of length $n - 1 - |\alpha|$ starting at p and let s be the endpoint of β. There is a path γ of length $\exp_0(D)$ from s to r. Now the path $\beta\gamma\alpha$ leads from p to q (through s and r) and $|\beta\gamma\alpha| = n - 1 + \exp_0(D)$. Thus, the digraph D is primitive, and moreover, we have the following inequality:

$$\exp_0(D) + n - 1 \geq \exp(D). \tag{2}$$

The reader who may wonder why we need such a slight variation of the standard notion will see that this variation fits better into a more general framework that we will present below.

First, however, we demonstrate how to construct Eulerian automata with a relatively large reset threshold on the basis of the notion of 0-primitivity. For this, we need Eulerian digraphs with the largest possible exponent (or 0-exponent) among all primitive Eulerian digraphs with n vertices. Such digraphs have been classified by Shen [12].

For every odd $n \geq 5$, consider the automaton $\mathscr{D}_n$ with the state set $Q = \{1, 2, \ldots, n\}$ and the input letters a and b acting on Q as follows: $1 \cdot a = 2$, $1 \cdot b = 3$; $(n - 1) \cdot a = 2$, $(n - 1) \cdot b = 1$; $n \cdot a = 1$, $n \cdot b = 3$; and for every $1 < k < n - 1$

$$k \cdot a = \begin{cases} k + 2 & \text{if } k \text{ is even,} \\ k + 1 & \text{if } k \text{ is odd;} \end{cases} \qquad k \cdot b = \begin{cases} k + 3 & \text{if } k \text{ is even,} \\ k + 2 & \text{if } k \text{ is odd.} \end{cases}$$

The automaton $\mathscr{D}_n$ is shown in Fig. 2. We denote the underlying graph of $\mathscr{D}_n$ by $\mathcal{D}_n$.

Proposition 1. *If $\mathcal{G}$ is a primitive Eulerian graph with outdegree 2 and n vertices, $n \geq 8$, then $\exp(\mathcal{G}) \leq \frac{(n-1)^2}{4} + 1$. The equality holds only for the graph $\mathcal{D}_n$.*

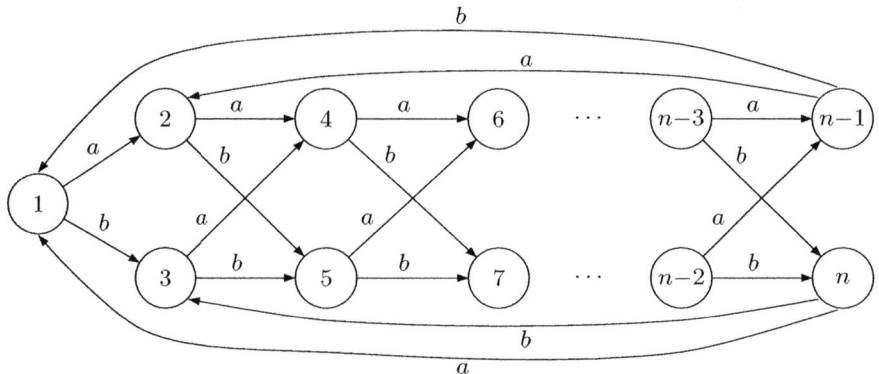

Fig. 2. The automaton $\mathcal{D}_n$

Proposition 1 and the inequalities (1) and (2) guarantee that every synchronizing coloring of the graph $\mathcal{D}_n$ has reset threshold of magnitude $\frac{n^2}{4} + o(n^2)$. In particular, we can prove the following result using the technique developed in [1].

Proposition 2. *The reset threshold of the automaton $\mathcal{D}_n$ is equal to $\frac{n^2-4n+11}{4}$.*

Proof. We start with estimating $\exp_0(\mathcal{D}_n)$. Observe that for every $\ell \geq \exp_0(\mathcal{D}_n)$ there is a cycle of length ℓ in $\mathcal{D}_n$. Indeed, let r be a state such that for every $p \in Q$ there is a path of length $\exp_0(\mathcal{D}_n)$ from p to r. Now take an arbitrary path α of length $\ell - \exp_0(\mathcal{D}_n)$ starting at r and let s be the endpoint of α. By the choice of r, there is a path β of length $\exp_0(\mathcal{D}_n)$ from s to r. Thus, the path $\alpha\beta$ is a cycle of length exactly ℓ.

Now consider the partition π of the set Q into $\frac{n+1}{2}$ classes V_i, $0 \leq i \leq \frac{n-1}{2}$, where $V_0 = 1$ and $V_i = \{2i, 2i+1\}$ for every $0 < i \leq \frac{n-1}{2}$. We define a graph $\mathcal{G}_n$ with the quotient set Q/π as the vertex set and with the edges induced by the edges of $\mathcal{D}_n$ as follows: there is an edge e' in $\mathcal{G}_n$ with $t(e') = V_i$ and $h(e) = V_j$ if and only if there is an edge e in $\mathcal{D}_n$ with $t(e) \in V_i$ and $h(e) \in V_j$. Then every cycle in $\mathcal{D}_n$ induces a cycle of the same length in $\mathcal{G}_n$. In particular, for every $\ell \geq \exp_0(\mathcal{D}_n)$ there is a cycle of length ℓ in $\mathcal{G}_n$. It is easy to see that the graph $\mathcal{G}_n$ has precisely two simple cycles: one of length $\frac{n-1}{2}$ and one of length $\frac{n+1}{2}$. We conclude that every $\ell \geq \exp_0(\mathcal{D}_n)$ is expressible as a non-negative integer combination of $\frac{n-1}{2}$ and $\frac{n+1}{2}$.

Here we invoke the following well-known and elementary result from arithmetic:

Lemma 1 ([9, Theorem 2.1.1]). *If k_1, k_2 are relatively prime positive integers, then $k_1 k_2 - k_1 - k_2$ is the largest integer that is not expressible as a non-negative integer combination of k_1 and k_2.*

Applying lemma 1 we conclude that $\exp_0(\mathcal{D}_n) \geq \frac{n^2-4n+3}{4}$ and there is no cycle of length $\frac{n^2-4n-1}{4}$ in $\mathcal{G}_n$. The inequality 1 implies that $\mathrm{rt}(\mathcal{D}_n) \geq \frac{n^2-4n+3}{4}$, and it remains to exclude two cases: $\text{rt}(\mathscr{D}_n) = \frac{n^2-4n+3}{4}$ and $\text{rt}(\mathscr{D}_n) = \frac{n^2-4n+7}{4}$. This is easy.

Suppose that w is a shortest reset word for $\mathscr{D}_n$ which leaves $\mathscr{D}_n$ in some state $r \in V_i$. Note that $i \neq 0$ (otherwise the word obtained by removing the last letter from w would be a shorter reset word, and this is impossible).

If $|w| = \frac{n^2-4n+3}{4}$, we write $w = xw'$ for some letter x and apply the word w to some state from V_{i-1}. We conclude that w' induces a cycle from V_i to V_i in $\mathcal{G}_n$. This cycle would be of length $\frac{n^2-4n-1}{4}$, which is impossible.

Finally suppose that the length of w is $\frac{n^2-4n+7}{4}$. If $i \neq 1$, then the same argument as in the previous paragraph leads to a contradiction. (We just apply w to a state from V_{i-2}.) If $i = 1$, let $w = xyw'$ for some letters x and y. Depending on x, either $n \cdot xy \in V_1$ or $(n-1) \cdot xy \in V_1$. In both cases w' induces a cycle from V_1 to V_1 in $\mathcal{G}_n$ of length $\frac{n^2-4n-1}{4}$, which is impossible.

We thus see that the reset threshold of the automaton $\mathscr{D}_n$ is at least $\frac{n^2-4n+11}{4}$. Since the word $aa(ba^{\frac{n-1}{2}})^{\frac{n-5}{2}}bb$ resets $\mathscr{D}_n$, we conclude that this bound is tight.

Our computational experiments show that the largest reset threshold among all synchronizing colorings of $\mathcal{D}_n$ is equal to $\frac{(n-1)^2}{4} + 1$. Therefore, $\frac{n^2}{4} + o(n^2)$ is the best lower bound on the reset threshold of synchronizing Eulerian automata with n states that can be obtained by a direct encoding of Eulerian graphs with large exponent. However, our main result (see Theorem 1 below) shows that for every odd n there is a synchronizing Eulerian automaton with n states and reset threshold $\frac{n^2-3n+4}{2}$. This indicates that the notion of 0-exponent is too weak to be useful for isolating synchronizing Eulerian automata with maximal reset threshold. The reason for this is that we have discarded too much information when passing from synchronizability to 0-primitivity—we forget everything about paths labelled by reset words except their length. Thus, we use another notion in which more information is preserved, namely, the Parikh vectors of the paths are taken into account.

Consider a DFA $\mathscr{A} = \langle Q, \Sigma \rangle$ with $|\Sigma| = k$ and fix some ordering of the letters in Σ. We define a subset $\text{E}_1(\mathscr{A})$ of $\mathbb{N}_0^k$ as follows: a vector $\mathbf{v} \in \mathbb{N}_0^k$ belongs to $\text{E}_1(\mathscr{A})$ if and only if there is state $r \in Q$ such that for every $p \in Q$, there exists a $\mathbf{v}$-path from p to r. If the set $\text{E}_1(\mathscr{A})$ is non-empty, then the automaton $\mathscr{A}$ is called 1-*primitive*. The minimum value of the sum $i_1 + i_2 + \cdots + i_k$ over all k-tuples $(i_1, i_2, \ldots, i_k)$ from $\text{E}_1(\mathscr{A})$ is called the 1-*exponent* of $\mathscr{A}$ and denoted by $\exp_1(\mathscr{A})$. We would like to note that a very close concept for colored multigraphs has been studied in [11,7]. Clearly, every synchronizing automaton $\mathscr{A}$ is 1-primitive and

$$\text{rt}(\mathscr{A}) \geq \exp_1(\mathscr{A}). \tag{3}$$

In order to illustrate how the notion of 1-exponent may be utilized, we prove a statement concerning the Černý automata $\mathscr{C}_n$ (this statement will be used in

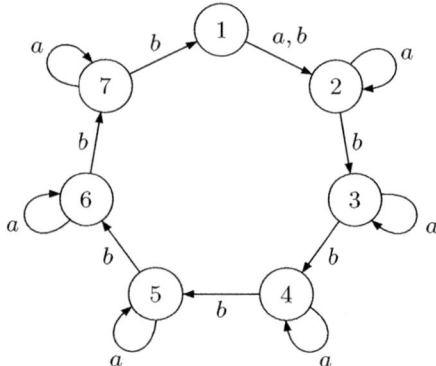

Fig. 3. The automaton $\mathscr{C}_n$ for $n = 7$

the proof of our main result). Recall the definition of $\mathscr{C}_n$. The state set of $\mathscr{C}_n$ is $Q = \{1, 2, \ldots, n\}$ and the letters a and b act on Q as follows:

$$i \cdot a = \begin{cases} 2 & \text{if } i = 1, \\ i & \text{if } 1 < i; \end{cases} \qquad i \cdot b = \begin{cases} i + 1 & \text{if } i < n, \\ 1 & \text{if } i = n. \end{cases}$$

The automaton $\mathscr{C}_n$ for $n = 7$ is shown in Fig. 3. Here and below we adopt the convention that edges bearing multiple labels represent bunches of edges sharing tails and heads. In particular, the edge $1 \xrightarrow{a,b} 2$ in Fig. 3 represents the two parallel edges $1 \xrightarrow{a} 2$ and $1 \xrightarrow{b} 2$.

Proposition 3. *Every reset word of the automaton $\mathscr{C}_n$ contains at least $n^2 - 3n + 2$ occurrences of the letter b and at least $n - 1$ occurrence of the letter a.*

Proof. Since the automaton $\mathscr{C}_n$ is synchronizing, the set $\mathrm{E}_1(\mathscr{C}_n)$ is non-empty. We make use of the following simple property of $\mathrm{E}_1(\mathscr{C}_n)$: if $\mathsf{v} = (\alpha, \beta) \in \mathrm{E}_1(\mathscr{C}_n)$, then for every $t \in \mathbb{N}$ we have $(\alpha, \beta + t) \in \mathrm{E}_1(\mathscr{C}_n)$. Indeed, let r be a state such that for every $p \in Q$ there is a v-path from p to r. We aim to show that there is also an $(\alpha, \beta + t)$-path from an arbitrary state p to r. Let $q = p \cdot b^t$, then by definition of r there is a v-path from q to r. Augmenting this path in the beginning by the path staring at p and labeled b^t, we obtain an $(\alpha, \beta + t)$-path from p to r.

Now observe that there is a v-path from r to r. This path is a cycle and it can be decomposed into simple cycles of the automaton $\mathscr{C}_n$. The simple paths in $\mathscr{C}_n$ are loops labeled a with the Parikh vector $(1, 0)$, the cycle

$$1 \xrightarrow{a} 2 \xrightarrow{b} 3 \xrightarrow{b} \ldots \xrightarrow{b} n - 1 \xrightarrow{b} n \xrightarrow{b} 1$$

with the Parikh vector $(1, n - 1)$ and the cycle

$$1 \xrightarrow{b} 2 \xrightarrow{b} 3 \xrightarrow{b} \ldots \xrightarrow{b} n - 1 \xrightarrow{b} n \xrightarrow{b} 1$$

with the Parikh vector $(1, n - 1)$. Thus, there are some $x, y, z \in \mathbb{N}_0$ such that the following equality holds true:

$$(\alpha, \beta) = x(1, 0) + y(1, n - 1) + z(0, n).$$

It readily implies that $\beta = y(n - 1) + zn$. Since for every $t \in \mathbb{N}$ the vector $(\alpha, \beta + t)$ also belongs to $\mathrm{E}_1(\mathscr{C}_n)$, we conclude that $\beta + t$ is also expressible as a non-negative integer combination of n and $n - 1$. Lemma 1 implies that $\beta \geq n(n-1) - n - (n-1) + 1 = n^2 - 3n + 2$. If w is a reset word of the automaton $\mathscr{C}_n$, then the Parikh vector of w belongs to $\mathrm{E}_1(\mathscr{C}_n)$, whence w contains at least $n^2 - 3n + 2$ occurrences of the letter b.

It remains to prove that w contains at least $n - 1$ occurrences of the letter a. Note that for every set S of states, we have $|S \cdot b| = |S|$ and $|S \cdot a| \geq |S| - 1$. Hence, to decrease the cardinality from n to 1, one has to apply a at least $n - 1$ times, and any word w such that $|Q \cdot w| = 1$ must contain at least $n - 1$ occurrences of a.

As a corollary we immediately obtain Černý's result [2, Lemma 1] that $\mathrm{rt}(\mathscr{C}_n) = (n-1)^2$. Indeed, Proposition 3 implies that the reset threshold is at least $(n-1)^2$, and it is easy to check that the word $(ab^{n-1})^{n-2}a$ of length $(n-1)^2$ resets $\mathscr{C}_n$. Also we see that a reset word w of minimal length for $\mathscr{C}_n$ is unique. Indeed, w cannot start or end with b because b acts as a cyclic permutation. Thus, $w = aua$ and the word u has $n^2 - 3n + 2$ occurrences of b and $n - 3$ occurrences of a. Note that b^n cannot occur as a factor of u since b^n acts is an identity mapping. Clearly, there is only one way to insert $n - 3$ letters a in the word b^{n^2-3n+2} such that the resulting word contains no factor b^n. Though the series $\mathscr{C}_n$ is very well studied, to the best of our knowledge the uniqueness of the shortest reset word for $\mathscr{C}_n$ has not been explicitly stated in the literature.

Observe that $\exp_0(\mathscr{C}_n) = n - 1$ and we could not extract any strong lower bound for $\mathrm{rt}(\mathscr{C}_n)$ from the inequality (2). In [1] a tight lower bound for $\mathrm{rt}(\mathscr{C}_n)$ has been obtained in an indirect way, via relating $\mathscr{C}_n$ to graphs with largest possible 0-exponent from [16]. In contrast, Proposition 3 implies that $\exp_1(\mathscr{C}_n)$ is close to $(n - 1)^2$ so the inequality (3) gives a stronger lower bound.

Now we are ready to present main result of this paper. We define the automaton $\mathscr{M}_n$ (from Matricaria) on the state set $Q = \{1, 2, \ldots, n\}$, where $n \geq 5$ is odd, in which the letters a and b act as follows:

$$k \cdot a = \begin{cases} k & \text{if } k \text{ is odd,} \\ k + 1 & \text{if } k \text{ is even;} \end{cases} \qquad k \cdot b = \begin{cases} k + 1 & \text{if } k \neq n \text{ is odd,} \\ k & \text{if } k \text{ is even,} \\ 1 & \text{if } k = n. \end{cases}$$

Observe that $\mathscr{M}_n$ is Eulerian. The automaton $\mathscr{M}_n$ for $n = 7$ is shown in Fig. 4 on the left.

Theorem 1. *If $n \geq 5$ is odd, then the automaton $\mathscr{M}_n$ is synchronizing and its reset threshold is equal to $\frac{n^2 - 3n + 4}{2}$.*

Proof. Let w be a reset word of minimum length for $\mathscr{M}_n$. Note that the action of aa is the same as the action of a. Therefore aa could not be a factor of w.

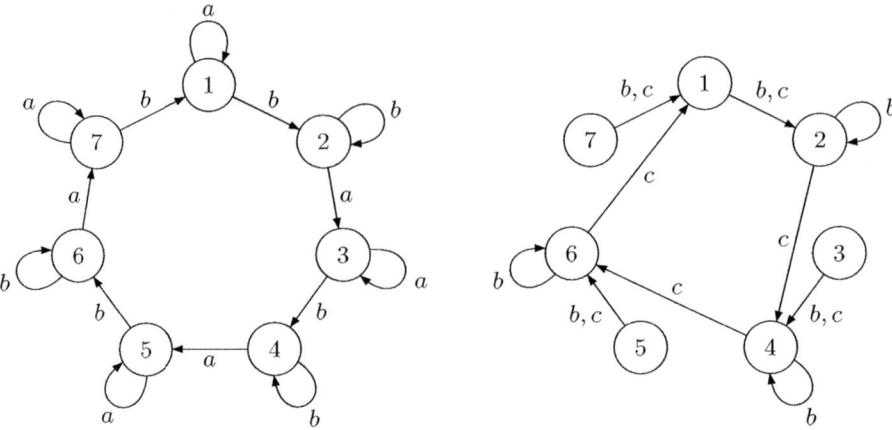

Fig. 4. The automaton $\mathscr{M}_n$ for $n = 7$ and the automaton induced by the actions of b and $c = ab$

(Otherwise reducing this factor to just a results in a shorter reset word.) So every occurrence of a, maybe except the last one, is followed by b. If we let $c = ab$, then either w or wb (if w ends with a) could be rewritten into a word u over the alphabet $\{b, c\}$. The actions of b and c induce a new automaton on the state set of $\mathscr{M}_n$ (this induced automaton is shown in Fig. 4 on the right). It is not hard to see that in both cases u is a reset word for the induced automaton. After applying the first letter of u it remains to synchronize the subautomaton on the set of states $S = \{1\} \cup \{2k \mid 1 \le k \le \frac{n-1}{2}\}$, and this subautomaton is isomorphic to $\mathscr{C}_{\frac{n+1}{2}}$.

Suppose $u = u'c$ for some word u' over $\{b, c\}$. Since the action of c on any subset of S cannot decrease its cardinality, we conclude that u' is also a reset word for the induced automaton. But c is the last letter of u only if $w = w'a$ and w' was rewritten into u'. Thus, w' also is a reset word for $\mathscr{M}_n$, which is a contradiction.

If $u = xu'$ for some letter x, then by Proposition 3 we conclude that u' has at least $(\frac{n+1}{2})^2 - 3(\frac{n+1}{2}) + 2 = \frac{n^2 - 4n + 3}{4}$ occurrences of c and at least $\frac{n-1}{2}$ occurrences of b. Since each occurrence of c in u' corresponds to an occurrence of the factor ab in w, we conclude that the length of w is at least $1 + 2\frac{n^2 - 4n + 3}{4} + \frac{n-1}{2} = \frac{n^2 - 3n + 4}{2}$.

One can verify that the word $b(b(ab)^{\frac{n-1}{2}})^{\frac{n-3}{2}}b$ is a reset word for $\mathscr{M}_n$ whence the above bound is tight.

It is not hard to see that $\exp_0(\mathscr{M}_n) = n - 1$ and also $\exp_1(\mathscr{M}_n)$ is linear in n. Thus, both 0-exponent and 1-exponent are far too weak to give a good lower bound for the reset threshold of $\mathscr{M}_n$. That is why we have obtained a tight lower bound for $\mathrm{rt}(\mathscr{M}_n)$ in an indirect way, via relating $\mathscr{M}_n$ to an automaton with a large 1-exponent (namely, to $\mathscr{C}_{\frac{n+1}{2}}$). Now we are going to develop a notion that can give a good bound in a more direct way.

Observe that the most important part of the proof of Theorem 1 deals with estimating the number of occurrences of the factor ab in a reset word. In fact, a rough estimation can be done directly. Let w be a reset word that leaves $\mathscr{M}_n$ in the state 2 and $k = |w|_{ab}$. Consider a path from 2 to 2 in which the state 2 does not occur in the middle. Words labeling such paths come from the language $L = b^*(a^+b^+)^{\frac{n-1}{2}}ba^*b$. Thus, w can be divided into several blocks from L. Since every block has either $\frac{n-1}{2}$ or $\frac{n+1}{2}$ occurrences of the factor ab, we conclude that k is expressible as a non-negative integer combination of the numbers $\frac{n-1}{2}$ and $\frac{n+1}{2}$. Note that $(ab)^t w$, where $t \in \mathbb{N}$, is a reset word that leaves $\mathscr{M}_n$ in the state 2. Since ab occurs $k+t$ times as a factor in $(ab)^t w$, we see that $k+t$ also is expressible as a non-negative integer combination of $\frac{n-1}{2}$ and $\frac{n+1}{2}$. Applying lemma 1 we conclude that $k \geq \frac{n^2-4n+3}{4}$. Thus, the length of w is at least $\frac{n^2-4n+3}{2}$.

The above reasoning suggests the following generalization. Let $\mathscr{A} = \langle Q, \Sigma \rangle$ be a DFA with $Q = \{1, 2, \ldots, n\}$ and let k be a non-negative integer. We say that the automaton $\mathscr{A}$ is k-primitive if there exist words $u_1, u_2, \ldots, u_n$ such that $1 \cdot u_1 = 2 \cdot u_2 = \cdots = n \cdot u_n$ and for every word v of length at most k we have $|u_1|_v = |u_2|_v = \ldots = |u_n|_v$. Note that the last condition implies that all words $u_1, u_2, \ldots, u_n$ have the same length. The minimal length of words that witness k-primitivity of $\mathscr{A}$ is called the k-exponent of $\mathscr{A}$ and is denoted by $\exp_k(\mathscr{A})$. Observe that the rough estimation in the previous paragraph shows that $\exp_2(\mathscr{M}_n)$ is close to $\mathrm{rt}(\mathscr{M}_n)$.

Consider now an arbitrary synchronizing automaton $\mathscr{A}$. It is clear that $\mathscr{A}$ is k-primitive for every k and $\mathrm{rt}(\mathscr{A}) \geq \exp_k(\mathscr{A})$. Thus, we have the following non-decreasing sequence:

$$\exp_0(\mathscr{A}) \leq \exp_1(\mathscr{A}) \leq \cdots \leq \exp_k(\mathscr{A}) \leq \exp_{k+1}(\mathscr{A}) \leq \ldots . \qquad (4)$$

At every next step we require that words $u_1, u_2, \ldots, u_n$ get more similar to each other than they were in previous step. Thus, eventually these words converge to a reset word and the sequence stabilizes at $\mathrm{rt}(\mathscr{A})$. So we hope that studying the sequence (4) may lead to a new approach to the Černý conjecture.

References

1. Ananichev, D.S., Gusev, V.V., Volkov, M.V.: Slowly synchronizing automata and digraphs. In: Hliněný, P., Kučera, A. (eds.) MFCS 2010. LNCS, vol. 6281, pp. 55–65. Springer, Heidelberg (2010)
2. Černý, J.: Poznámka k homogénnym eksperimentom s konečnými automatami. Matem.-fyzikalny Časopis Slovensk. Akad. Vied 14(3), 208–216 (1964) (in Slovak)
3. Dubuc, L.: Sur les automates circulaires et la conjecture de Černý. RAIRO Inform. Théor. Appl. 32, 21–34 (1998) (in French)
4. Dulmage, A.L., Mendelsohn, N.S.: Gaps in the exponent set of primitive matrices. Ill. J. Math. 8, 642–656 (1964)
5. Eppstein, D.: Reset sequences for monotonic automata. SIAM J. Comput. 19, 500–510 (1990)
6. Kari, J.: Synchronizing finite automata on Eulerian digraphs. Theoret. Comput. Sci. 295, 223–232 (2003)

7. Olesky, D.D., Shader, B., van den Driessche, P.: Exponents of tuples of nonnegative matrices. Linear Algebra Appl. 356, 123–134 (2002)
8. Pin, J.-E.: On two combinatorial problems arising from automata theory. Ann. Discrete Math. 17, 535–548 (1983)
9. Ramírez Alfonsín, J.L.: The diophantine Frobenius problem. Oxford University Press, Oxford (2005)
10. Sandberg, S.: Homing and synchronizing sequences. In: Broy, M., Jonsson, B., Katoen, J.-P., Leucker, M., Pretschner, A. (eds.) Model-Based Testing of Reactive Systems. LNCS, vol. 3472, pp. 5–33. Springer, Heidelberg (2005)
11. Shader, B.L., Suwilo, S.: Exponents of nonnegative matrix pairs. Linear Algebra Appl. 363, 275–293 (2003)
12. Shen, J.: Exponents of 2-regular digraphs. Discrete Math. 214, 211–219 (2000)
13. Trahtman, A.N.: The Černý conjecture for aperiodic automata. Discrete Math. Theor. Comput. Sci. 9(2), 3–10 (2007)
14. Volkov, M.V.: Synchronizing automata and the Černý conjecture. In: Martín-Vide, C., Otto, F., Fernau, H. (eds.) LATA 2008. LNCS, vol. 5196, pp. 11–27. Springer, Heidelberg (2008)
15. Volkov, M.V.: Synchronizing automata preserving a chain of partial orders. Theoret. Comput. Sci. 410, 2992–2998 (2009)
16. Wielandt, H.: Unzerlegbare, nicht negative Matrizen. Math. Z. 52, 642–648 (1950) (in German)

Parametric Verification and Test Coverage for Hybrid Automata Using the Inverse Method

Laurent Fribourg and Ulrich Kühne

LSV - ENS Cachan & CNRS, 94235 Cachan, France
{kuehne,fribourg}@lsv.ens-cachan.fr

Abstract. Hybrid systems combine continuous and discrete behavior. Hybrid Automata are a powerful formalism for the modeling and verification of such systems. A common problem in hybrid system verification is the good parameters problem, which consists in identifying a set of parameter valuations which guarantee a certain behavior of a system. Recently, a method has been presented for attacking this problem for Timed Automata. In this paper, we show the extension of this methodology for hybrid automata with linear and affine dynamics. The method is demonstrated with a hybrid system benchmark from the literature.

1 Introduction

Hybrid systems combine continuous and discrete behavior. They are especially useful for the verification of embedded systems, as they allow the unified modeling and the interaction of discrete control and the continuous environment or system state such as position, temperature or pressure.

There are several classes of formal models for hybrid systems. In general, there is a trade-off between the expressivity of the model and the complexity of the algorithmic apparatus that is needed for its formal analysis. Linear Hybrid Automata (LIIA) provide a good compromise. In contrast to more general hybrid automata models, which allow arbitrary dynamics of the continuous state variables, LHA are restricted to linear dynamics. This allows the use of efficient algorithms based on convex polyhedra. Furthermore, more complex dynamics – like hybrid automata with affine dynamics (AHA) – can easily be approximated conservatively by LHA. Although reachability is undecidable for LHA [12], practically relevant results have been obtained using this formalism [11].

For the modeling of embedded systems it is handy to use *parameters* either to describe uncertainties or to introduce tuning parameters that are subject to optimization. Instead of setting these parameters manually and then verifying the resulting concrete system, parameterized models are used to perform automatic *parameter synthesis*. A common assumption is the existence of a set of bad states that should never be reached. Then the parameter synthesis can be solved by treating the parameters as additional state variables and computing the reachable states of the parameterized system in a standard manner[11]. However, this standard approach is not feasible except for very simple cases. It is therefore

G. Delzanno and I. Potapov (Eds.): RP 2011, LNCS 6945, pp. 191–204, 2011.

essential to dynamically prune the search space. The method presented in [9] is based on the CEGAR approach, iteratively refining a constraint over the parameters by discarding states that violate a given property. A similar refinement scheme has already been used for (non-parameterized) reachability problems of hybrid systems (see e.g. [14]), starting with an abstraction and refining until the property has been proved or a counterexample has been found.

While these traditional approaches to parameter synthesis are based on the analysis of bad states or failure traces, a complementary – or *inverse* – method has been proposed in [4]. It uses a parameter instantiation that is known to guarantee a *good* behavior in order to derive a constraint on the parameters that leads to the same behavior. While the algorithm in [4] is restricted to Timed Automata (TA), we present its extension to LHA in this paper.

There are different scenarios for the application of the presented approach. If a given parameter instantiation is known to guarantee certain properties, the inverse method can be used to derive an enlarged area of the parameter space that preserves these properties, while possibly allowing for enhanced performance of the system. In the The inverse method can also be used to obtain a measure of *coverage* of the parameter space by computing the zones of equivalent behavior for each point. This approach is also known as *behavioral cartography* [5] and will be discussed in this paper. While the natural extension of these algorithms works well for simple LHA, it does not scale well to LHA models that approximate more complex dynamics. Therefore, we present an enhanced algorithm that can be applied on affine hybrid automata.

The presented algorithms are implemented in a tool called IMITATOR (*Inverse Method for Inferring Time AbstracT behaviOR*) [3]. The tool has originally been developed for the analysis of TA. The new version IMITATOR 3 implements the semantics of LHA as presented in Sect. 3. The manipulation of symbolic states is based on the polyhedral operations of the Parma Polyhedra Library [6].

Throughout the paper, we will use a running example – a distributed temperature control system – to illustrate the presented concepts. Further examples can be found in [10].

2   Related Work

The presented approach exhibits the same general differences with the CEGAR-based approach of [9] at the LHA level as formerly at the TA level. First, the input of CEGAR-based methods is a bad location to be avoided while the input of our inverse method is a good reference valuation for the parameters; second, the constraint in CEGAR-based methods guarantees the avoidance of bad locations while the constraint generated by the inverse method guarantees the same behavior (in terms of discrete moves) as under the reference valuation.

Additionally, our inverse method based approach for LHA is comparable to the symbolic analysis presented in [1] for improving the simulation coverage of hybrid systems. In their work, Alur et al. start from an initial state x and a discrete-time simulation trajectory, and compute a constraint describing those initial states that are guaranteed to be equivalent to x, where two initial states are considered to be equivalent if the resulting trajectories contain the same locations at each discrete step of execution. The same kind of constraint can be generated by our inverse method when initial values of the continuous variables are defined using parameters. The two methods are however methodologically different. On the one hand, the generalization process done by the inverse method works, using forward analysis, by refining the current constraint over the parameters that repeatedly discards the generated states that are incompatible with the initial valuation of x; on the other hand, the method of Alur et al. generalizes the initial value of x by performing a backward propagation of sets of equivalent states. This latter approach can be practically done because the system is supposed to be *deterministic*, thus making easy the identification of transitions between discrete states during the execution. Our inverse method, in contrast, can also treat nondeterministic systems.

The approach presented in [15] shares a similar goal, namely identifying for single test cases a robust environment that leads to the same qualitative behavior. Instead of using symbolic reachability techniques, their approach is based on the stability of the continuous dynamics. By using a bisimulation function (or contraction map), a robust neighborhood can be constructed for each test point. As traditional numeric simulation can be used, this makes the technique computationally effective. But, for weakly stable systems, a lot of test points have to be considered in order to achieve a reasonable coverage. For some of the examples in [15], we achieve better or comparable results (see [10]).

3   Hybrid Automata with Parameters

3.1   Basic Definitions

In the sequel, we will refer to a set of continuous variables $X = x_1, \ldots, x_N$ and a set of parameters $P = p_1, \ldots, p_M$. Continuous variables can take any real value. We define a valuation as a function $w : X \to \mathbb{R}$, and the set of valuations over variables X is denoted by $\mathcal{V}(X)$. A valuation w will often be identified with the point $(w(x_1), \ldots, w(x_N)) \in \mathbb{R}^N$. A parameter valuation is a function $\pi : P \to \mathbb{R}$ mapping the parameters to the real numbers.

Given a set of variables X, a linear inequality has the form $\sum_{i=1}^{N} \alpha_i x_i \bowtie \beta$, where $x_i \in X$, $\alpha_i, \beta \in \mathbb{Z}$ and $\bowtie \in \{<, \leq, =\}$. A convex linear constraint is a finite conjunction of linear inequalities. The set of convex linear constraints over X is denoted by $\mathcal{L}(X)$. For a constraint $C \in \mathcal{L}(X)$ satisfied by a valuation $w \in \mathcal{V}(X)$, we write $w \models C$. For a constraint over continuous variables and parameters $C \in \mathcal{L}(X \cup P)$ satisfied by a valuation w and a parameter valuation π, we write $\langle w, \pi \rangle \models C$. By convention, we also write $w \models C$ for partial valuations. For example, a valuation $w \in \mathcal{V}(X)$ is said to satisfy a constraint $C \in \mathcal{L}(X \cup P)$ iff it can be extended with at least one parameter valuation π such that $\langle w, \pi \rangle \models C$.

Sometimes we will refer to a variable domain X', which is obtained by renaming the variables in X. Explicit renaming of variables is denoted by the substitution operation. Here, $(C)_{[X/Y]}$ denotes the constraint obtained by replacing in C the variables of X by the variables of Y.

A convex linear constraint can also be interpreted as a set of points in the space $\mathbb{R}^N$, more precisely as a convex polyhedron. We will use these notions synonymously. In this geometric context, a valuation satisfying a constraint is equivalent to the polyhedron containing the corresponding point, written as $w \in C$. Also here, for a partial valuation w (i.e. a point of a subspace of C), we write $w \in C$ iff w is contained in the projection of C on the variables of w.

Definition 1. *Given a set of continuous variables X and a set of parameters P, a (parameterized) hybrid automaton is a tuple $\mathcal{A} = (\Sigma, Q, q_0, I, D, \rightarrow)$, consisting of the following*

- *a finite set of actions Σ*
- *a finite set of locations Q*
- *an initial location $q_0 \in Q$*
- *a convex linear invariant $I_q \in \mathcal{L}(X \cup P)$ for each location q*
- *an activity $D_q : \mathbb{R}^n \rightarrow \mathbb{R}^n$ for each location q*
- *discrete transitions $q \xrightarrow{g,a,\mu} q'$, with guard condition $g \in \mathcal{L}(X \cup P)$, action $a \in \Sigma$ and a jump relation $\mu \in \mathcal{L}(X \cup P \cup X')$.*

Given a parameter constraint $K \in \mathcal{L}(P)$, the automaton $\mathcal{A}$ with the parameters restricted to K is denoted by $\mathcal{A}(K)$. Given a parameter valuation π, the automaton $\mathcal{A}$ with all parameters instantiated as in π is denoted by $\mathcal{A}[\pi]$.

Without loss of generality, it is assumed here that all continuous variables x are initialized with $x = 0$. Arbitrary initial values can be modeled by adding a transition with appropriate variable updates. Parameters can be seen as additional state variables which do not evolve in time (null activity).

The activities D_q describe how the continuous variables evolve within each location q. In order to obtain automata models which can be symbolically analyzed, restrictions have to be made to these activities. This leads to the following classes of hybrid automata.

Definition 2. *We define the following subclasses of hybrid automata.*

(1) A linear hybrid automaton *(LHA) is a hybrid automaton, where in each location q, the activity is given by a convex linear constraint $D_q \in \mathcal{L}(\dot{X})$ over the time derivatives of the variables.*

(2) An affine hybrid automaton *(AHA) is a hybrid automaton, where in each location q, the activity is given by a convex linear constraint $D_q \in \mathcal{L}(X \cup \dot{X})$ over the variables and the time derivatives.*

The class of timed automata can be obtained by restricting the derivatives to $\dot{x} = 1$ and limiting the jump relations to either $x' = x$ or $x' = 0$ (clock reset) for all variables $x \in X$. In total, the automata models defined above form a hierarchy $TA \subset LHA \subset AHA$.

The reachable states of LHA can be efficiently represented by convex polyhedra. Due to the more complex dynamics, this is not true for AHA. In the following, we consider linear hybrid automata with parameters. But, AHA can be approximated by LHA with arbitrary precision by partitioning the state space, as e.g. described in [8]. In Sect. 4.3 it is discussed, how these techniques can be adapted to suit our methods. In the following, we give an example of a hybrid system, that will later on be used to illustrate the approaches proposed here.

Example 1. The *room heating benchmark* (RHB) has been described in [7]. It models a distributed temperature control system. There are m movable heaters for $n > m$ rooms. The temperature x_i in each room i is a continuous variable that depends on the (constant) outside temperature u, the temperature of the adjacent rooms, and whether there is an activated heater in the room.

Depending on the relations between the temperatures measured, the heaters will be moved. If there is no heater in room i, a heater will be moved there from an adjacent room j, if the temperature has reached a threshold $x_i \leq get_i$ and there is a minimum difference of the temperatures $x_j - x_i \geq dif_i$. Note that in contrast with the RHB modeled in [1], the heater move from a room to another one is nondeterministic, since multiple guard conditions can be enabled simultaneously (in [1], the nondeterminism is resolved by moving only the heater with the smallest index). The dynamics is given by equations of the form:

$$\dot{x}_i = c_i h_i + b_i(u - x_i) + \sum_{i \neq j} a_{i,j}(x_j - x_i) \tag{1}$$

where $a_{i,j}$ are constant components of a symmetric adjacency matrix, constants b_i and c_i define the influence of the outside temperature and the effectiveness of the heater for each room i, and $h_i = 1$ if there is a heater in room i and $h_i = 0$ otherwise. Here, we will study an instantiation of RHB as given in [1] with $n = 3, m = 2$, outside temperature $u = 4$, the constants $b = (0.4, 0.3, 0.4), c = (6, 7, 8)$. The adjacency matrix $a_{i,j}$ is given as $\begin{pmatrix} 0.0 & 0.5 & 0.0 \\ 0.5 & 0.0 & 0.5 \\ 0.0 & 0.5 & 0.0 \end{pmatrix}$ and the thresholds are set to $get = 18$ and $dif = 1$ for all rooms.

The system can be modeled as an AHA, as shown in Fig. 1. There are three control modes, corresponding to the positions of the two heaters. The automaton has four variables, the temperatures $X = \{x_1, x_2, x_3\}$ and a variable t acting as clock. In this example, the temperatures are sampled at a constant rate $\frac{1}{h}$, where h is a parameter of the automaton. This sampling scheme is used in the models of sampled-data hybrid systems of [16] and simulink/stateflow models [1].

3.2   Symbolic Semantics

The symbolic semantics of a LHA $\mathcal{A}(K)$ are defined at the level of constraints, a symbolic state is a pair (q, C) of a location q and a constraint C over variables and parameters. The corresponding operations are therefore performed on convex polyhedra rather than on concrete valuations. One necessary operation is the progress of time within a symbolic state, modeled by the *time-elapse* operation.

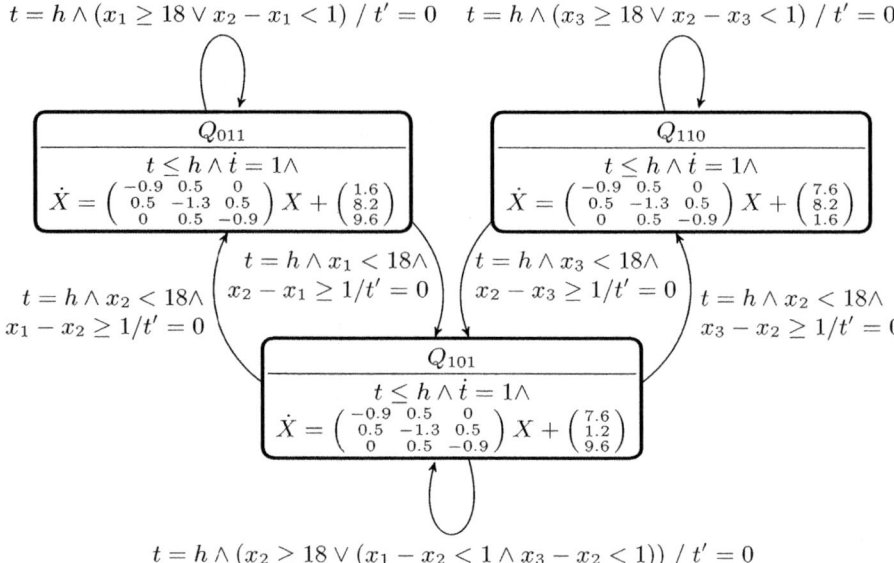

$$t = h \wedge (x_1 \geq 18 \vee x_2 - x_1 < 1) \,/\, t' = 0 \qquad t = h \wedge (x_3 \geq 18 \vee x_2 - x_3 < 1) \,/\, t' = 0$$

Fig. 1. Automaton model for room heating benchmark

Definition 3. *Given a symbolic state (q, C), the states reached by letting t time units elapse, while respecting the invariant of q, are characterized as follows:*

$$w' \in C \uparrow_q^t \quad iff \quad \exists w \in C, v \in D_q : w' = w + t \cdot v \wedge w' \in I_q.$$

We write $w' \in C \uparrow_q$ if $w' \in C \uparrow_q^t$ for some $t \in \mathbb{R}_+$.

Note that due to the convexity of the invariants, if $C \subseteq I_q$ and $C \uparrow_q^t \subseteq I_q$, then also $\forall t' \in [0, t] : C \uparrow_q^{t'} \subseteq I_q$. The operator preserves the convexity of C. Furthermore, the operator $C \downarrow_X$ denotes the projection of the constraint C on the variables in X. Based on these definitions, the symbolic semantics of a LHA $\mathcal{A}(K)$ is given by a labeled transition system (LTS).

Definition 4. *A labeled transition system over a set of symbols Σ is a triple $(S, S_0, \Rightarrow)$ with a set of states S, a set of initial states $S_0 \subseteq S$ and a transition relation $\Rightarrow \subseteq S \times \Sigma \times S$. We write $s \xrightarrow{a} s'$ for $(s, a, s') \in \Rightarrow$. A run of length m is a finite alternating sequence of states and symbols of the form $s_0 \xrightarrow{a_0} s_1 \xrightarrow{a_1} \ldots \xrightarrow{a_{m-1}} s_m$, where $s_0 \in S_0$.*

Definition 5. *The symbolic semantics of LHA $\mathcal{A}(K)$ is a LTS with*

- *states $S = \{(q, C) \in Q \times \mathcal{L}(X \cup P) \mid C \subseteq I_q\}$*
- *initial state $s_0 = (q_0, C_0)$ with $C_0 = K \wedge [\bigwedge_{i=1}^{N} x_i = 0] \uparrow_{q_0}$*
- *discrete transitions $(q, C) \xrightarrow{a} (q', C')$ if exists $q \xrightarrow{a, g, \mu} q'$ and $C' = ([C(X) \wedge g(X) \wedge \mu(X, X')] \downarrow_{X'} \wedge I_{q'}(X'))_{[X'/X]}$*

- *delay transitions* $(q, C) \xrightarrow{t} (q, C')$ with $C' = C \uparrow_q^t$
- *transitions* $(q, C) \xrightarrow{a} (q', C')$ if $\exists t, C'' : (q, C) \xrightarrow{a} (q', C'') \xrightarrow{t} (q', C')$

The *trace of a symbolic run* $(q_0, C_0) \xrightarrow{a_0} \ldots \xrightarrow{a_{m-1}} (q_m, C_m)$ is obtained by projecting the symbolic states to the locations, which gives: $q_0 \xrightarrow{a_0} \ldots \xrightarrow{a_{m-1}} q_m$. Two runs are said to be *equivalent*, if their corresponding traces are equal.

The set of states reachable from any state in a set S in exactly i steps is denoted as $Post^i_{\mathcal{A}(K)}(S) = \{s' \mid \exists s \in S : s \xrightarrow{a_0} \ldots \xrightarrow{a_{i-1}} s'\}$. Likewise, the set of all reachable states from S is defined as $Post^*_{\mathcal{A}(K)}(S) = \bigcup_{i \geq 0} Post^i_{\mathcal{A}(K)}$. The reachable states of an automaton $\mathcal{A}(K)$ are defined as $Reach_{\mathcal{A}(K)} = Post^*_{\mathcal{A}(K)}(\{s_0\})$, where s_0 is the initial state of $\mathcal{A}(K)$.

Note that during a run of $\mathcal{A}(K)$, the parameter constraints associated to the reachable states can only get stronger, since the parameters do not evolve under the time elapse operation, and can only be further constrained by invariants or guard conditions. This gives rise to the following observation.

Lemma 1. *For any reachable state* $(q, C) \in Reach_{\mathcal{A}(K)}$, *it holds that* $(\exists X : C) \subseteq K$. *This implies that for each parameter valuation* $\pi \models C$, *also* $\pi \models K$.

The lemma follows directly from the definition of the symbolic semantics. We say that a state (q, C) is *compatible* with a parameter valuation π, or just π-*compatible*, if $\pi \models C$. Conversely, it is π-*incompatible* if $\pi \not\models C$. These observations are the basis for the *Inverse Method*, is described in next section.

4 Algorithm

4.1 Inverse Method

The Inverse Method for LHA attacks the good parameters problem by generalizing a parameter valuation π that is known to guarantee a good behavior. Thereby, the valuation π is relaxed to a constraint K such that the *discrete behavior* – i.e. the set of traces – of $\mathcal{A}[\pi]$ and $\mathcal{A}(K)$ is identical. The algorithm has first been described for parametric timed automata in [4]. It has been applied for the synthesis of timing constraints for memory circuits [2].

Algorithm 1 describes the Inverse Method for LHA. The overall structure is similar to a reachability analysis. In the main loop, the reachable states with increasing depth i are computed. In parallel, the constraint K is derived. It is initialized with **true**. Each time a π-incompatible state (q, C) is reached, K is refined such that the incompatible state is unreachable for $\mathcal{A}(K)$. If C is π-incompatible, then there must be at least one inequality J in its projection on the parameters $(\exists X : C)$, which is incompatible with π. The algorithm selects one such inequality and adds its negation $\neg J$ to the constraint K. Before continuing with the search, the reachable states found so far are updated to comply with the new constraint K (line 1). If there are no more π-incompatible states, then i is increased and the loop continues.

Algorithm 1. $IM(\mathcal{A}, \pi)$

---

   input : Parametric linear hybrid automaton $\mathcal{A}$
   input : Valuation π of the parameters
   output: Constraint K_0 on the parameters
1 $i \leftarrow 0$; $K \leftarrow$ **true** ; $S \leftarrow \{s_0\}$
2 **while true do**
3 $\quad$ **while** *there are π-incompatible states in S* **do**
4 $\quad\quad$ Select a π-incompatible state (q, C) of S (i.e., s.t. $\pi \not\models C$) ;
5 $\quad\quad$ Select a π-incompatible inequality J in $(\exists X : C)$ (i.e., s.t. $\pi \not\models J$) ;
6 $\quad\quad$ $K \leftarrow K \wedge \neg J$;
7 $\quad\quad$ $S \leftarrow \bigcup_{j=0}^{i} Post_{\mathcal{A}(K)}^{j}(\{s_0\})$;
8 $\quad$ **if** $Post_{\mathcal{A}(K)}(S) \sqsubseteq S$ **then return** $K_0 \leftarrow \bigcap_{(q,C) \in S}(\exists X : C)$
9 $\quad$ $i \leftarrow i + 1$;
10 $\quad$ $S \leftarrow S \cup Post_{\mathcal{A}(K)}(S)$

The algorithm stops as soon as no new states are found (line 1). The output of the algorithm is then a parameter constraint K_0, obtained as the intersection of the constraints associated with the reachable states. The resulting constraint can be characterized as follows.

Proposition 1. *Suppose that the algorithm $IM(\mathcal{A}, \pi_0, k)$ terminates with the output K_0. Then the following holds:*

- *$\pi_0 \models K_0$*
- *For all $\pi \models K_0$, $\mathcal{A}[\pi_0]$ and $\mathcal{A}[\pi]$ have the same sets of traces.*

A proof along the lines of [13] can be found in [10]. We obtain a (convex) constraint K_0 including the initial point π_0, that describes a set of parameter valuations for which the same set of traces is observable. In particular, if $\mathcal{A}[\pi_0]$ is known to avoid a set of (bad) locations for π_0, so will $\mathcal{A}[\pi]$ for any $\pi \models K_0$.

The algorithm IM is not guaranteed to terminate[1]. Note also that the presented algorithm involves nondeterminism. In Algorithm 1 in lines 1 and 1, one can possibly choose among several incompatible states and inequalities. This may lead to different – nevertheless correct – results. This implies in particular that the resulting constraint K_0 is not maximal in general. (In order to overcome this limitation, the *behavioral cartography* method will be proposed in Section 4.2).

Example 2. In order to enable the application of the inverse method as described above to the RHB from example 1, the AHA automaton is converted to a LHA. This is done using the method described in [8]. The space is partitioned into regions, and within each region, the activity field is overapproximated using linear sets of activity vectors. For each region R delimiting a portion of the partitioned state space, the activities are statically overapproximated as

$$\dot{x}_i \in [min\{f_i(x) \mid x \in R\}, max\{f_i(x) \mid x \in R\}],$$

---

[1] Termination of such a general reachability-based procedure cannot be guaranteed due to undecidability of reachability for TA with parameters and LHA [12].

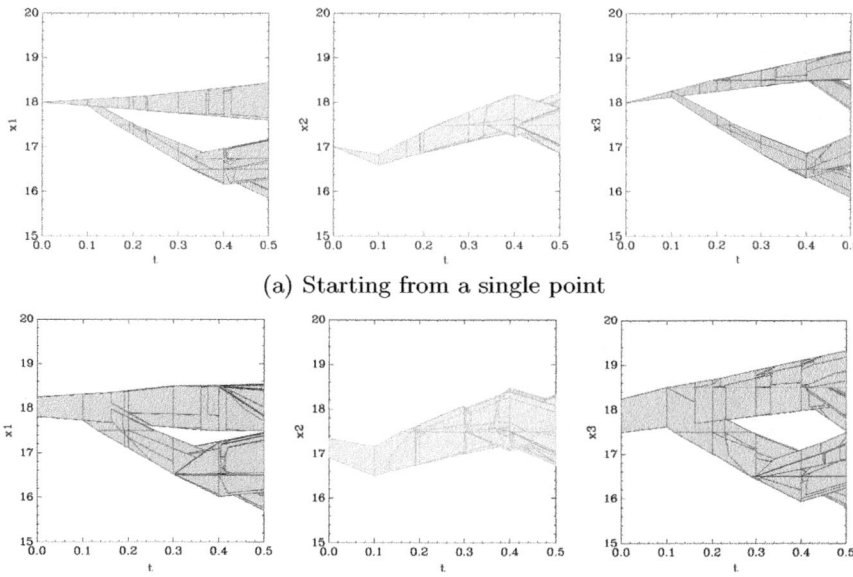

(a) Starting from a single point

(b) Starting from a tile synthesized by the Inverse Method

Fig. 2. Reachable states for room heating benchmark

where $f_i(x)$ corresponds to the right-hand side in (1). The approximation can be made arbitrarily accurate by approximating over suitably small regions of the state space. Here, each region R corresponds to a unit cube (of size 1 degree Celsius) in the dimensions x_1, x_2, x_3.

We now consider the following (bounded liveness) property:

Prop1: At least one of the heaters will be moved within a given time interval $[0, t_{max}]$ with $t_{max} = \frac{1}{2}$ and a sampling time $h = \frac{1}{10}$.

The upper bound t_{max} plays here the role of the maximal number of discrete transitions that are used in the method of [1]. In the automaton model, a violation of the property is modeled by a transition to a location q_{bad}. To check the property *Prop1* for varying initial conditions, we add the parameters a_1, a_2, a_3 and constrain the initial state with $x_1 = a_1 \wedge x_2 = a_2 \wedge x_3 = a3$. For initial point $(a_1, a_2, a_3) = (18, 17, 18)$, the reachable states for the variables x_1, x_2 and x_3 are shown in Fig. 2(a). The bad location is not reached from this point. Using the Inverse Method (Algorithm 1), the initial point can be generalized to a larger region around the starting point $(18, 17, 18)$, resulting in the constraint

$$a_1 \geq a_2 + \tfrac{181}{200} \wedge a_1 < \tfrac{a_3}{2} + \tfrac{37}{4} \wedge a_2 > \tfrac{3381}{200} \wedge a_2 < \tfrac{35}{2} \wedge a_3 > \tfrac{35}{2} \wedge a_3 < \tfrac{456}{25}.$$

The symbolic runs starting from this enlarged initial region are depicted in Fig. 2(b). The sets of traces of the two figures coincide, i.e. the sequence of discrete transitions of every run represented in Fig. 2(b) is identical to the sequence of discrete transitions of some run in Fig. 2(a).

Algorithm 2. BC

---

    input   : Parametric linear hybrid automaton $\mathcal{A}$
    input   : Parameter bounds $min_1 \ldots min_M$ and $max_1 \ldots max_M$
    input   : Step sizes $\delta_1 \ldots \delta_M$
    output: Set of constraints Z on the parameters

  1   $Z \leftarrow \varnothing$
  2   $V \leftarrow \{\pi \mid \pi_i = min_i + \ell_i \cdot \delta_i, \ \pi_i \leq max_i, \ \ell_1, \ldots, \ell_M \in \mathbb{N}\}$
  3   **while true do**
  4 | Select point $\pi \in V$ with $\forall K \in Z : \pi \not\models K$
  5 | $K \leftarrow IM(\mathcal{A}, \pi)$
  6 | $Z \leftarrow Z \cup \{K\}$
  7 | **if** $\forall \pi \in V : \exists K \in Z : \pi \models K$ **then**
  8 | | **return** Z

4.2 Behavioral Cartography

The inverse method works efficiently in many cases, since large parts of the state space can effectively be pruned by refining the parameter constraint K. In this way, many bad states never have to be computed, in contrast to the traditional approach to parameter synthesis. A drawback of the inverse method is that the notion of equivalence of the traces may be too strict for some cases. If e.g. one is interested in the non-reachability of a certain bad state, then there may exist several admissible regions in the parameter space that differ in terms of the discrete behavior or trace-sets. In order to discover these regions, the inverse method needs to be applied iteratively with different starting points.

The systematic exploration of the parameter space using the inverse method is called *behavioral cartography* [5]. It works as shown in Algorithm 2. For each parameter p_i, the interval $[min_i, max_i]$, possibly containing a single point, specifies the region of interest. This results in a rectangular zone $v_0 = [min_1, max_1] \times \cdots \times [min_M, max_M]$. Furthermore, step sizes $\delta_i \in \mathbb{R}$ are given. The algorithm selects (yet uncovered) points defined by the region v_0 and the step sizes and calls the inverse method on them. The set Z contains the tiles (i.e. parameter constraints) computed so far. The algorithm proceeds until all starting points are covered by some tile $K \in Z$.

By testing the inclusion in some computed tile, repeated computations are avoided for already covered points. The result of the cartography is a set of tiles of the parameter space, each representing a distinct behavior of the LHA $\mathcal{A}$. Note that the computed tiles do not necessarily cover the complete region v_0. On the other hand, it is possible that v_0 be covered by very few calls to the inverse method. Note also that, compared to the algorithm in [1], this is a stronger result, as each tile corresponds to a *set* of traces that exploits all possible behavior for the covered parameter valuations, including nondeterminism.

Example 3. The cartography is illustrated by a further experiment on the RHB model from example 2. Again, we check *Prop1*. The initial point is varied for

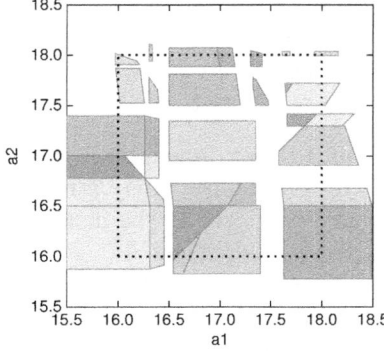

Fig. 3. Cartography of the initial states of RHB

the initial values a_1 and a_2, while fixing $a_3 = 18$. Therefore, the cartography procedure is used, iterating the initial point within the rectangle $[16, 18]^2$ (i.e, $min_1 = min_2 = 16$ and $max_1 = max_2 = 18$) with a step size of $\delta_1 = \delta_2 = \frac{1}{3}$. This leads to a total of 32 tiles, shown in Fig. 3. By analyzing the cartography, one gets a quantitative measure of the coverage of the considered region (shown as a dashed rectangle in the figure). In this case, the computed tiles cover 56% of the rectangle. All tiles in the figure have been classified as good tiles.

4.3   Enhancement of the Method for Affine Dynamics

It can be observed that for some systems there are areas in the parameter space, where slight variations of the initial conditions lead to many different traces. In this case, a good coverage based the cartography approach will be very costly, since many points have to be considered. In general, the inverse method and the behavioral cartography is quite limited when applied to LHA models that were obtained from AHA by static partitioning.

As described in [8], AHA can be approximated by LHA with arbitrary precision. This is done by partitioning the invariant of a location, usually into a set of small rectangular regions. For each region R, the affine dynamics are over-approximated by linear dynamics. In this way, the locations are split up until the desired precision is obtained.

Due to this partitioning, the resulting LHA will have more locations than the original AHA, leading also to more different traces for each parameter instantiation. This renders the inverse method ineffective for AHA, as the region around a parameter valuation π that corresponds to the same trace set, will generally be very small. This is because the traces contain a lot of information on the transitions between partitions that are irrelevant wrt. the system's behavior.

These limitations can be overcome by grouping reachable states that only represent different partitions of the same invariant of a location q. In our algorithm, this is done as an extension of the time-elapse operator. Each time that the time-elapse $C \uparrow_q$ needs to be computed for a location with affine dynamics D_q, the following steps are performed:

1. Build local partitions P of the invariant I_q
2. Compute a linear over-approximation $\hat{D}_P$ of D_q for each partition P
3. Compute the locally reachable states S wrt. partitions P and dynamics $\hat{D}_P$
4. Compute the convex hull of the states S

Here, the number of partitions Δ per dimension is chosen by the user. Note that cost and precision of the overall analysis may strongly depend on the chosen value for Δ. In practice, one would iterate the methods presented in this paper in order to refine the analysis by increasing Δ.

Given this variant of the time-elapse for affine dynamics, the computed reachable states are an over-approximation due to the piecewise linearization of the dynamics and the convex hull operation. Thus, the trace equivalence is no longer valid. But, as we compute an over-approximation of the possible runs, non-reachability is preserved.

Proposition 2. *Given an AHA $\mathcal{A}$, suppose that the algorithm $IM(\mathcal{A}, \pi_0, k)$ terminates with the output K_0. Then the following holds:*

- *$\pi_0 \models K_0$*
- *If for $\mathcal{A}[\pi_0]$, a location q_{bad} is unreachable, then it is also unreachable for all $\mathcal{A}[\pi]$ with $\pi \models K_0$*

Example 4. The adapted algorithm is applied to the RHB. With the discussed techniques, we can apply the inverse method and thus the cartography directly on the AHA model, without statically partitioning the state space in order to obtain a LHA. Again, by repeating the inverse method, a large part of the system's initial state space is decomposed into tiles of distinct discrete behavior. The reachability analysis for the AHA model is quite costly. Therefore, we will try to cover large parts of the parameter space using a very coarse linearization, given by a small number Δ of partitions. This is illustrated in the following. As reported in Example 3, applying the cartography on the statically linearized

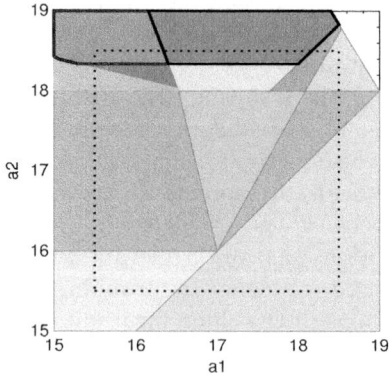

Fig. 4. Enhanced cartography for room heating benchmark

RHB model delivers a coverage of only 56% when fixing $a_3 = 18$. Instead, we apply the enhanced method directly on the AHA model, again regarding property *Prop1*. Here, the initial values a_1 and a_2 are varied within the rectangle $[15.5, 18.5]^2$ (i.e, $min_1 = min_2 = 15.5$ and $max_1 = max_2 = 18.5$) with a step size of $\delta_1 = \delta_2 = \frac{1}{2}$. In the first step, the invariants will be uniformly linearized, i.e. we set $\Delta = 1$. The resulting cartography in Fig. 4 consists of 12 tiles, where the good ones are shown in green, while the tiles corresponding to a bad behavior are shown in red (and outlined in bold). Note that the whole rectangular region is covered and that already with a coarse linearization, most of the tiles could be proved good.

5 Final Remarks

In this paper, we present a method to derive parameter constraints for LHA, that guarantee the same behavior as for a reference valuation of the parameters. This method has been recently introduced for deriving timing constraints for timed automata. Here, we provide the extension of the method to LHA. Furthermore, it is shown how the reachability procedure can be adapted to enable the analysis of systems with affine dynamics. By early pruning of invalid states, the method is more efficient than the parameter synthesis based on standard reachability analysis. Repeated analysis for different starting points yields a "behavioral cartography". This allows to cover large parts of the initial state space of nondeterministic hybrid systems, and provides an alternative tool to the symbolic simulation method of [1], which gives sometimes better results.

References

1. Alur, R., Kanade, A., Ramesh, S., Shashidhar, K.: Symbolic analysis for improving simulation coverage of simulink/stateflow models. In: EMSOFT, pp. 89–98 (2008)
2. André, É.: IMITATOR: A tool for synthesizing constraints on timing bounds of timed automata. In: Leucker, M., Morgan, C. (eds.) ICTAC 2009. LNCS, vol. 5684, pp. 336–342. Springer, Heidelberg (2009)
3. André, É.: IMITATOR II: A tool for solving the good parameters problem in timed automata. In: INFINITY. EPTCS, vol. 39, pp. 91–99 (September 2010)
4. André, É., Chatain, T., Encrenaz, E., Fribourg, L.: An inverse method for parametric timed automata. IJFCS 20(5), 819–836 (2009)
5. André, É., Fribourg, L.: Behavioral cartography of timed automata. In: Kučera, A., Potapov, I. (eds.) RP 2010. LNCS, vol. 6227, pp. 76–90. Springer, Heidelberg (2010)
6. Bagnara, R., Hill, P., Zaffanella, E.: Applications of polyhedral computations to the analysis and verification of hardware and software systems. Theoretical Computer Science 410(46), 4672–4691 (2009)
7. Fehnker, A., Ivancic, F.: Benchmarks for hybrid systems verification. In: Alur, R., Pappas, G.J. (eds.) HSCC 2004. LNCS, vol. 2993, pp. 326–341. Springer, Heidelberg (2004)
8. Frehse, G.: PHAVer: algorithmic verification of hybrid systems past HyTech. STTT 10(3), 263–279 (2008)

9. Frehse, G., Jha, S., Krogh, B.: A counterexample-guided approach to parameter synthesis for linear hybrid automata. In: Egerstedt, M., Mishra, B. (eds.) HSCC 2008. LNCS, vol. 4981, pp. 187–200. Springer, Heidelberg (2008)
10. Fribourg, L., Kühne, U.: Parametric verification of hybrid automata using the inverse method. Research Report LSV-11-04, LSV, ENS Cachan, France (2011)
11. Henzinger, T., Ho, P.-H., Wong-Toi, H.: HyTech: A model checker for hybrid systems. STTT 1, 110–122 (1997)
12. Henzinger, T., Kopke, P., Puri, A., Varaiya, P.: What's decidable about hybrid automata? In: JCSS, pp. 373–382 (1995)
13. Hune, T., Romijn, J., Stoelinga, M., Vaandrager, F.: Linear parametric model checking of timed automata. JLAP 52-53, 183–220 (2002)
14. Jha, S., Krogh, B., Weimer, J., Clarke, E.: Reachability for linear hybrid automata using iterative relaxation abstraction. In: Bemporad, A., Bicchi, A., Buttazzo, G. (eds.) HSCC 2007. LNCS, vol. 4416, pp. 287–300. Springer, Heidelberg (2007)
15. Julius, A., Fainekos, G., Anand, M., Lee, I., Pappas, G.: Robust test generation and coverage for hybrid systems. In: Bemporad, A., Bicchi, A., Buttazzo, G. (eds.) HSCC 2007. LNCS, vol. 4416, pp. 329–342. Springer, Heidelberg (2007)
16. Silva, B., Krogh, B.: Modeling and verification of sampled-data hybrid systems. In: ADPM (2000)

A New Weakly Universal Cellular Automaton in the $3D$ Hyperbolic Space with Two States

Maurice Margenstern

Université Paul Verlaine – Metz,
LITA EA 3097, UFR MIM, and CNRS, LORIA,
Campus du Saulcy,
57045 METZ Cédex 1, France
margens@univ-metz.fr
http://www.lita.sciences.univ-metz.fr/~margens

Abstract. – In this paper, we show a construction of a weakly universal cellular automaton in the $3D$ hyperbolic space with two states. Moreover, based on a new implementation of a railway circuit in the dodecagrid, the construction is a truly $3D$-one. This result under the hypothesis of weak universality and in this space cannot be improved.

Keywords: cellular automata, weak universality, hyperbolic spaces, tilings.

1 Introduction

In this paper, we construct a weakly universal cellular automaton in the $3D$ hyperbolic space with two states. Moreover, based on a new implementation of a railway circuit in the dodecagrid,the construction is a truly $3D$-one.

The dodecagrid is the tiling $\{5, 3, 4\}$ of the $3D$ hyperbolic space, and we refer the reader to [15,6] for an algorithmic approach to this tiling. We remind the reader that in the just mention denotation of the tiling, 5 represents the number of sides of a face, 3 the number of faces meeting at a vertex and 4 the number of dodecahedra around an edge. We also refer the reader to [5,7] for an implementation of a railway circuit in the dodecagrid which yields a weakly universal cellular automaton with 5 states. The circuit is the one used in other papers by the author, alone or with co-authors, inspired by the circuit devised by Ian Stewart, see [18]. The notion of weak universality is discussed in previous papers, see for instance [20,3,9,5] and comes from the fact that the initial configuration is infinite. However, it is not an arbitrary configuration: it has to be regular at large according to what was done previously, see [5,17,16,11].

Due to the small room left for this paper, we very briefly mention the new features of the implementation of the railway circuit in Section 2. In Section 3, we define the elements of the new implementation and, in Section 4 we describe the scenario of the simulation. We refer the reader to [14] for a detailed study and, in particular, for the rules of the automaton whose correctness was checked by a computer program, allowing us to state the following result:

G. Delzanno and I. Potapov (Eds.): RP 2011, LNCS 6945, pp. 205–217, 2011.

Theorem 1. *There is a weakly universal cellular automaton in the dodecagrid which is weakly universal and which has two states exactly, one state being the quiescent state. Moreover, the cellular automaton is rotation invariant and the set of its cells changing their state is a truly 3D-structure.*

In Section 5 we look at the remaining tasks.

2   The Railway Circuit

As initially devised in [18] and then mentioned in [4,2,17,16,7], the circuit uses tracks represented by lines and quarters of circles and switches. There are three kinds of switches: the **fixed**, the **memory** and the **flip-flop** switches. They are represented by the schemes given in Figure 1.

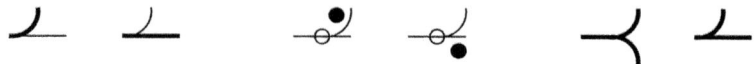

Fig. 1. The three kinds of switches. From left to right: fixed, flip-flop and memory switches.

A switch is an oriented structure: on one side, it has a single track u and, on the the other side, it has two tracks a and b. This defines two ways of crossing a switch. Call the way from u to a or b **active**. Call the other way, from a or b to u **passive**. The names comes from the fact that in a passive way, the switch plays no role on the trajectory of the locomotive. On the contrary, in an active crossing, the switch indicates which track, either a or b will be followed by the locomotive after running on u: the new track is called the **selected** track. The **fixed switch** is left unchanged by the passage of the locomotive. It always remains in the same position: when actively crossed by the locomotive, the switch always sends it onto the same track. The flip-flop switch is assumed to be crossed actively only. Now, after each crossing by the locomotive, it changes the selected track. The memory switch can be crossed by the locomotive actively and passively. In an active passage, the locomotive is sent onto the selected track. Now, the selected track is defined by the track of the last passive crossing by the locomotive. Of course, at initial time, the selected track is fixed.

With the help of these three kinds of switches, we define an **elementary circuit** as in [18], which exactly contains one bit of information. The circuit is illustrated by Figure 2, below and it is implemented in the Euclidean plane. It can be remarked that the working of the circuit strongly depends on how the locomotive enters it. If the locomotive enters the circuit through E, it leaves the circuit through O_0 or O_1, depending on the selected track of the memory switch which stands near E. If the locomotive enters through U, the application of the given definitions shows that the selected track at the switches near E and U are both changed: the switch at U is a flip-flop which is changed by the actual active passage of the locomotive and the switch at E is a memory one which

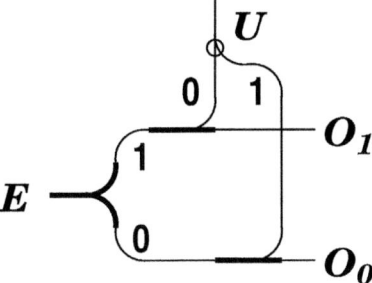

Fig. 2. The elementary circuit

is changed because it is passively crossed by the locomotive and through the non-selected track. The actions of the locomotive just described correspond to a **read** and a **write** operation on the bit contained by the circuit which consists of the configurations of the switches at E and at U. It is assumed that the write operation is triggered when we know that we have to change the bit which we wish to rewrite.

For a detailed use of this element, the reader is referred to [4,2]. For an implementation in the hyperbolic plane and in the hyperbolic 3D space, the reader is referred to [10,14].

3   A New Implementation in the Dodecagrid

Remember that the first weakly universal cellular automaton in the dodecagrid had five states, see [5]. The second result of the same type in the dodecagrid, see [10] is a cellular automaton with 3 states. As usual in this type of research, this important improvement requires something new in the scenario. The difference between these results relies on the implementation of the tracks. In the former paper, the track is materialized by a state which is different from the quiescent state and the locomotive consists of two contiguous cells of the track where one of them is another colour. In the latter paper, the track is a more complex structure: the locomotive always occupies cells which are quiescent when the locomotive is not there. The track is not materialized, but it is indicated by milestones, cells in another state, one of the two ones required by the locomotive. These milestones are placed close to the track in a way which allows the motion of the locomotive in both directions along the track, see [10]. As a consequence, the configuration of the switches is very different in the latter paper from what it is in the former.

This new important improvement, the last one as the present result cannot be improved for what is the number of states, is also based on a new implementation of the tracks with deep consequences on the tracks.

In order the reader could better understand this new implementation, we remind a few features on the representation of the dodecagrid.

3.1   Representation of the Dodecagrid

Due to the small room of this paper, it is not possible to provide the reader with a short introduction to hyperbolic geometry and to the tiling which we are using now. Here, we assume that the reader knows the pentagrid, the tiling $\{5, 4\}$ of the hyperbolic plane and we refer him/her to the papers of the author, see [6,7] as an example.

If we fix a face F of a dodecahedron Δ of the dodecagrid, the plane which supports F also supports faces of infinitely many dodecahedra. We shall use this property to project parts of the dodecagrid on planes supporting faces of its tiles as illustrated by Figure 3. For each of these dodecahedra, we shall say that its face on the plane of F is its **back**. Its face, opposite to its face is its **top**.

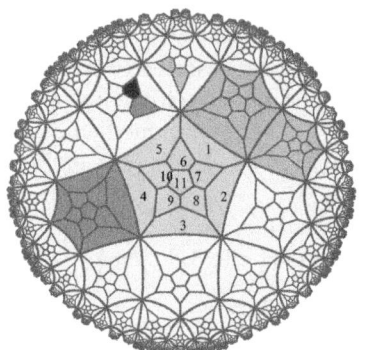 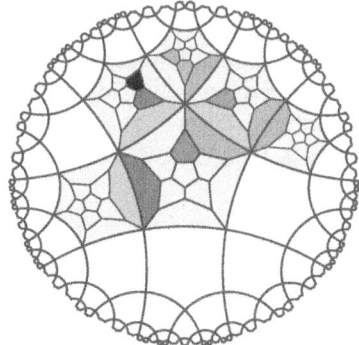

Fig. 3. Two different ways for representing a pseudo-projection on the pentagrid. On the left-hand side: the tiles have their colour. On the right-hand side: the colour of a tile is reflected by its neighbours only. Note the numbering of the central tile on the left-hand side picture. Face 0 of the tile is on the plane of projection.

On these figures, a dodecahedron is represented by its central projection on the plane Π of its back from a point which is on the axis joining the back and the top. Consider two neighbouring dodecahedra Δ_1 and Δ_2 with their backs on Π. Let σ be their common edge lying on Π. Then, in the projection, the common face of Δ_1 and Δ_2 is represented by two faces of the projections having in common σ. We shall take advantage of this property to represent the state of a cell by its colour on the projections of its outer faces, those in contact with Π on the projections of the same spatial face in the neighbour of the cell.

When the projection on one plane will not be enough, we shall use a projection on another plane, perpendicular to the first one.

3.2   The New Tracks

In this new setting, the locomotive is reduced to one cell and it will be called the **particle**. We have two colours this time: white and black. White is the quiescent state and the particle is always black. The implementation of the milestones around a cell of the track is not symmetric with respect to the plane of the back of the cell. For this reason, we decide that a track is one way.

Accordingly, when the return motion is needed, tracks are defined by pairs, as this is mostly the case in real life where they are parallel. Now, in the hyperbolic plane, it is very difficult to construct 'parallel' lines. Thanks to the third dimension, we can find a more efficient solution, illustrated by Figure 4.

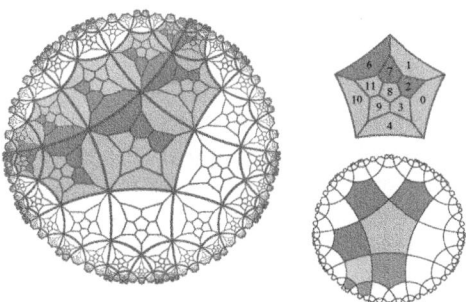

Fig. 4. Left-hand side: The new vertical ways with two tracks. In yellow, one direction; in brown, the opposite direction. Right-hand side: up: the numbering of the faces in a dodecagrid of the tracks; down: a cut of the tracks in the plane of the face 10 of the central cell.

The idea is to put the track one upon the other on both sides of the plane *Π*, taking advantage of the third dimension. On the figure, especially in the cut performed according to a plane orthogonal to *Π*, we can see that the milestones are organized as a kind of catenary above the back faces of the cells of the tracks.

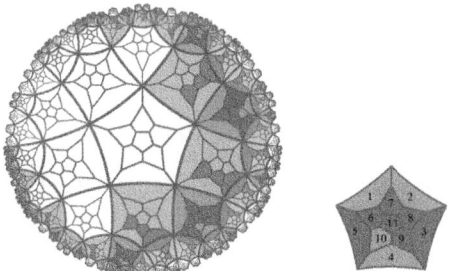

Fig. 5. The new horizontal ways with two tracks. In yellow and green, one direction. In brown and purple, the opposite direction. Note that here we have straight elements and corners. We also have two kinds of horizontal elements as well as two kinds of corners.

The track represented in Figure 4 represent one kind of tracks which play here the role of the verticals in the Euclidean representation. We need another kind of tracks, playing the role of the horizontals, which are represented by Figure 5.

Now, as in the previous papers, we take advantage of the third dimension to remove the crossings: they are replaced by bridges.

The bridges require some care but we have no room here to detail this issue. We just mention the following feature. Let Π be the plane separating the direct and the return tracks. As the return track is below Π, the part concerning this track is a bridge below Π too: it is the reflection of the bridge over Π on which the direct track travels. The reader is referred to [14] for the exact implementation.

Now, we can turn to the implementation of the switches in this new setting.

3.3   The New Switches

For this analysis, we shall again use the tracks u, a and b defined in Section 2. We remind that the active passage goes from u to either a or b and that the passive crossing goes from either a or b to u. As we split the ways into two tracks, we shall denote them by u_d, u_r, a_d, a_r, b_d and b_r respectively, where the subscript d indicates the active direction and r indicates the return one. *A priori*, this defines two switches: the first one from u_d to a_d or b_d and the second one from a_r or b_r to u_r. We shall call the first one the **active switch** and the second one the **passive switch**. Note that each of these new switches deals with one-way tracks only. This can be illustrated by the right-hand side picture of Figure 6.

As the flip-flop switch is used in an active passage only, there is no return track and, consequently no passive switch. A flip-flop makes use of single tracks only, in the direction to the switch for the way u defined in Section 2.

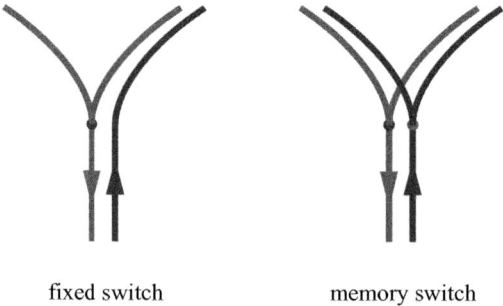

fixed switch   memory switch

Fig. 6. The new switches. On the left-hand side, the fixed switch; on the right-hand side: the memory switch.

And so, we remain with the fixed and the memory switches.

First, let us look at the fixed switch. As the switch is fixed, we may assume that u always goes to a. This means that the track b_d is useless. Now, it is plain that u_d and a_d constitute the two rays of a line issued from a point of the line. Consequently, there is no active fixed switch. The switch is concerned by the return tracks only: u_r, a_r and b_r, it is a passive switch. It is easy to see that it works as a **collector**: it collects what comes from both a_r and b_r and send this onto u_r. Now, as at any time there is at most one particle arriving at the switch from the union of a_r and b_r, the collector receives the particle from a_r or b_r alternately, never at the same time, see the left-hand side picture of Figure 6.

Now, let us look at the memory switch. It is clear that, in this case, we have both an active and a passive switch. However, a closer look at the situation shows that the passive memory switch has a tight connection with the fixed switch and that the active memory switch has a tight connection with the flip-flop switch. We shall return to this point in Section 4 to which we turn now.

4   The Scenario of the Simulation

4.1   The Motion of the Particle

We briefly mention the general way of motion. On a vertical segment, where the cells have their back on the same plane Π, the particle enters a cell through its face 10 and it exits through its face 1, see the cell with numbered faces in Figure 4. Other directions, as an example from face 4 to face 1 are possible. On a horizontal segment, the straight elements work in a similar way. But there are also corners, see Figure 5. As they are not symmetric, there are two kinds of them. One is used for one way, from face 1 to face 2 and the other is used for the opposite direction, from face 2 to face 1, again see Figure 5.

We are now ready to study the implementation of the switches.

We shall examine the three kinds of switch successively. We start with the trivial case of the fixed switch. We go on with the rather easy case of the flip-flop switch and we complete the study by the rather involved case of the memory switch.

4.2   Fixed Switches

As known from Section 2, we have a passive switch only to implement. The definition of the motion of the particle by a passage from face 4 or 10 to face 1 shows that there is nothing to do but abutting the two arriving tracks to the exiting one. This is illustrated by Figure 7. It is enough to make the central cell

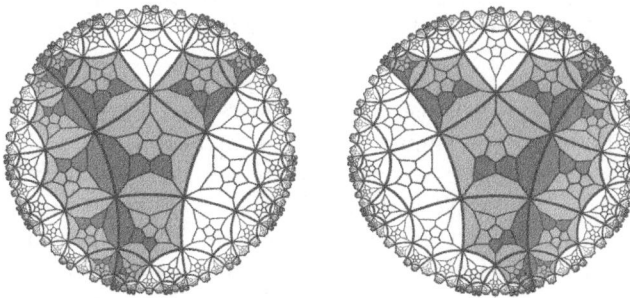

Fig. 7. Two fixed switch: left-hand side and right-hand side. Note the symmetry of the figure. Also note that this requires straight elements only.

symmetric, which is easy to realize: it is enough to rotate a straight element around its face 1 in such a way that the face lying on Π_0 is now face 0.

Now, it is easy to put the return track, either along one arriving track or along the other: Figure 7 shows that the orientation of the milestones presented in the figure makes both constructions possible.

We shall see that this easy implementation will help us to realize a rather simple implementation of the passive memory switch.

4.3 Flip-Flop Switches

Now, we have to realize the flip-flop switch. We can easily see that it is not enough to reverse the straight elements with respect to the previous figure in order to realize the switch. The central cell must have a specific pattern. We decide to append just one additional milestone and we change a bit the pattern, making it symmetric and significantly different from that of the fixed switch, see Figure 8.

This fixes the frame but now, we have to implement the mechanism which first, forces the particle to go to one side and not to the other one and then, to change the selected track.

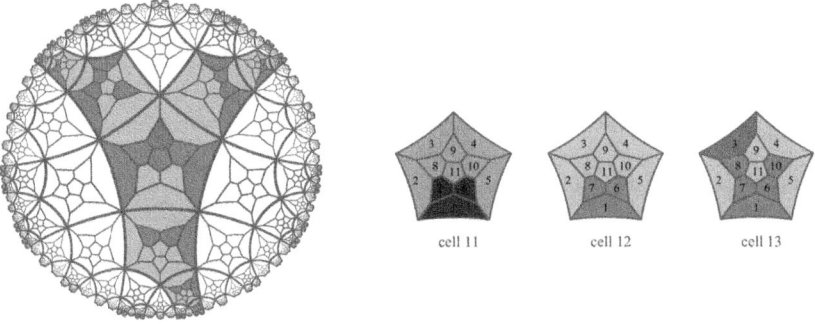

Fig. 8. The flip-flop switch: Left-hand side: here the selected track is the right-hand side one. Three cells have a particular patter: the central one and those which the ways a and b abut. Right-hand side: the control device; cell 13 is the controller, cells 11 and 12 are mere signals. Here cell 11 is black, signaling the forbidden track.

Let us number a few cells using the numbers they are given in the computer program which checks the simulation: the central cell of the picture is 4, and its neighbours on the tracks a and b are 5 and 8 respectively. All these cells have their back as face 0. We distinguish cells 11, 12 and 13 on cells 5, 8 and 4 respectively which are on the face 9 of these cells. The numbering of the faces of cells 11, 12 and 13 in the right-hand side of Figure 8 allows to locate the face 0 of cells 4, 5 and 8 respectively. Note that cells 11 and 13 are in contact: the face 4 of the former coincides with the face 3 of the latter. Cells 12 and 13 are in contact symmetrically. Cells 11 and 12 are **sensors** and cell 13 is the **controller**.

Cells 11 and 12 signalize the selected track: one of them is white and the other is black, the black cell being on the forbidden track. The controller is usually white and when it flashes to black, cells 11 and 12 simultaneously change their state to the opposite one. Cell 13 is black just for this action: after that it goes back to the white state.

How can cell 13 detect the situation? As already indicated, cells 11 and 12 are in contact with cell 13. It is also in contact with cell 4. When the particle is in cell 4, cell 13 detects it and flashes. So, at the next step, the particle is on the expected track, as the black cell 11 or 12 bars the corresponding track. Now, cells 11 and 12 can see cell 13 in black: they both change their state. This is visible at the next time but cell 13 is again white and the particle has gone further so that cells 11 and 12 have the opposite state with respect to the one they had before the particle arrived to cell 4.

4.4   Memory Switches

The memory switch is the most complex structure in our implementation. This is not due to the fact that we have to implement two single-track switches: a passive one and an active one. It is mainly because these two single-track switches must be connected. This comes from the definition of the memory switch: the selected track of the active switch is defined by the last crossing of the passive switch.

The passive switch is not as inactive as its name would mean. If the particle happens to cross the passive switch through the non-selected track, it changes the selected track in the passive switch and, also, it sends a signal which triggers the change of the selected track in the active switch. This can be performed by carefully mixing the characteristics of the fixed and the flip-flop switches.

Figure 9 represents the passive memory switch. Here, the selected track is the left-hand side one.

As in the case of the flip-flop switch, we number the cells involved in the passive switch by taking the number they received in the computer program.

The central cell again receives number 4 and cells 5 and 8 are its neighbours belonging to the parts a and b of the tracks. Cell 4 can see cells 5 and 8 through its faces 3 and 4 respectively and, conversely, cells 5 and 8 can see cell 4 through their face 4 and 3 respectively. Cells 11, 12 and 13 are very different from the cells with the same numbers in a flip-flop switch. Here, cell 13 can be characterized as follows: let A be the common vertex of cells 4, 5 and 8. Above Π_0, there are four dodecahedra sharing A. We just mentioned three of them. The fourth one is cell 13, which is obtained from cell 4 by reflection in the plane orthogonal to Π_0 which passes through A and which cuts the faces 1 of cell 5 and 8 perpendicularly. Now, we can number the faces of cell 13 in such a way that cell 13 can see cells 5 and 8 through its faces 4 and 3. Cells 11 and 12 are put on cell 13, on its faces 10 and 8. Cells 11 and 12 are the reflection of cells 5 and 8 respectively in the edge which is shared by both the concerned dodecahedra. If the selected track is b, cell 12 is black and cell 11 is white. If the selected track is a, cell 11 is black and cell 12 is white. Now we can describe what happens more clearly.

As required by the working of a memory switch, if the particle crosses the passive memory switch through the selected track, nothing happens.

And so, consider the case when the particle crosses the passive switch through the non-selected track. When the particle is in cell 8, cell 13 can see the particle and, as its cell 12 is black, it knows that it must flash. This means that it becomes black for the next time and then returns to the white state at the following time.

When cell 13 flashes, cells 11 and 12 exchange their states: if it was black it becomes white and conversely.

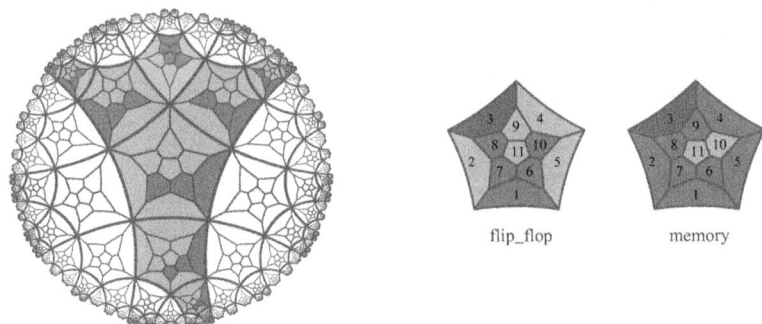

flip_flop   memory

Fig. 9. Left-hand side: the passive memory switch, here the left-hand side one. Nothing happens if the particle comes through this track. If it comes through the other one, this will trigger the change of selection, both in the passive and in the active switch. Right-hand side: the difference between the patterns of the controller of the flip-flop and that of the passive memory switch.

But this is not enough. The change must also happen in the active switch. To this purpose, a track conveys the particle created by the flash of cell 13 to the active memory switch. We can see this as a kind of branching which is ready and which captures the particle which consists of cell 13 when it is black.

This track leads the new particle to the active switch while the previous particle, that which plays the role of the locomotive goes on along the way defined by its track. And so, as long as the new particle does not reach its goal in the active memory switch, we have two particles in the circuit. The new particle can be seen as a temporary copy of the first one.

The active switch is implemented as a flip-flop switch with its cells with the same number as previously. However, here cell 13 is very different from the cell 13 in a flip-flop switch. This can be seen on the patterns illustrated by the right-hand side picture of Figure 9. Now, the active switch is put on a plane Π_1 which has a common perpendicular δ with the plane Π_0 of the passive memory switch. We can choose this common perpendicular in such a way that the top faces of the cell 13 in both switches are also perpendicular to δ. The track leading the new particle arrives on the top of the cell 13 of the active memory switch. Now the active switch works as a fixed switch: nothing happens when it is crossed by the particle playing the role of the locomotive. But when the temporary particle arrives to cell 13, this makes this cell to flash as it becomes black, but at the next time it becomes again white so that the new particle vanishes. Now, like in the case of the flip-flop switch, when cell 13 flashes, cells 11 and 12 exchange their states which triggers the change of the selected track in the active memory switch too.

We have not the room here to give the rules of the cellular automaton which complete the proof of Theorem 1. We refer the reader to [14] for a precise account about the rules.

4.5   A Word about the Computer Program

The computer program is based on the same one which was used to check the rules of the weakly universal cellular automaton on the dodecagrid with three states which was used in [10]. However, it was adapted to this automaton in the part computing the initial configuration and, also in the main function as the control steps are a bit different from those of the previous program. All the steps of the scenario were checked by this program which also makes sure that the rules are rotation invariant: they are unchanged if the states of the neighbour are changed according to a rotation leaving invariant the dodecahedron on which the cell is based.

This allows us to say that with this checking, the proof of Theorem 1 is now complete.

5   Conclusion

With this result, we reached the frontier between decidability and weak universality for cellular automata in hyperbolic spaces: starting from 2 states there are weakly universal such cellular automata, with 1 state, there are none, which is trivial. Moreover, the set of cells run over by the particle is a true spatial structure. We can see that the third dimension is much more used in this implementation than in the one considered in [10].

What can be done further?

In fact, the question is not yet completely closed. In [12] we proved that it is possible to implement a $1D$-cellular automaton with n states into the pentagrid, the heptagrid and the dodecagrid and also a whole family of tilings of the hyperbolic plane with the same number of states. For the pentagrid, it was needed to append an additional condition which is satisfied, in particular, by the elementary cellular automaton defined by rule 110, for which the reader is referred to [1,19]. Consequently, as stated in [12]:

Theorem 2. (M. Margenstern, [12]). There is a weakly universal rotation invariant cellular automaton with two states in the pentagrid, in the heptagrid and in the dodecagrid.

However, this is a general theorem based on the very complicate proof of a deep result involving a number of computations in comparison with which those of this paper are quasi-nothing. Also, the implementation provides a structure which is basically a linear one. This is why, it seems to me that the construction of the present paper is worth of interest: it is a truly spatial construction. Moreover, the construction is very elementary.

Now, there are still a few questions. What can be done in the plane with a true planar construction? At the present moment, the smaller number of states is 4, in the heptagrid, see [11], while it is 9 in the pentagrid, see [17].

Moreover, the question of the number of states in order to achieve strong universality is almost completely open. Remember that strong universality means that we start the computations from a finite initial configuration: all cells are question except finitely many of them. There is a result in [13] which constructs a strongly universal cellular automaton in the pentagrid with 9 states only. Most probably, this can also be performed in the dodecagrid. But, again, the automaton relies on a structure which is mainly a linear one, so that the question is still relevant.

And so, there is some definite effort before closing this question.

References

1. Cook, M.: Universality in elementary cellular automata. Complex Systems 15(1), 1–40 (2004)
2. Herrmann, F., Margenstern, M.: A universal cellular automaton in the hyperbolic plane. Theoretical Computer Science 296, 327–364 (2003)
3. Margenstern, M.: Frontier between decidability and undecidability a survey. Theoretical Computer Science 231(2), 217–251 (2000)
4. Margenstern, M.: Two railway circuits: a universal circuit and an NP-difficult one. Computer Science Journal of Moldova 9, 1–35 (2001)
5. Margenstern, M.: A universal cellular automaton with five states in the 3D hyperbolic space. Journal of Cellular Automata 1(4), 315–351 (2006)
6. Margenstern, M.: Cellular Automata in Hyperbolic Spaces. Theory, vol. 1, 422 p. OCP, Philadelphia (2007)
7. Margenstern, M.: Cellular Automata in Hyperbolic Spaces. Implementation and computations, vol. 2, 360 p. OCP, Philadelphia (2008)
8. Margenstern, M.: Surprising Areas in the Quest for Small Universal Devices. Electronic Notes in Theoretical Computer Science 225, 201–220 (2009)
9. Margenstern, M.: Turing machines with two letters and two states. Complex Systems (2010) (accepted)
10. Margenstern, M.: A weakly universal cellular automaton in the hyperbolic 3D space with three states. arXiv:1002.4290[cs.FL], 54 (2010)
11. Margenstern, M.: A universal cellular automaton on the heptagrid of the hyperbolic plane with four states. Theoretical Computer Science (2010) (to appear)
12. Margenstern, M.: About the embedding of one dimensional cellular automata into hyperbolic cellular automata. arXiv:1004.1830 [cs.FL], 19 (2010)
13. Margenstern, M.: An upper bound on the number of states for a strongly universal hyperbolic cellular automaton on the pentagrid. In: JAC 2010, TUCS Proceedings, Turku, Finland (2010)
14. A new weakly universal cellular automaton in the 3D hyperbolic space with two states. arXiv:1005.4826v1[cs.FL], 38 (2010)

15. Margenstern, M., Skordev, G.: Tools for devising cellular automata in the hyperbolic 3D space. Fundamenta Informaticae 58(2), 369–398 (2003)
16. Margenstern, M., Song, Y.: A universal cellular automaton on the ternary heptagrid. Electronic Notes in Theoretical Computer Science 223, 167–185 (2008)
17. Margenstern, M., Song, Y.: A new universal cellular automaton on the pentagrid. Parallel Processing Letters 19(2), 227–246 (2009)
18. Stewart, I.: A Subway Named Turing, Mathematical Recreations. Scientific American, 90–92 (1994)
19. Wolfram, S.: A new kind of science. Wolfram Media, Inc., Champaign (2002)
20. Neary, T., Woods, D.: Small Semi-Weakly Universal Turing Machines. Fundamenta Informaticae 91(1), 179–195 (2009)

A Fully Symbolic Bisimulation Algorithm

Malcolm Mumme and Gianfranco Ciardo

University of California, Riverside CA 92521, USA
{mummem,ciardo}@cs.ucr.edu

Abstract. We apply the saturation heuristic to the bisimulation problem for deterministic discrete-event models, obtaining the fastest to date symbolic bisimulation algorithm, able to deal with large quotient spaces. We compare performance of our algorithm with that of Wimmer et al., on a collection of models. As the number of equivalence classes increases, our algorithm tends to have improved time and space consumption compared with the algorithm of Wimmer et al., while, for some models with fixed numbers of state variables, our algorithm merely produced a moderate extension of the number of classes that could be processed before succumbing to state-space explosion. We conclude that it may be possible to solve the bisimulation problem for systems having only visible deterministic transitions (e.g., Petri nets where each transition has a distinct label) even if the quotient space is large (e.g., 10^9 classes), as long as there is strong event locality.

Keywords: bisimulation, symbolic methods, algorithms, decision diagrams, saturation, locality, verification

1 Introduction

The bisimulation problem has applications in verification of model equivalence and model minimization in preparation for analysis or composition. An algorithm by Paige and Tarjan [9] finds the largest bisimulation of an explicitly-represented transition system with N states and M transitions using $O(M \log N)$ time. The only major performance improvement over that algorithm, for explicitly represented systems with cycles, is the linear time algorithm [10] for the *single function coarsest partition problem*, the special case of the bisimulation problem where there is a single transition relation and that relation is a function.

In practice, it is desirable to minimize transition systems with large state spaces which cannot conveniently be represented explicitly. *Symbolic* encodings can be used, in some cases, to store state spaces many orders of magnitude larger than possible with explicit representations. These encodings have been used for bisimulation with varying degrees of success. The empirically fastest prior symbolic algorithms [6,14] represent bisimulations as partitions corresponding to the minimized state space, and use data structures with size at least linear in the number of partition classes. Those algorithms may be efficient when there are relatively few large partition classes, but become infeasible when the minimized transition system still has many states, since each class is symbolically represented while the collection of classes itself is instead explicitly represented.

G. Delzanno and I. Potapov (Eds.): RP 2011, LNCS 6945, pp. 218–230, 2011.

It was noted [2] that an *interleaved* representation (Section 2.3) has the potential to efficiently handle the desired partitions. However, the bisimulation algorithm [2] employing this representation performed relatively poorly on large problems. We consider the case where the transition relation of a globally asynchronous system is disjunctively partitioned into multiple transition relations according to the high-level system description. Our algorithm handles the special case where each of these transition relations is a partial function, as is the case with Petri nets with unique transition labels. We reduce the functional case of bisimulation to a transitive closure problem, then show that, usually, this problem may be solved efficiently using a saturation-based algorithm with interleaved partition representation.

2   Background

2.1   Deterministic Colored Labeled Transition Systems (DCLTSs)

A *deterministic colored labeled transition system* is a tuple $\langle \mathcal{S}, \mathcal{S}_{init}, \mathcal{E}, \mathcal{T}_\mathcal{E}, \mathcal{C}, c \rangle$, where $\mathcal{S}$ is a set of *states*; $\mathcal{S}_{init} \subseteq \mathcal{S}$ are the *initial states*; $\mathcal{E}$ is a set of *events* (transition labels); $\mathcal{T}_\mathcal{E} \subseteq \mathcal{S} \times \mathcal{E} \times \mathcal{S}$ is a labeled set of partial *transition functions* over $\mathcal{S}$, so that, for $\alpha \in \mathcal{E}$, we have $\mathcal{T}_\alpha : \mathcal{S} \nrightarrow \mathcal{S}$, where the notation "$\nrightarrow$" stresses that the function might be undefined for some domain elements; $\mathcal{C}$ is a set of colors; and $c : \mathcal{S} \to \mathcal{C}$ is a state coloring function. We write $\langle s_1, \alpha, s_2 \rangle \in \mathcal{T}_\mathcal{E}$ as $s_2 = \mathcal{T}_\alpha(s_1)$ or $s_1 \xrightarrow{\alpha} s_2$ for convenience. If the DCLTS is in state $s_1 \in \mathcal{S}$, its state may nondeterministically change to s_2, for any transition $\langle s_1, \alpha, s_2 \rangle \in \mathcal{T}_\mathcal{E}$.

In the context of bisimulation, a DCLTS may be thought of as a black box, since the current state is not directly visible, only its color is. Upon instantiation, a DCLTS has its state set to one of the initial states in $\mathcal{S}_{init}$. Beyond instantiation and querying the current state color, the only other operation permitted on the DCLTS is requesting to take the transition corresponding to $\alpha \in \mathcal{E}$ in the current state (the operation cannot be completed if $\mathcal{T}_\alpha$ is undefined in the current state).

2.2   Bisimulation [8,11]

A bisimulation is an equivalence relation among the states of a colored labeled transition system (CLTS). For example, the states of a minimized finite-state automaton (FSA) are given by the quotient of the original set of states over the largest bisimulation of the original FSA. States s_1 and s_2 are extensionally equivalent iff $s_1 \sim s_2$, where $\sim$ is the largest bisimulation of the automaton. Equivalent (bisimilar) states of the original FSA are merged into a single state in the minimized FSA. Formally, the largest bisimulation $\sim$ of a DCLTS $\langle \mathcal{S}, \mathcal{S}_{init}, \mathcal{E}, \mathcal{T}_\mathcal{E}, \mathcal{C}, c \rangle$, is the largest equivalence relation $\mathcal{B} \subseteq \mathcal{S} \times \mathcal{S}$ where:

1. Bisimilar states (pairs in $\mathcal{B}$) have the same color: $\langle p, q \rangle \in \mathcal{B} \Rightarrow c(p) = c(q)$.
2. Bisimilar states have only matching transitions, to bisimilar states:
   $\langle p, q \rangle \in \mathcal{B} \Rightarrow \forall \alpha \in \mathcal{E}, \forall p' \in \mathcal{S}, (p \xrightarrow{\alpha} p' \Rightarrow \exists q' \in \mathcal{S}, q \xrightarrow{\alpha} q' \wedge \langle p', q' \rangle \in \mathcal{B})$.

2.3 Quasi-Reduced Ordered Multi-way Decision Diagrams

Given a sequence of K finite sets $\mathcal{S}_{K:1} \stackrel{\text{def}}{=} (\mathcal{S}_K, ..., \mathcal{S}_1)$ (w.l.o.g. we assume $\mathcal{S}_k = \{0, 1, ..., n_k - 1\}$), we encode a non-empty set of K-tuples $\mathcal{X} \subseteq \hat{\mathcal{S}} = \mathcal{S}_K \times \cdots \times \mathcal{S}_1$ as a Quasi-reduced ordered Multi-way Decision Diagram (MDD) of depth K. An MDD is a uniform-depth, acyclic, finite, single-rooted, directed graph, where each non-leaf node r at level $level(r) = k \in K, ..., 1$ (the distance from r to a leaf node) has at most n_k outgoing edges uniquely labeled from $\mathcal{S}_k$, is itself an MDD root (of the subgraph induced by the nodes reachable from r), and is *canonical* (no two nodes are roots of MDDs encoding identical sets).

We write $r[i]$ to denote the MDD node reached by following the i-labeled edge from r, if it exists. We use an MDD d to encode a (non-empty) set with characteristic function $Map(d)$ from K-tuples $\hat{\mathcal{S}}$ to booleans, defined as follows:

- $Map(d) = \langle \to true \rangle$ where $\langle \to b \rangle$ is the nullary map to b, (for a leaf MDD).
- $Map(d) = \bigcup_{s \in \mathcal{S}_{level(d)}} s \times \left\{ \begin{array}{ll} Map(d[s]) & \text{if } d[s] \text{ exists} \\ \langle \mathcal{S}_{level(d-1)} \times \cdots \times \mathcal{S}_1 \to false \rangle & \text{otherwise} \end{array} \right\}$ (otherwise).

Thus, MDD a has a path with arcs labeled $s_{K:1}$ from the root a to some leaf labeled b iff $Map(a)$ contains the tuple $\langle s_{K:1} \to b \rangle$. Depending on the context, we use a single symbol (say $\mathcal{Q}$) to represent either the MDD, the root node of the MDD, or the set whose characteristic function is $Map(\mathcal{Q})$.

The ability to combine tuples provides for encoding of arbitrary finite relations over the elements of a tuple domain. Both concatenation and interleaving are used to combine tuples. In *concatenated* representation, we represent a binary relation $\mathcal{R}$ on $\hat{\mathcal{S}}$ as a set of tuples from the product domain $\hat{\mathcal{S}} \times \hat{\mathcal{S}}$, by concatenation of the elements in each pair of $\mathcal{R}$, so that the pair $\langle \langle s_{K:1} \rangle, \langle s'_{K:1} \rangle \rangle$ becomes $\langle s_K, ..., s_1, s'_K, ..., s'_1 \rangle$. In *interleaved* representation, the same pair in $\mathcal{R}$ is instead represented as $\langle s_K, s'_K, ..., s_1, s'_1 \rangle$. Interleaved representations often produce more compact MDD encodings than concatenated representations, presumably due to the proximity of correlated variables in the former.

We encode sets of states in DCLTSs, as well as sets of transitions and partitions, as MDDs, typically in interleaved representation. Our SMART MDD library provides efficient implementations of set operations, such as union, intersection, difference, symmetric difference, relation composition, relational product, cartesian product, and quantification, in interleaved or concatenated representation. In SMART, a *unique-table* mechanism maintains canonicity. A newly-constructed node might be the root of an MDD that coincidentally encodes the same map as some other existing MDD, thus is submitted to the unique-table mechanism which provides a node at the root of an MDD encoding with the desired map while preserving the canonicity of the collection of stored MDDs. Mutation of existing data structures is not allowed, resulting in a functional-like style of programming where operations on maps encoded as MDDs tend to be written recursively, employing concurrent DFS-like searches through homologous parts of input MDDs. Function caching is used to avoid exponential runtimes.

2.4   Saturation

The transitive closure of a labeled set of relations $\mathcal{T}_\mathcal{E}$ over $\hat{S}$ starting from a set of initial states $\mathcal{S}_{init}$ is the least fixpoint $\mathcal{S}$ of the union $\bigcup_{\alpha \in \mathcal{E}} \mathcal{T}_\alpha$ of those relations containing $\mathcal{S}_{init}$ and may be written as $\mathcal{S} = (\bigcup_{\alpha \in \mathcal{E}} \mathcal{T}_\alpha)^*(\mathcal{S}_{init})$. The *saturation* [13] heuristic is a state-of-the-art symbolic algorithm to efficiently explore large state spaces of asynchronous systems by evaluating $(\bigcup_{\alpha \in \mathcal{E}} \mathcal{T}_\alpha)^*(\mathcal{S}_{init})$. In this work, we employ saturation to produce a similar fixpoint, on relations.

A breadth-first algorithm for this iterates $\mathcal{S}_{temp} \leftarrow (\mathcal{S}_{temp}) \cup (\bigcup_{\alpha \in \mathcal{E}} \mathcal{T}_\alpha(\mathcal{S}_{temp}))$ on a variable $\mathcal{S}_{temp}$ initialized to $\mathcal{S}_{init}$, until no change occurs, leaving the final result in $\mathcal{S}_{temp}$. A chaining heuristic relies on the decomposition of $\mathcal{T}_\mathcal{E}$ into the individual $\mathcal{T}_\alpha$'s, for $\alpha \in \mathcal{E}$, to break each iteration into a sequence of smaller iterations, such as $\mathcal{S}_{temp} \leftarrow \mathcal{S}_{temp} \cup \mathcal{T}_\alpha(\mathcal{S}_{temp})$ for all $\alpha \in \mathcal{E}$. Saturation uses the following decomposition, which may be assumed w.l.o.g.: a state is described as a K-tuple of simple variables, $\langle s_{K:1} \rangle \in \hat{S}$. The transition relation $\mathcal{T} = \bigcup_{\alpha \in \mathcal{E}} \mathcal{T}_\alpha$ is partitioned in such a way that $\mathcal{E} = \{1, ..., K\}$ and the support of $\mathcal{T}_\alpha$ is limited to variables $\{1, ..., \alpha\}$. That is, there exists an "equivalent" transition relation $\mathcal{T}'_\alpha$, involving only variables $1, ..., \alpha$ of the state tuple, so that variables $\alpha + 1 ... K$ do not influence the enabling of $\mathcal{T}'_\alpha$ and $\mathcal{T}'_\alpha$ acts on them as the identity function: $\langle s_{K:1}, s'_{K:1} \rangle \in \mathcal{T}_\alpha \Leftrightarrow (s_{K:\alpha+1} = s'_{K:\alpha+1} \wedge \langle s_{\alpha:1}, s'_{\alpha:1} \rangle \in \mathcal{T}'_\alpha)$.

Thus, $\mathcal{T}'_\alpha$ contains the same information as $\mathcal{T}_\alpha$. Since the MDD encoding of $\mathcal{T}_\alpha$ has depth $2K$, saturation instead uses $\mathcal{T}'_\alpha$, with encoded depth 2α.

Saturation requires that a transition $\mathcal{T}_\alpha$ is applied (1) only to sets already closed under all the transitions $\mathcal{T}_{\alpha-1:1}$, and (2) only at level α of an encoding of $\mathcal{S}_{temp}$, by applying $\mathcal{T}'_\alpha$ (with interleaved encoding) instead of $\mathcal{T}_\alpha$. Requirement (1) has a tendency to keep the encodings of various values of $\mathcal{S}_{temp}$ compact, while requirement (2) tends to result in efficient application of the $\mathcal{T}_\alpha$'s. Both requirements can be implemented with a recursive algorithm as shown by the pseudocode in Fig. 1, where *SatClos* uses saturation to compute the smallest closure of input $\mathcal{S}_{in}$ over the relations $\mathcal{T}_{k:1}$ corresponding to the input MDDs encoding $\mathcal{T}'_{k:1}$ (in interleaved representation). This algorithm returns the MDD in working variable $\mathcal{S}$. In line 7, the children of working variable $\mathcal{S}$ are initialized to versions of children of the input $\mathcal{S}_{in}$, closed under relations $\mathcal{T}_{k-1:1}$. Consequently, $\mathcal{S}$ is also closed under relations $\mathcal{T}_{k-1:1}$, and $\mathcal{S}_{in} \subseteq \mathcal{S}$. Property $\mathcal{S}_{in} \subseteq \mathcal{S}$ continues to hold, since $\mathcal{S}$ is only increased in lines 11 and 15, and its elements are never removed. Line 14 guarantees the value of $\mathcal{S}'$ is closed under relations $\mathcal{T}_{k-1:1}$. The purpose of the code in lines 6-16 is to iteratively close a working variable $\mathcal{S}$ over transition $\mathcal{T}'_k$, using the assignment $\mathcal{S} \leftarrow \mathcal{S} \cup \mathcal{T}'_k(\mathcal{S})$, while maintaining closure over transitions $\mathcal{T}_{k-1:1}$. It must therefore assure that $\mathcal{T}'_k(\mathcal{S}) \subseteq \mathcal{S}$ and $\mathcal{S}_{in} \subseteq \mathcal{S}$. This is guaranteed by line 11 and the termination condition, together with the fact that no element is ever removed from $\mathcal{S}$. Loop termination can be shown by considering the termination condition together with the monotonically increasing nature of $\mathcal{S}$. Consequently, after line 14, $\mathcal{S} = \mathcal{T}^*(\mathcal{S}_{in})$. The recursive calls always terminate, as the leaf case is trivially handled in the first line while the remaining cases involve recursive calls, always with parameters of lower rank.

```
MDD SatClos(in MDD T'_{k:1}, in MDD S_{in}) is
  local   MDD S                                            • converges to output
  local   MDD S_{prev}
  1  if k = 0 or S_{in} = ∅                                • leaf or empty
  2    then return S_{in}
  3  if SatClos(T'_{k:1}, S_{in}) is in cache then
  4    return cached result
  5  S ← new MDD node of size |S_k|
  6  foreach i ∈ S_k do
  7    S[i] ← SatClos(T'_{k-1:1}, S_{in}[i])
  8  S ← unique(S)                                         • make canonical
  9  repeat                                  • invariant: S closed on T'_{k-1:1}
  10   S_{prev} ← S
  11   S ← S ∪ T'_k(S)                                     • break invariant
  12   S' ← new MDD node of size |S_k|
  13   foreach i ∈ S_k do                                  • reestablish invariant
  14     S'[i] ← SatClos(T'_{k-1:1}, S[i])
  15   S ← unique(S')                                      • canonical
  16  until S = S_{prev}
  17  place into cache:
  18    S = SatClos(T'_{k:1}, S_{in})
  19  return S
```

Fig. 1. Least set closure computation

The basic saturation algorithm described above may be enhanced in various ways, such as addition of *fine-grained chaining* [3]. Although some enhancements may produce additional improvement, our work uses the basic saturation algorithm described above to obtain the largest bisimulation of a DCLTS.

2.5   Symbolic Bisimulation

Several symbolic algorithms exist for largest bisimulation (see Section 5). Among them, the ones with faster observed performance are limited by the fact that their runtime is always at least linear in the number of states of the minimized automaton. Our algorithm overcomes this limitation by focusing on the special case where the automaton has deterministic transitions, as in a Petri net with unique transition labels. In this case, the bisimulation problem can be reduced to an equivalent transitive closure problem, to which saturation applies well.

Symbolic bisimulation algorithms are typically based on iterative partition refinement, using *splitting* techniques. Generally, a working variable P initially holds a partition of the state space, either the partition containing a single equivalence class, or *block* (i.e., the entire state space) or, when there is more than one color, the partition containing a block for each color (i.e., each block contains all states of the corresponding color). Ultimately, the output is a partition of the state space with the equivalence classes based on bisimilarity. After initialization, P is iteratively refined by a splitting operation, which may split one or more blocks to increase compliance with clause 2 of the definition of bisimilarity: if a block B_1 contains states x and y and, for some transition α, there is a block B_2 containing state x' such that $x \xrightarrow{\alpha} x'$, but B_2 contains no y' such that $y \xrightarrow{\alpha} y'$, then states x and y are not bisimilar. Consequently, we may use the *splitter* B_2 to split block B_1 into blocks $B_1' = \{x \in B_1 | \exists x' \in B_2 : x \xrightarrow{\alpha} x'\}$ and $B_1'' = B_1 \setminus B_1' = \{y \in B_1 \mid \neg\exists y' \in B_2 : y \xrightarrow{\alpha} y'\}$.

Repeated iterations produce the partition corresponding to the maximum bisimulation. More advanced algorithms combine many splitting operations into one step, or chain them in an organized manner to improve efficiency [6,14].

3   Main Results

3.1   Bisimulation Partition Refinement Relations

Given a DCLTS with state space $\mathcal{S}$ and transition relations $\{\mathcal{T}_\alpha : \alpha \in \mathcal{E}\}$, we define a set of *bisimulation partition refinement relations* (BPRR) $\{\mathcal{P}_\alpha : \alpha \in \mathcal{E}\}$:

$$\forall \alpha \in \mathcal{E} : \mathcal{P}_\alpha \subseteq \hat{\mathcal{S}}^4 = \{\langle\langle s_1, s_2\rangle, \langle s_3, s_4\rangle\rangle : s_1 = \mathcal{T}_\alpha(s_3) \wedge s_2 = \mathcal{T}_\alpha(s_4)\},$$

where we may write $\langle s_3, s_4 \rangle \in \mathcal{P}_\alpha(\langle s_1, s_2 \rangle)$ to mean $\langle\langle s_1, s_2\rangle, \langle s_3, s_4\rangle\rangle \in \mathcal{P}_\alpha$. The complement $\approx$ of the maximum bisimulation of a transition system is closed under these BPRRs, since $\approx$ contains exactly all pairs of non-bisimilar states $\langle s_1, s_2 \rangle$ where $s_1 \not\sim s_2$. To prove this, consider a specific tuple $\langle\langle s_1, s_2\rangle, \langle s_3, s_4\rangle\rangle$ from a specific bisimulation partition refinement relation $\mathcal{P}_\alpha$. We need to show that $\langle s_1, s_2 \rangle \in \approx$ implies $\langle s_3, s_4 \rangle \in \approx$. Thus, according to our definition of BPRR, we must prove that $\langle \mathcal{T}_\alpha(s_3), \mathcal{T}_\alpha(s_4)\rangle \in \approx$ implies $\langle s_3, s_4 \rangle \in \approx$. This is the same as saying that $s_3 \sim s_4$ implies $\mathcal{T}_\alpha(s_3) \sim \mathcal{T}_\alpha(s_4)$, and the latter is a direct consequence of the definition of bisimulation together with $\mathcal{T}_\alpha$ being a function. This closure of $\approx$ over BPRRs suggests that computation of $\approx$ is reducible to a set closure problem for systems with deterministic transitions.

3.2   Reduction of Deterministic Bisimulation to Set Closure

In systems with deterministic transitions, the generic splitting step can be simplified by considering three cases, based on the partial/full nature of a transition relation under consideration. Given transition relation $\mathcal{T}_\alpha$ and a candidate pair $\langle s_3, s_4 \rangle$, for which we wish to determine whether $s_3 \sim s_4$, we have:

Case 1: If neither s_3 nor s_4 is in the domain of $\mathcal{T}_\alpha$, the pair cannot be eliminated using $\mathcal{T}_\alpha$ with any splitter.

Case 2: If exactly one member, say s_4, is in the domain of $\mathcal{T}_\alpha$ and if there is an $s_2 = \mathcal{T}_\alpha(s_4)$ for which there is no $s_1 = \mathcal{T}_\alpha(s_3)$ such that $\langle s_1, s_2 \rangle \in \sim$, then, the pair $\langle s_3, s_4 \rangle$ will be eliminated by splitting.

Case 3: If both s_3 and s_4 are in the domain of $\mathcal{T}_\alpha$, there is a pair $\langle s_1, s_2 \rangle$, where $s_1 = \mathcal{T}_\alpha(s_3) \wedge s_2 = \mathcal{T}_\alpha(s_4)$, so that $\langle s_3, s_4 \rangle \in \mathcal{P}_\alpha\langle s_1, s_2 \rangle$ as discussed above. Then, splitting will eventually eliminate $\langle s_3, s_4 \rangle$ iff $\langle s_1, s_2 \rangle$ is absent or eliminated.

```
BPRRclosure:
 1  foreach α ∈ ℰ do                                          • 𝒫_α =BPRR for 𝒯_α
 2      𝒫_α ← {⟨⟨s₁,s₂⟩⟨s₃,s₄⟩⟩ : s₁=𝒯'_α(s₃),s₂=𝒯'_α(s₄)}
 3  let ℒ = {level(𝒫_α) : α ∈ ℰ}
 4  foreach β ∈ ℒ do
 5      𝒫'_β ← ⋃_{α∈ℰ|level(𝒫_α)=β} 𝒫_α
 6  foreach β ∉ ℒ do
 7      𝒫'_β ← ∅                                               • 𝒫' is 𝒫 merged by top level
 8  foreach α ∈ ℰ do
 9      𝒮_α ← {s∈𝒮: ∃s', s →^α s'}                            • states enabling α
10  B̄₀ ← {⟨s₁,s₂⟩ : c(s₁)≠c(s₂)}                             • initialize B̄₀
11  foreach α ∈ ℰ do                                          • augment B̄₀
12      B̄₀ ← B̄₀∪(𝒮_α×(𝒮\𝒮_α))∪((𝒮\𝒮_α)×𝒮_α)
13  B̄ ← SatClos(𝒫'_{4K:1}, B̄₀)                              • B̄ ← 𝒫*(B̄₀)
14  return ℬ ← (𝒮 × 𝒮) \ B̄
```

Fig. 2. Saturation for DCLTS bisimulation

Thus, we may restructure the generic splitting algorithm as follows. First, we initialize variable $\overline{B}$ (complement of a partition) to the set of pairs $\langle s_3,s_4 \rangle$ where, for some $\alpha \in \mathcal{E}$, exactly one member of $\langle s_3,s_4 \rangle$ is in the domain of $\mathcal{T}_\alpha$. This eliminates all pairs discussed in Case 2 above. Then, we iterate the following until $\overline{B}$ is stable: for each $\alpha \in \mathcal{E}$, augment $\overline{B}$ with $\mathcal{P}_\alpha(\langle s_1,s_2 \rangle)$ for each $\langle s_1,s_2 \rangle \in \overline{B}$. This eventually eliminates pairs as required in Case 3. At the end of these steps, $\overline{B} = \approx$, so we compute $\sim$ as $\mathcal{S}^2 \setminus \overline{B}$.

This algorithm may be stated more formally as follows. Consider a DCLTS $\langle \mathcal{S}, \mathcal{S}_{init}, \mathcal{E}, \mathcal{T}_\mathcal{E}, \mathcal{C}, c \rangle$, where the relations $\mathcal{T}'_\mathcal{E}$, corresponding to the deterministic transition relations $\mathcal{T}_\mathcal{E}$, as described in Section 2.4, are also available. To compute the largest bisimulation $\mathcal{B}$ of this DCLTS, use algorithm *BPRRclosure* of Fig. 2.

4   Experimental Results

We implemented our algorithm, as well as the algorithm of [14], in the SMART verification tool using the SMART MDD library. We then measured the performance of each implementation by computing the largest bisimulation of many cases of various Petri net models including the four models described below.

Flexible Manufacturing System (Kanban). This net describes a kanban-style flexible manufacturing system [12], having fixed size and topology, and is parameterized by manufacturing station capacity. No two states are equivalent, so there are always as many equivalence classes as there are states.

This model comprises four connected identical manufacturing stations, each having the same capacity given by the model parameter N. Each task held by a station is either in work, in a waiting-for-rework state, or ready for output. The first station may begin processing any number of jobs up to its capacity. A task in the first station that is ready for output may spawn a task in both

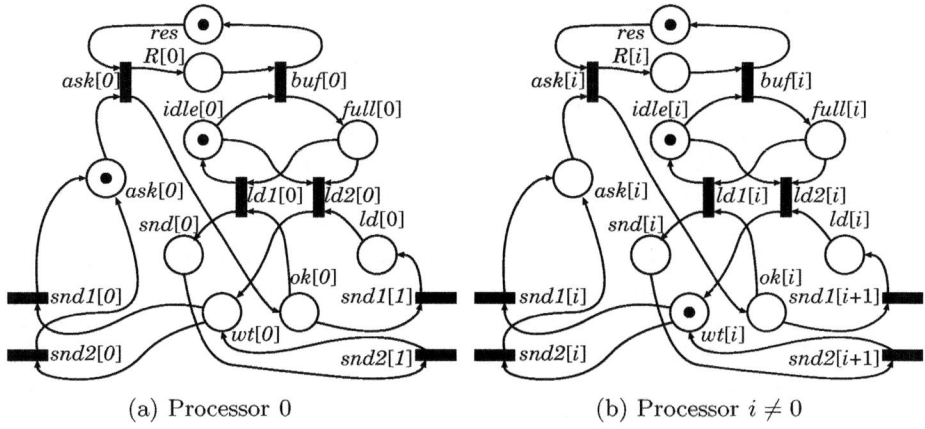

(a) Processor 0   (b) Processor $i \neq 0$

Fig. 3. Petri net model: robin with initial marking

the second and third stations, when those stations have sufficient capacity; this
event releases capacity in the first station. A task finishing in both the second
and third stations produce a task in the fourth station, releasing capacity in
the second and third station. Finally, a task finishing in the fourth station is
considered finished and immediately releases capacity in the fourth station.

Scheduler (Round-Robin). This round-robin scheduler model [4,7] is similar
to Milner's scheduler [8], with some additional complications. An extra place,
shared by all processors, is used as a lock to disallow states where more than one
processor is in a "start" local state. The case where a processor finishes before
its successor finishes a previous task is distinguished from the case where the
processor finishes after its successor finished a previous task, by the occurrence
of a separate transition sequence. This model is variable-sized, parameterized
in the number N of processors. As with the previous model, no two states are
equivalent, so there are always as many equivalence classes as there are states.

The model comprises a ring of N processors (numbered from 0 to $N - 1$)
which must be scheduled so that, for $i \in 0...N - 1$, processor $(i + 1)$ mod N may
not start a new task until processor i starts a new task, with the exception that
a specific processor, 0, may initially start. Each processor executes a single task
to completion, and the tasks may finish in any order. The Petri net shown in
Fig. 3(b) corresponds to a single typical processor in the ring, showing the initial
marking. Fig. 3(a) shows the initial marking for the Petri net modeling processor
0, which starts the first task. In these (sub)Petri nets, place *res* is common to
all processors, while the places and transitions having names with brackets (i.e.,
"$[i]$") have one copy per processor. Also, transitions $snd1[i]$ and $snd2[i]$, at the
left and right of each subnet, are shared between adjacent processors.

The presence of a token in place $ask[i]$ indicates that processor i may start a
new task, while a token in $wt[i]$ means processor i is waiting to start a new task.
Firing of $ask[i]$ places processor i exclusively in a "start" state, as this requires

the only token from the shared place *res*. Processor i moves from the start state to a fully running state by firing $buf[i]$, returning the token to *res*, and moving the token in $idle[i]$ to $full[i]$. Since firing $ask[i]$ also places a token in $ok[i]$, transition $snd1[(i + 1) \bmod N]$ is enabled if processor $(i + 1) \bmod N$ is waiting. Then, $snd1[(i + 1) \bmod N]$ may fire, permitting processor $(i + 1) \bmod N$ to possibly start and also enabling $ld2[i]$. Firing of $ld2[i]$ indicates task completion, moving the token in $full[i]$ back to $idle[i]$, placing a token in $wt[i]$, indicating waiting. If processor $(i+1) \bmod N$ is busy (not waiting, as indicated by absence of a token in $wt[(i+1) \bmod N]$), processor i may indicate completion by firing $ld1[i]$, moving the token in $full[i]$ back to $idle[i]$, and also moving the token from $ok[i]$ to $snd[i]$. If processor $(i+1) \bmod N$ subsequently waits (as indicated by presence of a token in $wt[(i+1) \bmod N]$), enabling $snd2[(i+1) \bmod N]$, then $snd1[(i+1) \bmod N]$ is disabled, distinguishing the two cases. Firing $snd2[(i+1) \bmod N]$ puts processor i into a waiting state, and enables processor $(i + 1) \bmod N$ to start a new task.

Extrema Finding (Leader). This model simulates the distributed extrema finding algorithm of [5]. Numeric tokens are passed unidirectionally around a ring of processes in a sequence of phases resulting in the recognition of the largest token. This model is variable-sized, parameterized in the number N of processes. There is significant event locality. Some states are equivalent, resulting in about 10% fewer equivalence classes than states.

This model comprises a ring of processes joined by unidirectional buffers. Initially, each process holds a unique numeric token, and is considered "active". In the initial phase, each process sends this token to its successor. Between phases, the state of each process includes three items: (1) the token it currently holds, (2) the value of the largest token it has previously held, and (3) whether the process is active or inactive. In subsequent phases, before the largest token is recognized, each active process receives from its predecessor a token which it will hold next. Inactive processes simply forward messages. An active process not holding the largest token may become inactive as follows. It becomes inactive if the value of the token it will hold next is less than the value of the largest token it has previously held. Each active process also forwards the value of the token it will hold to its successor process, and receives such a value from its predecessor. It then becomes inactive if the value received is greater than the value of the token it will hold next. Eventually exactly one process remains active, it receives the largest token and recognizes this as the same value remembered as the largest token it has previously held, and then enters a unique state, recognizing the maximum token.

Few Classes (Cascade). As ordinary models having only visible deterministic transitions (no "τ"-transitions) tend to have many equivalence classes, we developed a model with few equivalence classes, to evaluate algorithm performance in this important case. This model has a variable number of stages with three places each and three transitions between each stage of places. The initial marking, shown in Fig. 4, has one token in each place in the first stage and no other tokens. In any state other than the final state, all three transitions after one of

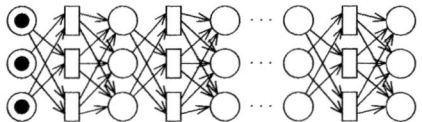

Fig. 4. Petri net model: cascade with initial marking

the stages are enabled and no other transition is enabled. The three transitions fire in such a way that one of them firing results in disabling the other two transitions and in depositing one token in each place of the next stage, enabling all transitions after the subsequent stage. Each firing also leaves one token in a place that records the firing, so that every state remembers the firing sequence that produced it. In this variable-size model, parameterized in the number of stages N, there are exponentially many states in the number of stages, while the number of classes equals the number of stages.

Performance. Table 1 lists the model sizes in the first set of columns, while the remaining columns summarize performance results for our saturation-based algorithm and for our implementation of the algorithm of Wimmer et al. The number of classes is shown as "=" when the number of classes is the same as the number of states. For the algorithm of Wimmer et al., column *iter* reports the number of iterations executed (including a final iteration resulting in no changes), while column *refine* gives the number of calls to the "*Refine*" subroutine of their algorithm, indicating the number of attempted block refinements [14]. The *runtime* and maximum *memory* to store the unique table for all runs are given in seconds and mega-bytes, respectively.

It is clear that, for models with strong event locality, the memory performance of our saturation-based algorithm quickly becomes superior as model size increases, resulting in vastly improved runtimes for those cases. For the kanban model, which has limited locality due to fixed model size (independent of N), the improvement does not appear to grow as rapidly with increasing N as it does with the other models. Nevertheless, the memory and runtime advantages with respect to Wimmer et al.'s algorithm is still enormous.

5   Related Work

Symbolic Bisimulation Minimisation[2]. describes an algorithm based on iterative partition refinement that, at each iteration, splits every block using every other block as a splitter, performing a constant number of advanced symbolic operations. It is among the earliest work on symbolic bisimulation methods. The purpose of their algorithm is to compute *weak bisimulation*, a form of bisimulation where some transitions are unobservable. Pre-processing the collection of transition relations transforms a weak bisimulation problem into the standard bisimulation problem. Their article compares their algorithm using interleaved decision diagrams vs. using non-interleaved diagrams and found that the relative

Table 1. Performance results summary for bisimulation algorithms

| N | states | classes | Wimmer et al. (no block ordering) | | | | Saturation | |
			iter	refine	runtime	memory	runtime	memory
kanban ($K = 21$)								
2	4600	=	4	5424	92.268	27.403	0.171	0.805
3	58400	=	5	71946	7700.186	307.898	0.646	1.816
4	454475	=	out of memory (> 1.5 GB)				1.527	3.660
9	384392800	=					231.041	41.267
10	1005927208	=					1501.159	75.444
11	2435541472	=					21720.048	112.244
robin ($K = N + 1$)								
9	10368	=	3	11208	141.732	28.965	1.092	3.368
11	50688	=	3	54861	2895.590	153.717	2.303	8.091
13	239616	=	3	259344	60823.424	699.049	4.338	14.027
15	1105920	=	timeout (> 1 day)				6.916	21.280
20	47185920	=					19.336	46.180
25	1887436800	=					52.838	123.950
leader ($K = 18N$)								
3	488	467	7	499	19.575	12.509	3.851	14.534
4	3576	3418	17	4139	291.322	76.391	15.433	38.786
5	26560	25329	28	34016	9579.915	389.100	25.185	45.286
6	197984	188322	out of memory (> 1.5 GB)				49.003	105.547
10	614229504	5.7×10^8					156.238	229.728
15	1.4×10^{13}	1.2×10^{13}					434.673	619.918
cascade ($K = 3N$)								
9	9841	9	2	8	15.206	36.120	0.101	0.630
10	29524	10	2	9	49.640	98.490	0.124	0.767
11	88573	11	2	10	191.265	302.936	0.158	0.882
20	1.7×10^9	20	out of memory (> 1.5 GB)				0.947	3.176
40	6.1×10^{18}	40					4.000	11.089
80	7.4×10^{37}	80					27.290	53.081
160	1.1×10^{76}	160					216.283	189.881

performance using these structures depended on the final partition. Interleaved decision diagrams perform better when the partition has many blocks, while non-interleaved diagrams perform better when it has very few blocks.

An Efficient Algorithm for Computing Bisimulation Equivalence[6]. produces a (probably) good initial partition of the state-space based on the computed *rank* of states. Their explicit implementation uses the set-theoretic notion of rank, where nodes in the transition graph correspond to sets (say, node c corresponds to set c'), and arcs between nodes correspond to membership between the respective sets (arc $a \to b$ corresponds to $b' \in a'$). This is reasonable because states with different rank cannot be equivalent. The possibility of cycles in transition graphs forces the use of non-well-founded-set theory [1]. Their symbolic implementation uses the following definition of rank: the rank of a non-well-founded node is one more than the highest rank of any well-founded node it reaches on any path, or $-\infty$ when there is no such path. They show that their algorithm always terminates in a number of symbolic steps linear in the number of states. After initial partitioning based on rank, some other algorithm must be used to complete the partitioning. The rank sequence of the initial partition can be used to efficiently direct the order of splitting operations.

Forwarding, Splitting, and Block Ordering to Optimize BDD-Based Bisimulation Computation[14]. describes several methods to accelerate symbolic bisimulation computation. The main algorithm is similar to that of [2], although blocks are represented explicitly, and assigned unique serial numbers. Aside from complications relating to weak bisimulation, their main optimizations are:

(1) Use of state *signatures*, as in [2], to compute block refinements. States with different signatures belong to different partition blocks. Their BDD encoding for signatures puts state variables toward the root and signature variables toward the leaves. This allows efficient partition refinement by substitution of block serial numbers into the BDD at the level of the signature, as the canonicity of the BDD guarantees that each node at that level corresponds to a distinct block. This technique is doomed to failure when there are many classes, as the encoding requires at least one BDD node per class.

(2) *Block forwarding* updates the current partition immediately after blocks are split. They split (and compute signatures for) only one block at a time. The partition is updated after each block splitting, allowing subsequent splittings to benefit immediately from the more refined partition.

(3) *Split-driven refinement* uses a *backward signature* (similar to the inverse of the transition relation) to determine which blocks may require splitting after the splitting of a given block, and skips the splitting of blocks whose elements signatures include no blocks split since the given block was created.

(4) *Block ordering* is the deliberate choice of which splitter to use at any given time that such a choice is available. They found that two heuristic orderings: "choose the block with a larger backward signature", and "choose the larger block", both produced improved run times compared to random choice.

6 Conclusions

Saturation-based bisimulation provides efficient bisimulation for DCLTS where strong event locality is present (the same models where saturation is ideal for state-space generation). Our algorithm is effective both when the bisimulation results in few classes and when it instead results in many classes, possibly even one per state. This is particularly important because we envision a tool chain where a bisimulation reduction is routinely attempted after state-space generation and prior to any further analysis. In this framework, it is then essential for the bisimulation algorithm to be efficient even (or perhaps, especially) when its application does not result in a reduction of the state space.

Techniques that produced improvement in saturation-based state-space generation may yet improve this algorithm. We believe that the saturation heuristic is also applicable to weak bisimulation and to non-deterministic CLTS bisimulation (nondeterminism may arise even in Petri nets, if multiple transitions share the same label). Our work is proceeding in that direction.

References

1. Aczel, P.: Non-Well-Founded Sets. CSLI, Stanford (1988)
2. Bouali, A., Simone, R.D.: Symbolic bisimulation minimisation. In: Probst, D.K., von Bochmann, G. (eds.) CAV 1992. LNCS, vol. 663, pp. 96–108. Springer, Heidelberg (1993)
3. Chung, M.Y., Ciardo, G., Yu, A.J.: A fine-grained fullness-guided chaining heuristic for symbolic reachability analysis. In: Graf, S., Zhang, W. (eds.) ATVA 2006. LNCS, vol. 4218, pp. 51–66. Springer, Heidelberg (2006)
4. Ciardo, G., et al.: SMART: Stochastic Model checking Analyzer for Reliability and Timing, User Manual, http://www.cs.ucr.edu/~ciardo/SMART/
5. Dolev, D., Klawe, M., Rodeh, M.: An $O(n \log n)$ unidirectional distributed algorithm for extrema finding in a circle. J. of Algorithms 3(3), 245–260 (1982)
6. Dovier, A., Piazza, C., Policriti, A.: An efficient algorithm for computing bisimulation equivalence. Theor. Comput. Sci. 311, 221–256 (2004)
7. Graf, S., Steffen, B., Lüttgen, G.: Compositional minimisation of finite state systems using interface specifications. Journal of Formal Aspects of Computing 8(5), 607–616 (1996)
8. Milner, R.: Communication and concurrency. Prentice-Hall, Inc., NJ (1989)
9. Paige, R., Tarjan, R.E.: Three partition refinement algorithms. SIAM Journal of Computing 16, 973–989 (1987)
10. Paige, R., Tarjan, R.E., Bonic, R.: A linear time solution to the single function coarsest partition problem. Theoretical Computer Science 40, 67–84 (1985)
11. Park, D.: Concurrency and automata on infinite sequences. In: Deussen, P. (ed.) GI-TCS 1981. LNCS, vol. 104, pp. 167–183. Springer, Heidelberg (1981)
12. Tilgner, M., Takahashi, Y., Ciardo, G.: SNS 1.0: Synchronized Network Solver. In: 1st Int. Workshop on Manuf. & Petri Nets, Osaka, Japan, pp. 215–234 (June 1996)
13. Wan, M., Ciardo, G.: Symbolic state-space generation of asynchronous systems using extensible decision diagrams. In: Nielsen, M., Kučera, A., Miltersen, P.B., Palamidessi, C., Tůma, P., Valencia, F. (eds.) SOFSEM 2009. LNCS, vol. 5404, pp. 582–594. Springer, Heidelberg (2009)
14. Wimmer, R., Herbstritt, M., Becker, B.: Forwarding, splitting, and block ordering to optimize BDD-based bisimulation computation. In: Haubelt, C., Teich, J. (eds.) Proceedings of the 10th GI/ITG/GMM-Workshop "Methoden und Beschreibungssprachen zur Modellierung und Verifikation von Schaltungen und Systemen" (MBMV), pp. 203–212. Shaker Verlag, Erlangen (2007)

Reachability for Finite-State Process Algebras Using Static Analysis

Nataliya Skrypnyuk and Flemming Nielson

Technical University of Denmark, 2800 Kgs. Lyngby, Denmark
{nsk,nielson}@imm.dtu.dk

Abstract. In this work we present an algorithm for solving the reachability problem in finite systems that are modelled with process algebras. Our method uses Static Analysis, in particular, Data Flow Analysis, of the syntax of a process algebraic system with multi-way synchronisation. The results of the Data Flow Analysis are used in order to "cut off" some of the branches in the reachability analysis that are not important for determining, whether or not a state is reachable. In this way, it is possible for our reachability algorithm to avoid building large parts of the system altogether and still solve the reachability problem in a precise way.

Keywords: reachability, process algebra, static analysis.

1 Introduction

Process algebras describe systems in a compositional way: they separately describe subsystems and interactions between them, thus achieving concise syntactic descriptions of the systems. In order to solve the reachability problem for a system specified in some process algebra it is however necessary to "expand" the definition and to make all the interactions between subsystems explicit, thus computing the *semantics* of the system. This may lead to the infamous state space explosion problem [1].

In this work we shall make use of the conciseness of the syntactic description of a system while solving the reachability problem. The idea is to analyse the syntax with Static Analysis methods [7] first. These methods can be exponentially faster than the methods based on the semantics of the systems. On the other hand, Static Analysis methods return in general approximations (in one direction only) of the actual results [7].

In our algorithm we estimate state reachability, while possibly being "overoptimistic". This means that in case we assess the state in question as not reachable, then it is definitely not reachable. If we estimate that the state is reachable then it is only possibly reachable. We use this kind of estimation in order not to build transitions out of states from which the state in question definitely cannot be reached. This allows us in some case to avoid building large parts of the systems and still obtain the correct answer to the reachability problem.

This work is a modification and extension of the reachability analysis of the algebra of IMC [2, 4] that has been developed in the PhD thesis [13]. In particular, the algorithms in Section 4 are new compared to [13].

G. Delzanno and I. Potapov (Eds.): RP 2011, LNCS 6945, pp. 231–244, 2011.
© Springer-Verlag Berlin Heidelberg 2011

Table 1. A process is in PA if it has a syntactic form of E, $a \in \mathbf{Act} \cup \{\tau\}$, $\ell \in \mathbf{Lab}$, $X \in \mathbf{Var}$, $A \subseteq \mathbf{Act}$

$$
\begin{aligned}
P ::= \ &\mathbf{0} & | \ &(1) \\
&a^{\ell}.X & | \ &(2) \\
&a^{\ell}.P & | \ &(3) \\
&P + P & | \ &(4) \\
&\underline{X := P} & &(5) \\
E ::= \ &P & | \ &(6) \\
&\text{hide } A \text{ in } E & | \ &(7) \\
&E \|_A\| E & &(8)
\end{aligned}
$$

2   The Process Algebra

Our reachability algorithm will be developed for a process algebra called PA. The algebra of PA has a CSP-style, i.e. multi-way, synchronisation model [5]. An important difference with CSP is that PA only allows for finite semantic models. This is required in order for the Data Flow Analysis developed by us to be applicable to it without using techniques like widening [7]. Finiteness of semantic models of PA processes is ensured on the syntactic level – i.e., if a process' syntactic definition complies with the PA syntax from Table 1, then it is ensured that its semantic model is finite.

The syntax of PA is defined using a set of *external actions* **Act**, a distinguished *internal action* τ, with $\tau \notin \mathbf{Act}$, a set of *labels* **Lab** and a set of *process identifiers* or *process variables* **Var**. The syntactic description of a particular PA process is always finite. The syntax of PA contains a number of standard process algebraic operators listed in Table 1. These are *prefixing* (2)-(3), *choice* (4), *process recursive definition* (5), *abstraction* or *hiding* (7) and *parallel composition* (8) operators. The *terminal process* (1) is also allowed in PA. If P is a (not necessary genuine) subexpression of a PA expression E, then we denote this fact by $P \preceq E$. Clearly $E \preceq E$ always holds.

An apparent difference between the syntax of PA and the syntax of the majority of other process algebras is that action names in PA are supplied with labels. This feature has been adopted from the variant of CCS process algebra [6] proposed in [9] where Data Flow Analysis has been developed for it. Labels do not influence the semantics of PA (we will see in Table 2 that they just decorate the transitions) but they serve as pointers into the syntax that make it easier to express the properties that we will be proving for PA.

The definition of the syntax of PA in Table 1, in particular, the application of abstraction (7) and parallel composition (8) operators only on top of process recursive definitions (5), has been inspired by the syntax of the algebra of Interactive Markov Chains (IMC) [2, 4]. The set of closed (i.e. no free process variables) PA expressions is actually a subset of $\mathrm{IMC}_{\mathrm{XL}}$ defined in [4]. It has been proved in [4] that $\mathrm{IMC}_{\mathrm{XL}}$ expressions have only finite semantic models, therefore the same is true for closed PA expressions. Guardedness of process variables in PA expressions (which is required in order to guarantee that their

Table 2. Structural Operational Semantics of PA: $a \in \mathbf{Act} \cup \{\tau\}$, $C \subseteq \mathbf{Lab}$, $C_1 \subseteq \mathbf{Lab}$, $C_2 \subseteq \mathbf{Lab}$, $\ell \in \mathbf{Lab}$, $X \in \mathbf{Var}$, $A \subseteq \mathbf{Act}$; E, E', F, F' are PA processes

$$a^\ell . E \xrightarrow[\{\ell\}]{a} E \qquad (1)$$

$$\frac{E \xrightarrow{a}_{C} E'}{E + F \xrightarrow{a}_{C} E'} \quad (2) \qquad\qquad \frac{F \xrightarrow{a}_{C} F'}{E + F \xrightarrow{a}_{C} F'} \quad (3)$$

$$\frac{E \xrightarrow{a}_{C} E' \quad a \notin A}{\text{hide } A \text{ in } E \xrightarrow{a}_{C} \text{hide } A \text{ in } E'} \quad (4) \qquad \frac{E \xrightarrow{a}_{C} E' \quad a \in A}{\text{hide } A \text{ in } E \xrightarrow{\tau}_{C} \text{hide } A \text{ in } E'} \quad (5)$$

$$\frac{E \xrightarrow{a}_{C} E' \quad a \notin A}{E \parallel_A \parallel F \xrightarrow{a}_{C} E' \parallel_A \parallel F} \quad (6) \qquad \frac{F \xrightarrow{a}_{C} F' \quad a \notin A}{E \parallel_A \parallel F \xrightarrow{a}_{C} E \parallel_A \parallel F'} \quad (7)$$

$$\frac{E \xrightarrow{a}_{C_1} E' \quad F \xrightarrow{a}_{C_2} F' \quad a \in A}{E \parallel_A \parallel F \xrightarrow{a}_{C_1 \cup C_2} E' \parallel_A \parallel F'} \quad (8) \qquad \frac{E\{X := E/X\} \xrightarrow{a}_{C} E'}{X := E \xrightarrow{a}_{C} E'} \quad (9)$$

semantics is well-defined) is ensured by the rule (2) in Table 1: all process identifiers are directly guarded in PA.

The semantics of PA is defined in the style of Structural Operational Semantics (SOS) introduced by G. Plotkin in [12]. We define the semantics of a PA expression as a Labelled Transition System (LTS) with the smallest set of transition relations that satisfy the rules in Table 2. Recursion unfolding in the rule (9) involves the substitution of each free X in E for its process definition $X := E$, denoted by $E\{X := E/X\}$.

Note that all the transitions in Table 2 are decorated by sets of labels. Labels in the set are those that are labelling actions which "participate" in the transition by giving rise to the rule (1) in Table 2. Due to the multi-way synchronisation ((8) in Table 2) there can be any number of actions "participating" in the transition. In this case all the corresponding labels will be contained in the set decorating the transition.

Assuming that E and E' are PA expressions, the notation $E \longrightarrow E'$ means that there exists a transition $E \xrightarrow{a}_{C} E'$ for some $a \in \mathbf{Act} \cup \{\tau\}$ and $C \subseteq \mathbf{Lab}$. The notation $E \xrightarrow{*} E'$ means that the pair (E, E') is contained in the reflexive and transitive closure of $\longrightarrow$.

We will make use of two auxiliary operators on PA expressions. The operator *Labs* returns all the labels occurring in its argument expression: for example, $Labs(\text{hide b in } a^{\ell_1}.b^{\ell_2}.\mathbf{0}) = \{\ell_1, \ell_2\}$. The operator *fl* ("free labels") returns all the labels that do not belong to τ or to an action that has been abstracted by the application of the rule (7) in Table 1: $fl(\text{hide b in } a^{\ell_1}.b^{\ell_2}.\mathbf{0}) = \{\ell_1\}$. Note that due to the application of the hiding operator ((7) in Table 1) solely on top of process definitions ((5) in Table 1) we avoid the situation where an originally free label can become hidden after a number of transitions. The last can namely occur only after a recursion unfolding, as in the expression $X := a^{\ell_1}.\text{hide a in } b^{\ell_2}.X$

but expressions of this kind are excluded from PA. The operators *Labs* and *fl* will be used in the Data Flow Analysis in Section 3.

3   Data Flow Analysis

The reachability algorithm that will be presented in Section 4.2 takes as input a PA expression and a state in the LTS induced by its semantics. The algorithm returns *true* in case the state is reachable from the start state of the LTS and *false* otherwise. While answering this question we will not build the whole LTS as a first step. Rather we will perform a particular kind of Static Analysis – Data Flow Analysis – on the input PA expression first.

Data Flow Analysis techniques have been originally developed for programming languages (see, for example, [7]). They have been transferred to the area of process algebras in [9, 10, 8, 11] etc. We have developed Data Flow Analysis for the algebra of IMC in the PhD thesis [13]. The description of the method and proved theoretical results will appear in [14]. The current section is based on the adaptation of the methods developed for IMC to the algebra of PA.

We will mostly do the Data Flow Analysis of so-called PA *programs*. These are closed PA expressions that are precisely defined: all the syntactically distinct actions are labeled with distinct labels and all the distinct process identifier definitions refer to distinct process variables.

The primal idea of the Data Flow Analysis is to represent each PA expression by a set of its *exposed labels*. These are labels that can decorate the transitions enabled for the expression according to its semantics. For example, exposed labels of $X := \mathsf{a}^{\ell_1}.X + \mathsf{b}^{\ell_2}.\mathbf{0}$ are ℓ_1 and ℓ_2: they decorate the transitions

$$X := \mathsf{a}^{\ell_1}.X + \mathsf{b}^{\ell_2}.\mathbf{0} \xrightarrow[\{\ell_1\}]{\mathsf{a}} X := \mathsf{a}^{\ell_1}.X + \mathsf{b}^{\ell_2}.\mathbf{0} \text{ and } X := \mathsf{a}^{\ell_1}.X + \mathsf{b}^{\ell_2}.\mathbf{0} \xrightarrow[\{\ell_2\}]{\mathsf{b}} \mathbf{0}.$$

The set of exposed labels can be computed inductively on the syntax of a PA expression by the operator $\mathcal{E}$ defined in Table 3. Note that the $\mathcal{E}$ is parametrised by the argument PA expression (i.e. we will compute $\mathcal{E}_F[\![F]\!]$ for a PA program F) in order to be able to determine the exposed labels of process identifiers (rule (8) in Table 3) which are in fact the exposed labels of the process identifier definitions.

Assume that there exists a transition decorated by a set containing only one label. We would like to determine the exposed labels in the state reachable through this transition. In order to be able to do this, we have defined two Data Flow Analysis operators on PA– the *generate* operator $\mathcal{G}$ and the *kill* operator $\mathcal{K}$. They are inspired by the definitions of the $\mathcal{G}$ and $\mathcal{K}$ operators in [9].

The $\mathcal{K}$ operator returns a mapping where each label in its argument PA expression is assigned to a set of labels that cease to be exposed in the PA expression reachable through the transition decorated by the label. In our example $X := \mathsf{a}^{\ell_1}.X + \mathsf{b}^{\ell_2}.\mathbf{0}$ both ℓ_1 and ℓ_2 will be mapped to $\{\ell_1, \ell_2\}$ by the $\mathcal{K}$ operator. The operator $\mathcal{G}$, on the other hand, returns a mapping where labels from its argument expression are assigned to sets of labels that will become newly exposed after the transition decorated by them. In our example ℓ_1 will be mapped to $\{\ell_1, \ell_2\}$ and ℓ_2 will be mapped to the empty set by the $\mathcal{G}$ operator.

Table 3. Definition of the exposed operator $\mathcal{E} : \mathrm{PA} \to 2^{\mathbf{Lab}}$, $F \in \mathrm{PA}$

$$\mathcal{E}_F[\![\mathbf{0}]\!] = \emptyset \tag{1}$$
$$\mathcal{E}_F[\![a^\ell.X]\!] = \{\ell\} \tag{2}$$
$$\mathcal{E}_F[\![a^\ell.P]\!] = \{\ell\} \tag{3}$$
$$\mathcal{E}_F[\![P_1 + P_2]\!] = \mathcal{E}_F[\![P_1]\!] \cup \mathcal{E}_F[\![P_2]\!] \tag{4}$$
$$\mathcal{E}_F[\![X := P]\!] = \mathcal{E}_F[\![P]\!] \tag{5}$$
$$\mathcal{E}_F[\![\text{hide } A \text{ in } P]\!] = \mathcal{E}_F[\![P]\!] \tag{6}$$
$$\mathcal{E}_F[\![P_1 \,\|\, A \,\|\, P_2]\!] = \mathcal{E}_F[\![P_1]\!] \cup \mathcal{E}_F[\![P_2]\!] \tag{7}$$
$$\mathcal{E}_F[\![X]\!] = \bigcup_{X := P \preceq F} \mathcal{E}_F[\![P]\!] \tag{8}$$

Table 4. Definition of the kill operator $\mathcal{K} : \mathrm{PA} \to (\mathbf{Lab} \to 2^{\mathbf{Lab}})$, $F \in \mathrm{PA}$

$$\mathcal{K}_F[\![\mathbf{0}]\!] = \bot_{\mathcal{K}} \tag{1}$$
$$\mathcal{K}_F[\![a^\ell.X]\!] = \bot_{\mathcal{K}}[\ell \mapsto \{\ell\}] \tag{2}$$
$$\mathcal{K}_F[\![a^\ell.P]\!](\ell') = \mathcal{K}_F[\![P]\!](\ell') \cup \bot_{\mathcal{K}}[\ell \mapsto \{\ell\}](\ell') \qquad \text{for all } \ell' \in \mathbf{Lab} . \tag{3}$$
$$\mathcal{K}_F[\![P_1 + P_2]\!](\ell) = \begin{cases} \mathcal{E}_F[\![P_1 + P_2]\!] & \text{if } \ell \in \mathcal{E}_F[\![P_1 + P_2]\!] , \\ \mathcal{K}_F[\![P_1]\!](\ell) \cup \mathcal{K}_F[\![P_2]\!](\ell) & \text{otherwise} . \end{cases} \tag{4}$$
$$\mathcal{K}_F[\![X := P]\!] = \mathcal{K}_F[\![P]\!] \tag{5}$$
$$\mathcal{K}_F[\![\text{hide } A \text{ in } P]\!] = \mathcal{K}_F[\![P]\!] \tag{6}$$
$$\mathcal{K}_F[\![P_1 \,\|\, A \,\|\, P_2]\!](\ell) = \mathcal{K}_F[\![P_1]\!](\ell) \cup \mathcal{K}_F[\![P_2]\!](\ell) \qquad \text{for all } \ell \in \mathbf{Lab} . \tag{7}$$

Table 5. Definition of the generate operator $\mathcal{G} : \mathrm{PA} \to (\mathbf{Lab} \to 2^{\mathbf{Lab}})$, $F \in \mathrm{PA}$

$$\mathcal{G}_F[\![\mathbf{0}]\!] = \bot_{\mathcal{G}} \tag{1}$$
$$\mathcal{G}_F[\![a^\ell.X]\!] = \bot_{\mathcal{G}}[\ell \mapsto \mathcal{E}_F[\![X]\!]] \tag{2}$$
$$\mathcal{G}_F[\![a^\ell.P]\!](\ell') = \mathcal{G}_F[\![P]\!](\ell') \cup \bot_{\mathcal{G}}[\ell \mapsto \mathcal{E}_F[\![P]\!]](\ell') \text{ for all } \ell' \in \mathbf{Lab} . \tag{3}$$
$$\mathcal{G}_F[\![P_1 + P_2]\!](\ell) = \mathcal{G}_F[\![P_1]\!](\ell) \cup \mathcal{G}_F[\![P_2]\!](\ell) \qquad \text{for all } \ell \in \mathbf{Lab} . \tag{4}$$
$$\mathcal{G}_F[\![X := P]\!] = \mathcal{G}_F[\![P]\!] \tag{5}$$
$$\mathcal{G}_F[\![\text{hide } A \text{ in } P]\!] = \mathcal{G}_F[\![P]\!] \tag{6}$$
$$\mathcal{G}_F[\![P_1 \,\|\, A \,\|\, P_2]\!](\ell) = \mathcal{G}_F[\![P_1]\!](\ell) \cup \mathcal{G}_F[\![P_2]\!](\ell) \qquad \text{for all } \ell \in \mathbf{Lab} . \tag{7}$$

The operators $\mathcal{K}$ and $\mathcal{G}$ are defined inductively on the syntax of PA in Tables 4 and 5. They make use of the mappings $\bot_{\mathcal{K}}$ and $\bot_{\mathcal{G}}$ that map all the labels from **Lab** to the empty set. The operators are parametrised by PA expressions which are given as arguments to the operator $\mathcal{E}$ whenever the last is called. At the top level we will be computing $\mathcal{K}_F[\![F]\!]$ and $\mathcal{G}_F[\![F]\!]$ for a PA program F.

We are now ready to explain the connection between the operators $\mathcal{E}$, $\mathcal{K}$ and $\mathcal{G}$. Assume that we are given a PA expression E. Then in case $E \xrightarrow[\{\ell\}]{a} E'$ for some a, ℓ and E', we have that $\mathcal{E}_{E'}[\![E']\!] = \mathcal{E}_E[\![E]\!] \backslash \mathcal{K}_E[\![E]\!](\ell) \cup \mathcal{G}_E[\![E]\!](\ell)$. The operators $\mathcal{G}$ and $\mathcal{K}$ thus "capture" the effect of the transition.

Assume now that we are given a PA program F and that $F \xrightarrow{*} E$. Then in the above equality we can in fact reuse the mappings computed on F. We can also use F as a parameter while computing the exposed labels of E because process

Table 6. Definition of the chain operator $\mathfrak{T} : \mathrm{PA} \to 2^{2^{\mathbf{Lab}}}$, $F \in \mathrm{PA}$

$$\mathfrak{T}_F[\![\mathbf{0}]\!] = \emptyset \tag{1}$$

$$\mathfrak{T}_F[\![\mathsf{a}^\ell.X]\!] = \{\{\ell\}\} \tag{2}$$

$$\mathfrak{T}_F[\![\mathsf{a}^\ell.P]\!] = \mathfrak{T}_F[\![P]\!] \ \cup \ \{\{\ell\}\} \tag{3}$$

$$\mathfrak{T}_F[\![P_1 + P_2]\!] = \mathfrak{T}_F[\![P_1]\!] \ \cup \ \mathfrak{T}_F[\![P_2]\!] \tag{4}$$

$$\mathfrak{T}_F[\![X := P]\!] = \mathfrak{T}_F[\![P]\!] \tag{5}$$

$$\mathfrak{T}_F[\![\text{hide } A \text{ in } P]\!] = \mathfrak{T}_F[\![P]\!] \tag{6}$$

$$\mathfrak{T}_F[\![P_1 \shortparallel A \shortparallel P_2]\!] = \{C | C \in \mathfrak{T}_F[\![P_1]\!], \exists \ell \in C, \exists \mathsf{a}^\ell.P_\ell \preceq P_1$$
$$\text{such that } (\mathsf{a} \notin A) \vee (\ell \notin \mathit{fl}(P_1))\} \ \cup$$

$$\{C | C \in \mathfrak{T}_F[\![P_2]\!], \exists \ell \in C, \exists \mathsf{a}^\ell.P_\ell \preceq P_2$$
$$\text{such that } (\mathsf{a} \notin A) \vee (\ell \notin \mathit{fl}(P_2))\} \ \cup$$

$$\{\{C_1 \cup C_2\} | C_1 \in \mathfrak{T}_F[\![P_1]\!], C_2 \in \mathfrak{T}_F[\![P_2]\!], \exists \ell_1 \in C_1,$$
$$\exists \ell_2 \in C_2, \exists \mathsf{a}_1{}^{\ell_1}.P_{\ell_1} \preceq P_1, \exists \mathsf{a}_2{}^{\ell_2}.P_{\ell_2} \preceq P_2,$$
$$\text{such that } (\mathsf{a}_1 = \mathsf{a}_2) \wedge (\mathsf{a}_1 \in A)$$
$$\wedge (\mathsf{a}_1 \in \mathit{fl}(P_1)) \wedge (\mathsf{a}_2 \in \mathit{fl}(P_2))\} \tag{7}$$

definitions in F and E do not differ. We thus obtain $\mathcal{E}_F[\![E']\!] = \mathcal{E}_F[\![E]\!] \backslash \mathcal{K}_F[\![F]\!](\ell) \cup \mathcal{G}_F[\![F]\!](\ell)$. In other words, knowing the mappings returned by the operators $\mathcal{G}$ and $\mathcal{K}$ on a PA program is enough to predict exposed labels after any number of transitions. The equality is unusual for Static Analysis methods – for example, the method in [9] is computing overapproximations of exposed labels – and is due to the finiteness of the semantic models of PA expressions.

Due to the multi-way synchronisation in PA, there can be more than one label decorating a transition. For example, for $\underline{X := \mathsf{a}^{\ell_1}.X} \shortparallel \shortparallel \underline{Y := \mathsf{a}^{\ell_2}.\mathsf{b}^{\ell_3}.Y}$ there is a transition decorated by $\{\ell_1, \ell_2\}$. In order to take care of this situation, we have introduced in Table 6 a so-called *chain* operator $\mathfrak{T}$ (inspired by the definition in [13]) on PA expressions. It returns all the sets of labels (that will be called *chains* in the following) that can decorate the transitions from an argument PA expression. In the above example it will return $\{\ell_1, \ell_2\}$ and $\{\ell_3\}$.

Similarly to the operators $\mathcal{K}$ and $\mathcal{G}$, the results returned by the $\mathfrak{T}$ on a PA program are applicable after any number of transitions. If a chain contains more than one label then in order to determine the effects of the $\mathcal{G}$ and $\mathcal{K}$ operators, the union of the mappings of all the labels in the chain should be taken.

The properties of the $\mathcal{E}$, $\mathcal{K}$, $\mathcal{G}$ and $\mathfrak{T}$ operators are expressed more formally in Theorem 1. It states that the results of the operators on a PA program are enough to reproduce its semantics. Note that due to the uniqueness of labels in PA programs, each label has its unique corresponding action name. All the labels in a chain have the same corresponding name according to the construction in Table 6. Theorem 1 will be used in the reachability algorithm in Section 4.2.

Theorem 1. *Given a* PA *program* F, *then for all* E *such that* $F \overset{*}{\longrightarrow} E$ *we have:*

$E \overset{\mathsf{a}}{\underset{C}{\longrightarrow}} E'$ *iff* $C \in \mathfrak{T}_F[\![F]\!]$, $C \subseteq \mathcal{E}_F[\![E]\!]$, $\mathcal{E}_F[\![E']\!] = (\mathcal{E}_F[\![E]\!] \backslash (\bigcup_{\ell \in C} \mathcal{K}_F[\![F]\!](\ell))) \cup \bigcup_{\ell \in C} \mathcal{G}_F[\![F]\!](\ell)$ *and* $\mathsf{a}^\ell.P_\ell \preceq F$ *for some* $P_\ell \preceq F$ *for all* $\ell \in C$.

Proof. The proof consists of several steps. We can first show that all the exposed labels in E are distinct. Therefore we can simply use label sets in our discussion and not multisets of labels as in, for example, [9]. We can then prove that all the labels decorating the transitions from any PA expression E_1 are exposed in it. Moreover, all the sets of labels decorating the transitions from E_1 are in $\mathfrak{T}_F[\![E_1]\!]$ and no further sets. We can then prove that in case $E_1 \xrightarrow{\ a\ } E_2$ for some action a, chain C and PA expression E_2, we have $\mathcal{E}_F[\![E_2]\!] = (\mathcal{E}_F[\![E_1]\!] \backslash (\bigcup_{\ell \in C} \mathcal{K}_F[\![E_1]\!](\ell))) \cup \bigcup_{\ell \in C} \mathcal{G}_F[\![E_1]\!](\ell)$.

After this we have to prove that the results of the generate, kill and chains operators on a PA program F are applicable after any number of transitions. In particular, if $F \xrightarrow{\ *\ } E$ and for some $C \in \mathfrak{T}_F[\![F]\!]$ all the labels in C are occurring in E, then $C \in \mathfrak{T}_F[\![E]\!]$ as well. A similar statement is true for the generate operator. For the kill operator the "killed" sets in $\mathcal{K}_F[\![E]\!]$ can in general be smaller than in $\mathcal{K}_F[\![F]\!]$, but the result of their subtraction from $\mathcal{E}_F[\![E]\!]$ is the same. Altogether these considerations prove the statement of the theorem.

Lemmas 3.14-3.15 and Theorem 3.16 in Chapter 3 of the PhD thesis [13] state similar facts for the variant of the calculus of IMC. See the proofs in the appendix of [13] for the details.   □

4   Reachability Algorithm

4.1   Determining Possible States

The reachability algorithm in Section 4.2 will determine whether a state represented by the exposed labels is reachable in the LTS induced by the semantics of a PA program. Before computing the state's reachability, we would like to determine whether the state in question actually represents a meaningful configuration of exposed labels. For example, for $X := a^{\ell_1}.b^{\ell_2}.X$ it does not make sense to ask whether the state represented by $\{\ell_1, \ell_2\}$ is reachable.

We will apply the operator *excl* to PA programs prior to conducting the reachability analysis. The operator *excl* is inductively defined on the syntax of PA in Table 7. It computes a set of label pairs: labels in a pair are those that mutually exclude the "exposedness" of each other. The operator makes use of the consideration that two labels cannot be exposed at the same time if one of them is in the prefix of the expression which contains the other one. Lemma 1 states the correctness of the *excl* operator.

Lemma 1. *Given a PA program F, then for all E such that $F \xrightarrow{\ *\ } E$ we have:*
$$((\ell_1, \ell_2) \in excl_F[\![F]\!]) \wedge (\ell_1 \in \mathcal{E}_F[\![E]\!]) \Rightarrow (\ell_2 \notin \mathcal{E}_F[\![E]\!]).$$

Proof. We prove the lemma by proving a more general statement that $((\ell_1, \ell_2) \in excl_F[\![F]\!]) \wedge (\ell_1 \in \mathcal{E}_F[\![E'']\!]) \Rightarrow (\ell_2 \notin \mathcal{E}_F[\![E'']\!])$ for any subexpression E'' of E. This will establish the lemma because $E \preceq E$ is true. We make our proof by induction on the number of steps in the derivation $F \xrightarrow{\ *\ } E$.

The statement holds for F: this can be shown by induction on the structure of F using the rules of the operator *excl* in Table 7. We have therefore to prove

Table 7. Definition of the operator $excl : \text{PA} \rightarrow 2^{\textbf{Lab} \times \textbf{Lab}}$

$$excl_F[\![\textbf{0}]\!] = \emptyset \tag{1}$$

$$excl_F[\![\textsf{a}^\ell.X]\!] = \emptyset \tag{2}$$

$$excl_F[\![\textsf{a}^\ell.P]\!] = excl_F[\![P]\!] \cup \bigcup_{\ell' \in Labs(P)}\{(\ell, \ell')\} \tag{3}$$

$$excl_F[\![P_1 + P_2]\!] = excl_F[\![P_1]\!] \cup excl_F[\![P_2]\!] \cup$$

$$\bigcup_{\ell_1 \in Labs(P_1)\setminus\mathcal{E}_F[\![P_1]\!]} \bigcup_{\ell_2 \in Labs(P_2)\setminus\mathcal{E}_F[\![P_2]\!]}\{(\ell_1, \ell_2)\} \tag{4}$$

$$excl_F[\![X := P]\!] = excl_F[\![P]\!] \tag{5}$$

$$excl_F[\![\text{hide } A \text{ in } P]\!] = excl_F[\![P]\!] \tag{6}$$

$$excl_F[\![P_1 \!\parallel\! A \!\parallel\! P_2]\!] = excl_F[\![P_1]\!] \cup excl_F[\![P_2]\!] \tag{7}$$

that in case $F \overset{*}{\longrightarrow} E \longrightarrow E'$ and the statement is true for E then it is also true for E'. We prove this by induction on the transition derivation according to the rules in Table 2.

The statement is clear for the rule (1) in Table 2 because in this case $E' \preceq E$. The statement easily follows from the induction hypothesis for the majority of the other rules (for the rules (6)-(8) it can be shown that $Labs(P_1) \cap Labs(P_2) = \emptyset$ holds for $P_1 \!\parallel\! A \!\parallel\! P_2 \preceq E$). For the rule (9) we need to additionally show that the statement of the lemma holds for $P\{X := P/X\}$ if it holds for $X := P$ – this is necessary for the induction hypothesis to become applicable.

We can deduce the statement for all $P'' \preceq P$ directly from the induction hypothesis for $X := P$. For subexpressions of the type $P''\{X := P/X\}$ it can be shown that $\mathcal{E}_F[\![P'']\!] = \mathcal{E}_F[\![P''\{X := P/X\}]\!]$ (this follows from the guardedness of the syntax of PA in Table 1). In this way, the induction hypothesis for the subexpressions of $X := P$ is applicable also to $P''\{X := P/X\}$ with $P'' \preceq P$. $\qquad\square$

4.2   Computing Reachable Labels

We will now describe the actual reachability algorithm in several steps. First, we define the algorithms *init* and *refine* in Table 8. The algorithm *init* initialises and returns the data structure *gchains*. This data structure maps the labels in a PA program (given as an input F) to the set of chains (computed on F) that contain a label which generates the label mapped to this set. For example, for $F \triangleq (\textsf{b}^{\ell_1}.\textsf{a}^{\ell_2}.\textsf{c}^{\ell_3}.\textbf{0} + \textsf{a}^{\ell_4}.\textsf{a}^{\ell_5}.\textsf{d}^{\ell_6}.\textbf{0}) \!\parallel\!\{\textsf{a}, \textsf{b}\}\!\parallel\! \textsf{a}^{\ell_7}.\textbf{0}$ we have $init(F) = \{\ell_1 \mapsto \emptyset, \ell_2 \mapsto \emptyset, \ell_3 \mapsto \{\{\ell_2, \ell_7\}\}, \ell_4 \mapsto \emptyset, \ell_5 \mapsto \{\{\ell_4, \ell_7\}\}, \ell_6 \mapsto \{\{\ell_5, \ell_7\}\}, \ell_7 \mapsto \emptyset\}$.

The algorithm *refine* computes a set of labels L which is an overapproximation of labels reachable from a state as represented by the exposed labels in an input S. The reachability is determined for the LTS induced by the semantics of an input PA program F. We call a label "reachable" if it is exposed in one of the reachable states.

The algorithm recursively deletes all those chains from the *gchains*'-mapping (initialised by the value of an input *gchains*) in which at least one constituting label is mapped to the empty set and is not in S, until no deleting is possible. Reachable labels are those that are either contained in S or are mapped to a

Table 8. Algorithms *init* and *refine*: $F \in$ PA, *gchains* $\in Labs(F) \rightarrow 2^{\mathfrak{T}_F[\![F]\!]}$, $S \subseteq Labs(F)$

Initialisation step:
proc *init(F)* is
for all $\ell \in Labs(F)$ do
    $gchains(\ell) := \{C \in \mathfrak{T}_F[\![F]\!] | \exists \ell' \in C$ such that $\ell \in \mathcal{G}_F[\![F]\!](\ell')\}$
return *gchains*
Refinement step:
proc *refine(F, S, gchains)* is
$L := S$; $gchains' := gchains$;
while $\exists \ell \in Labs(F)$ such that $(gchains'(\ell) = \emptyset) \wedge (\ell \notin S)$ do
    for all $\ell' \in Labs(F)$ do
        $gchains'(\ell') := gchains'(\ell') \backslash \{C \in \mathfrak{T}_F[\![F]\!] | \ell \in C\}$
for all $\ell \in Labs(F)$ do
    if $gchains'(\ell) \neq \emptyset$ then
        $L := L \cup \{\ell\}$;
return $L, gchains'$

---

non-empty set at the end of the algorithm's run. For the last there exists at least one chain generating it in which all the labels are considered to be reachable. We state in Lemma 2 the correctness of the *init* and *refine* procedures.

In our example, after computing $(L, gchains) = refine(F, \mathcal{E}_F[\![F]\!], init(F))$, we have $gchains = \{\ell_1 \mapsto \emptyset, \ell_2 \mapsto \emptyset, \ell_3 \mapsto \emptyset, \ell_4 \mapsto \emptyset, \ell_5 \mapsto \{\{\ell_4, \ell_7\}\}, \ell_6 \mapsto \{\{\ell_5, \ell_7\}\}, \ell_7 \mapsto \emptyset\}$ and therefore $L = \{\ell_1, \ell_4, \ell_5, \ell_6, \ell_7\}$. The mapping for the label ℓ_4 has been updated, therefore it will be correctly identified as unreachable. On the other hand, the label ℓ_6 will be considered reachable which is not the case, which shows that L is only an overapproximation of reachable labels.

Lemma 2. *Given a* PA *program F, we define a domain $D = \{d \in (Labs(F) \rightarrow 2^{\mathfrak{T}_F[\![F]\!]}) | d(\ell) \subseteq init(F)(\ell)$ for all $\ell \in Labs(F)\}$. Further we define a function $G_S : D \rightarrow D$ with $S \subseteq Labs(F)$ such that $G_S(d)(\ell) = \{C \in d(\ell) | \forall \ell' \in C \ (\ell' \in S) \vee (d(\ell') \neq \emptyset)\}$. For $(L, gchains) = refine(F, S, init(F))$ we have:*

1. *$gchains = GFP(G_S)$, where GFP stands for the greatest fixed point;*

2. *if $F \xrightarrow{*} E \xrightarrow{*} E'$ for some E and E' and $S = \mathcal{E}_F[\![E]\!]$ then $\mathcal{E}_F[\![E']\!] \subseteq L$.*

Proof. The domain D can be understood as a complete lattice with the order $d_1 \leq d_2$ iff $d_1(\ell) \subseteq d_2(\ell_2)$ for all $\ell \in Labs(F)$. The function G_S is monotone, therefore there exists the greatest fixed point of G_S on D according to the Knaster-Tarski theorem. The domain D satisfies the Descending Chain Condition because it is finite. Therefore the *GFP* of G_S can be computed by the repeated application of G_S to the greatest element of D until the descending chain stabilises. The algorithm *refine* in Table 8 can be understood as the repeated application of the function G_S with the initial argument $init(F)$ which is equal to the greatest element of D. Therefore it stabilises on $GFP(G_S)$.

Concerning the statement 2, it is clear that all the labels that are reachable from E are either exposed in E or are generated by a chain all the labels in which are reachable from E. We can define a mapping gch such that $gch(\ell) = \{C \in init(F)(\ell) | \exists E'' \text{ such that } (E \xrightarrow{*} E'') \wedge (C \subseteq \mathcal{E}_F[\![E'']\!])\}$ for all $\ell \in Labs(F)$. The mapping gch is a fixed point of $G_{\mathcal{E}_F[\![E]\!]}$, therefore it is smaller or equal to $GFP(G_{\mathcal{E}_F[\![E]\!]})$. All the labels reachable from E are either in $\mathcal{E}_F[\![E]\!]$ (and therefore in L) or are mapped to a non-empty set by gch – and therefore are mapped to a (larger or equal) non-empty set by $GFP(G_{\mathcal{E}_F[\![E]\!]})$, hence are in L.   □

We will usually apply the *init* once to a PA program F. On the contrary, we might need to compute the *refine* many times with different sets S. In case $F \xrightarrow{*} E \xrightarrow{*} E'$ we can, for example, reuse the *gchains*-parameter returned by $(L, gchains) = refine(F, \mathcal{E}_F[\![E]\!], init(F))$ as an input to the computation of $refine(F, \mathcal{E}_F[\![E']\!], gchains)$. We can safely reuse the previously computed mapping from labels to chains because the set of reachable labels can only become smaller after several transitions.

In the example above we have $F \xrightarrow[\{\ell_1, \ell_4\}]{a} E$ with $E \triangleq a^{\ell_2}.b^{\ell_3}.\mathbf{0} \| a \| \mathbf{0}$ and we can compute $refine(F, \mathcal{E}_F[\![E]\!], gchains)$ with *gchains* being the second output returned by $(L, gchains) = refine(F, \mathcal{E}_F[\![F]\!], init(F))$, which is faster than computing $refine(F, \mathcal{E}_F[\![E]\!], init(F))$. This consideration will be used in the reachability algorithm in Table 9.

4.3   Main Algorithm

The main reachability algorithm *reach* is described in Table 9. It takes as an input a PA program F and a state characterised by the exposed labels in $S_?$ and returns *true* if and only if the state is reachable from the start state in the semantics of F. The algorithm makes use of the Data Flow Analysis operators in determining states reachable after one transition (see Theorem 1).

Each newly computed state is checked for whether the set of its exposed labels is equal to $S_?$ (lines 1 and 17) or has been computed before, i.e. is in *States* (line 16) – in the first case the algorithm terminates returning *true* and in the second case it discards the state. Otherwise the *refine* algorithm is performed with the newly created state and the current value of *gchains* as input values (lines 7 and 18).

If all the labels in $S_?$ are in the set of reachable labels returned by *refine*, then all the transitions from the state are created (line 15). Otherwise the state is discarded (lines 8 and 19). In this way, we are "cutting off" branches in the LTS induced by the semantics of F that according to Lemma 2 do not lead to $S_?$. The algorithm terminates because the number of states is always finite. Theorem 2 states the correctness of the *reach* algorithm.

Theorem 2. *Given a PA program F, then $F \xrightarrow{*} E$ iff $reach(F, \mathcal{E}_F[\![E]\!]) = true$.*

Proof. It is easy to see from the algorithm *reach* in Table 9 that if for some $S_? \subseteq \mathbf{Lab}$ we have $reach(F, S_?) = true$ then there exists some E such that $F \xrightarrow{*} E$ and $\mathcal{E}_F[\![E]\!] = S_?$. We can namely use Theorem 1 in order to show

Table 9. Algorithm *reach* taking as arguments a PA program F and $S_? \subseteq \mathbf{Lab}$. The set *States* contains already encountered states, the set W is a working list.

```
proc reach(F, S?) is
 1:   if EF[[F]] = S? then
 2:      return true;
 3:   if S? ⊄ Labs(F) then
 4:      return false;
 5:   if ∃ℓ1 ∈ S?, ∃ℓ2 ∈ S? such that (ℓ1, ℓ2) ∈ exclF[[F]] then
 6:      return false;
 7:   gchains := init(F); (L', gchains') := refine(F, EF[[F]], gchains);
 8:   if S? ⊄ L' then
 9:      return false;
10:   States := {EF[[F]]};
11:   W := {(EF[[F]], gchains)};
12:   while (W ≠ ∅) do
13:      choose (S, gchains) from W; W := W\{(S, gchains)}
14:      for all C ∈ 𝔗F[[F]] such that C ⊆ S do
15:         S' := S\(⋃ℓ∈C 𝒦F[[F]](ℓ)) ∪ ⋃ℓ∈C 𝒢F[[F]](ℓ);
16:         if S' ∈ States then break;
17:         if (S' = S?) then return true;
18:         (L', gchains') := refine(F, S', gchains);
19:         if S? ⊄ L' then break;
20:         States := S ∪ {S'}; W := W ∪ {(S', gchains')};
21:   return false
```

that for each $(S, gchains)$ added to W holds $S = \mathcal{E}_F[[E]]$ for some $F \xrightarrow{\ *\ } E$. The algorithm returns *true* only in case $S = S_?$ for one of such $(S, gchains) \in W$.

It is left to show the other direction. The algorithm always terminates, therefore it is equivalent to showing that from $reach(F, S_?) = false$ follows that there does not exist any E such that $F \xrightarrow{\ *\ } E$ and $\mathcal{E}_F[[E]] = S_?$.

From the definition of *reach* in Table 9 and from Theorem 1, we can deduce that in case $reach(F, S_?) = false$ then for all $F \xrightarrow{\ *\ } E'$ we have either $\mathcal{E}_F[[E']] \neq S_?$ or there exists some E'' such that $F \xrightarrow{\ *\ } E'' \xrightarrow{\ *\ } E'$ and $S_? \not\subseteq L$ with L returned by $(L, gchains) = refine(F, \mathcal{E}_F[[E'']], init(F))$. According to Lemma 2, this means that at least one of the labels in $S_?$ is unreachable from E'', therefore $E' \neq S_?$. We are actually computing $refine(F, \mathcal{E}_F[[E'']], gchains)$ with the "current" value of *gchains* instead of $init(F)$ for efficiency reasons – it is "safe" to do because the set of reachable labels can only become smaller for E'' compared to F. □

4.4 Algorithmic Complexity

In this section we will assess the complexity of the proposed reachability analysis. We need first of all to assess the complexity of the Static Analysis operators.

Computations of the $\mathcal{E}$, $\mathcal{G}$ and $\mathcal{K}$ can be done in linear time in the syntax of a PA program F. We could save the exposed labels of the definitions of process variables while traversing F in order not to compute them anew according to the rule (8) in Table 3 each time the variables occur in F. Then the linear coefficient for the complexity of the $\mathcal{G}$ and $\mathcal{K}$ operators will be equal to the maximal number of components connected by the +-construct in the syntax of F. For example, for $F = a^{\ell_1}.(b^{\ell_2}.0 + c^{\ell_3}.0)$ the label ℓ_1 generates both ℓ_2 and ℓ_3 which are choice alternatives, similarly ℓ_2 "kills" both ℓ_2 and ℓ_3, therefore the $\mathcal{G}$ and $\mathcal{K}$ for this F can be computed in time complexity $2 * |F|$.

Computing the $\mathfrak{T}$ as it is represented in Table 6 is not very efficient, as the Cartesian product is taken in the rule (7). It is however possible to use a data structure for representing chains where each label is saved at most once. For example, we could save $[\ell_1, \{\ell_2, \ell_3\}]$ instead of both $\{\ell_1, \ell_2\}$ and $\{\ell_1, \ell_3\}$ for $a^{\ell_1}.0 \parallel_a \parallel (a^{\ell_2}.0 + a^{\ell_3}.0)$. We would need to "connect" for each $a \in A$ in each subexpression $P_1 \parallel A \parallel P_2$ of the analysed PA program the chains corresponding to a from P_1 and the chains corresponding to a from P_2. In this way, the computation of chains can be performed in linear time. The variant in Table 6 has been chosen for the sake of simplicity of presentation.

Computing the operator *excl* on a PA program F can be done in time quadratic in the syntax of the program: for each label ℓ there exists a set of labels (at most $|Labs(F)|$) such that they cannot be exposed together with ℓ. Checking whether a state is "valid" according to the *excl* can be done in time quadratic in the number of exposed labels in the state as we need to check whether there exists a pair of exposed labels in the state which cannot be exposed together.

The algorithm *reach* requires the execution of the procedure $init(F)$. Using the above mentioned concise data structure for chains, this can be performed in time linear in the number of labels in F. Each label will namely be linked with those labels in the chain data structure that generate it – then the number of such links is linear in the number of labels with linear coefficient equal to the maximal number of +-alternatives in F.

The main computational overhead in the *refine* procedure (which is performed in each new state created by *reach*) is the recomputation of the *gchains*-mapping. This can be done in time linear in the number of labels that were exposed before the transition into the state plus the number of labels that became unreachable after the transition. The reason is that each label that was previously exposed (and therefore reachable) should be checked for being reachable in the new state. If this is not the case then the links of all the labels generated by it into the chain data structure should be updated.

The *refine* procedure can be performed with the worklist W organised as a stack (the depth-first search). Then the maximal number of *gchains* structures saved in W at the same time will be equal to the maximal number of choice options encountered along one path in the LTS induced by the semantics of an input PA program. We do not need to save the whole *gchains* structure each time it is added to W but can just save changes compared to the previous *gchains*.

Altogether, the additional time overhead (compared to building the whole LTS) of our reachability analysis is at most quadratic in the size of the syntax. Additional space needed for performing the *reach* algorithm can also be quadratic in the worst case.

On the other hand, computing reachable labels beforehand in our method can bring considerable advantages in the reachability analysis. For example, in a program of the type $(a^{\ell_1}....b^{\ell_n}.\mathbf{0} + c^{\ell'_1}....d^{\ell'_n}.\mathbf{0})_{||\emptyset||} e^{\ell''_1}....f^{\ell''_n}.\mathbf{0}$ if we would like to know whether the state with $\{\ell_n\}$ exposed is reachable, we do not need to follow the choice branch $c^{\ell'_1}....d^{\ell'_n}.\mathbf{0}$ – we can determine that the label ℓ_n is not reachable along this choice branch with our *reach* algorithm. Moreover, we do not need to build the states reachable as the result of interleaving the processes $c^{\ell'_1}....d^{\ell'_n}.\mathbf{0}$ and $e^{\ell''_1}....f^{\ell''_n}.\mathbf{0}$, which leads to a considerable win in complexity.

We envision several possible enhancements of our algorithm. First, we can check the reachability of all/some of the states from a set by checking whether the labels of all/some of them are contained in the set of reachable labels. Second, we can not only check the reachability of a fully specified state but of a state defined by a subset of its exposed labels – this subset may represent some important property that holds or does not hold independently of other exposed labels. Finally, we can first determine which labels are "important" (i.e. influence the reachability of the state in question) and update the information only on them.

5   Conclusions

In this paper we have presented a reachability algorithm for a process algebra with finite state space. The algorithm is built upon the application of several Data Flow Analysis operators to the syntax of the system first. Data Flow Analysis is used in order to reduce the state space of the reachability analysis. In this way, we are combining Static Analysis and Model Checking methods in one algorithm, as the results of the Static Analysis are constantly refined until the correct answer to the reachability problem is obtained.

The algorithm requires some additional computational time and space in order to create and update auxiliary data structures, but it can also be considerably faster (see the example in Section 4.4) than an algorithm that builds the whole state space of the system. After slight modifications, our algorithm can also check the reachability of several states at the same time and of states which are only partially specified, i.e. with a subset of their exposed labels.

Future work includes the extension of the algorithm to other process algebras as well as to other formalisms specifying systems in a compositional way. We plan to make use of other proposed improvements in the study of reachability – for example, of identifying independent labels similar to the partial order reduction method [3, 1]. The order of independent labels is irrelevant for the computation of states reachable through the transitions decorated by them. This consideration may lead to further reduction of the state space of the reachability algorithm.

References

[1] Baier, C., Katoen, J.-P.: Principles of Model Checking (Representation and Mind Series). The MIT Press, Cambridge (2008)

[2] Brinksma, E., Hermanns, H.: Process Algebra and Markov Chains. In: Brinksma, E., Hermanns, H., Katoen, J.-P. (eds.) EEF School 2000 and FMPA 2000. LNCS, vol. 2090, pp. 183–231. Springer, Heidelberg (2001)

[3] Gerth, R., Kuiper, R., Peled, D., Penczek, W.: A partial order approach to branching time logic model checking. In: Proceedings of ISTCS 1995 (1995)

[4] Hermanns, H.: Interactive Markov Chains. LNCS, vol. 2428, pp. 129–154. Springer, Heidelberg (2002)

[5] Hoare, C.A.R.: Communicating Sequential Processes. Prentice-Hall, Englewood Cliffs (1985)

[6] Milner, R.: A Calculus of Communication Systems. LNCS, vol. 92. Springer, Heidelberg (1980)

[7] Nielson, F., Nielson, H.R., Hankin, C.L.: Principles of Program Analysis. Springer, Heidelberg (1999) Second printing, 2005

[8] Nielson, F., Nielson, H.R., Priami, C., Rosa, D.: Static analysis for systems biology. In: Proceedings of WISICT 2004. Trinity College Dublin (2004)

[9] Nielson, H.R., Nielson, F.: Data flow analysis for CCS. In: Reps, T., Sagiv, M., Bauer, J. (eds.) Wilhelm Festschrift. LNCS, vol. 4444, pp. 311–327. Springer, Heidelberg (2007)

[10] Nielson, H.R., Nielson, F.: A monotone framework for CCS. Comput. Lang. Syst. Struct. 35(4), 365–394 (2009)

[11] Pilegaard, H.: Language Based Techniques for Systems Biology. PhD thesis, Technical University of Denmark (2007)

[12] Plotkin, G.D.: A Structural Approach to Operational Semantics. Technical Report DAIMI FN-19, University of Aarhus (1981)

[13] Skrypnyuk, N.: Verification of Stochastic Process Calculi. PhD thesis, Technical University of Denmark (2011)

[14] Skrypnyuk, N., Nielson, F.: Pathway Analysis for IMC. Journal of Logic and Algebraic Programming (to appear)

Author Index

Abdulla, Parosh Aziz   125
Ábrahám, Erika   139
André, Étienne   31
Axelsson, Roland   45

Bersani, Marcello M.   58
Boichut, Yohan   72
Bonnet, Rémi   85
Bozzelli, Laura   96

Carioni, Alessandro   110
Cederberg, Jonathan   125
Chatterjee, Krishnendu   1
Chen, Taolue   2
Chen, Xin   139
Ciardo, Gianfranco   218
Courcelle, Bruno   26

Dao, Thi-Bich-Hanh   72

Eggermont, Christian E.J.   153

Fioravanti, Fabio   165
Frehse, Goran   139
Fribourg, Laurent   191
Frigeri, Achille   58

Ganty, Pierre   96
Ghilardi, Silvio   110
Gusev, Vladimir V.   180

Han, Tingting   2

Katoen, Joost-Pieter   2
Kühne, Ulrich   191

Lange, Martin   45

Margenstern, Maurice   205
Mereacre, Alexandru   2
Mumme, Malcolm   218
Murat, Valérie   72

Nielson, Flemming   231

Pettorossi, Alberto   165
Proietti, Maurizio   165

Ranise, Silvio   110
Raskin, Jean-François   28
Rossi, Matteo   58

San Pietro, Pierluigi   58
Senni, Valerio   165
Skrypnyuk, Nataliya   231
Soulat, Romain   31

Vojnar, Tomáš   125

Woeginger, Gerhard J.   153